How to use this book

This book has been written to help you develop the skills, knowledge and confidence needed for Paper 1 and Paper 2 options for AQA Level 3 Mathematical Studies (Core Maths).

Compulsory content

Chapters 1 – 3 cover the compulsory content required for Paper 1. You will also need to work through Chapter 4 as that contains compulsory content for each of the Paper 2 options.

Optional content

Chapters 5 – 7 cover Paper 2A: Statistical techniques.

Chapters 8 – 10 cover Paper 2B: Critical path and risk analysis.

Chapters 11 – 13 cover Paper 2C: Graphical techniques.

Practice questions

Chapter 14 contains plenty of practice questions which could be useful for revision.

How the chapters are structured

Each chapter begins with a **real-life introduction** to the topic to provide a relevant, interesting and understandable entry point. Each chapter is divided into the subject content as listed in the specification to make it easy for you to follow.

The **discussions** in the book are designed to allow you to discover the topic for yourself in real-life situations.

Detailed **mathematical explanations** of the topics are given which conclude with an **exercise** containing a variety of question styles and activities with an emphasis on real-world examples.

'**Mathematics in the real world**' sections use meaningful contexts and real-life data which complement your other post-16 studies.

Across each chapter **key terms** and their definitions are highlighted on the page and brought together in a handy glossary at the back of the book.

Each chapter concludes with a **real-life case study** and **project work** for you to extend your knowledge of the topic and further develop your critical thinking and problem-solving skills.

At the end of each chapter you can **check your progress** and determine if there are areas you need to review and revise.

We have tried to write a textbook that is different. Not for the sake of it but to make it real and relevant to you studying a fascinating subject in a world of digital media, soundbites and celebrity. We hope you appreciate the mathematics in everyday life just a little bit more after studying Level 3 Mathematical Studies using this book.

Helen Ball and Peter Ransom

1 Analysis of data

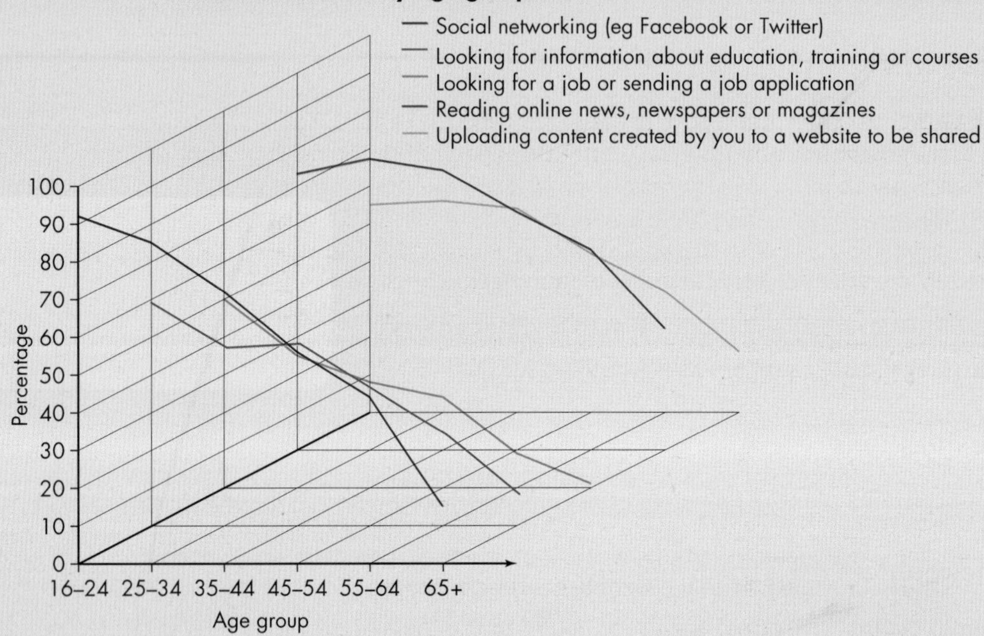

The chart shows the Internet activities of people in the UK during 2015.

Study the chart. What do you think of it?

Questions

Think about these questions.

1. How easy is it to interpret the chart? What other information would you like to see?
2. Why do the percentages for each age group, except the oldest age group, add up to more than 100%?
3. Do all the Internet activities decrease with age?
4. The Office for National Statistics reports that using the Internet for social networking has continued to grow, rising to 61% in 2015. This was an increase from 45% in 2011, and 54% in 2014. Social networking is widespread in all age groups, up to and including the 55–64 age group. In this group, 44% of adults reported using social networking. Of adults aged 65 and over, 15% use social networking. Social networking has become part of the daily lives of many adults. Of the 61% of adults who used social networking in the last three months, 79% did so more or less every day.
   - **a.** Do these statistics seem realistic to you?
   - **b.** How often do you use social networking?
5. What other trends and information can you interpret by looking at the chart?

Data analysis is important in many aspects of life. For example, teachers use data to see how students are progressing throughout the year. Supermarkets use data to regulate stock control. Manufacturers use data to monitor the efficiency of their machines. Entrepreneurs use data to gauge the success of their innovations.

Think about your daily life and some occasions when you collect and use data.

D1: Data

Learning objectives

You will learn how to:

- Identify qualitative, quantitative, primary and secondary data.
- Identify discrete and continuous data, independent and dependent variables.

Introduction

Data is classified to make it easier to process. Data can be collected in the form of measurements or observations of variables. Different kinds of data are represented in different diagrams, according to the type of data. For example, the way an apple wholesaler for a supermarket chain would represent the mass of each apple grown on a particular tree is very different from the way the wholesaler would represent the taste of each variety of apple, or the amount of organic material used in the orchard to grow the apples. The wholesaler must identify the type of data in order to choose the best diagram to use.

Mathematics in the real world

Website designers use qualitative data and quantitative data when gathering research for the design, redesign or improvement of a website. Here is an example of the kind of diagram that a website designer might use.

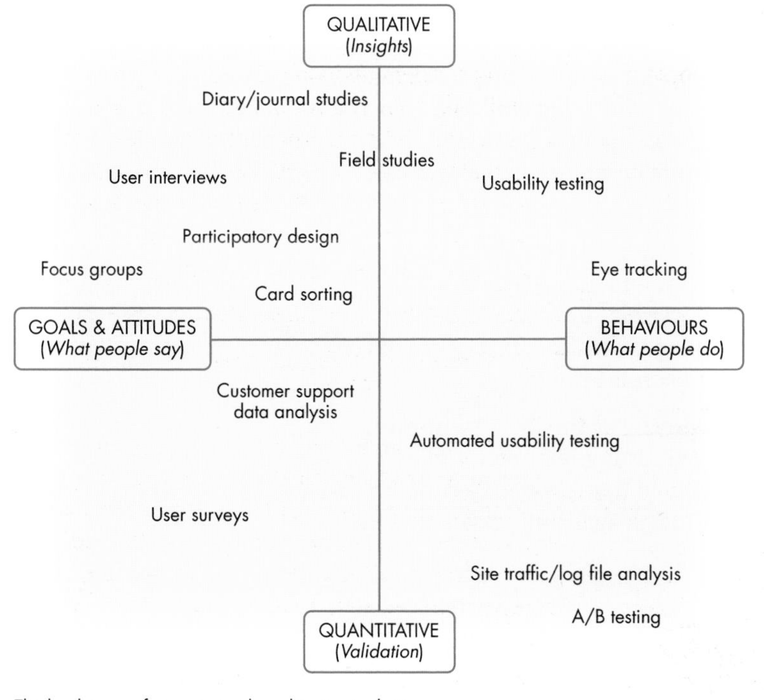

The landscape of user research and testing techniques.

Qualitative and quantitative data

- **Qualitative data** is non-numerical and cannot be ranked in a meaningful way. Qualitative data describes a quality that cannot be counted or measured by an instrument. For example, a brick manufacturer might describe a brick by its colour or its roughness.
- **Quantitative data** is data that is numerical and can be counted or measured. For example, the brick manufacturer might describe bricks by their mass or the number of bricks needed to build a house.

Qualitative data: non-numerical data that describes a quality.

Quantitative data: numerical data.

Discrete and continuous data

Quantitative data can be divided into **discrete** or **continuous** data.

- **Discrete data** is data that is counted, for example, the number of days you walk to college in a week.
- **Continuous data** is data that is measured, for example, the time it takes you to walk to college.

So, if the brick manufacturer describes bricks by their mass, this is an example of continuous data, because the mass can come as any number on a continuous scale. The number needed to build a house, however, is discrete data because the bricks must be provided as a whole number.

Discrete data: data that is counted.

Continuous data: data that is measured.

Discussion

Choose one of the following scenarios and discuss it with your peers. In each scenario, identify the data that you would collect; then explain why and whether this data would be qualitative or quantitative, discrete or continuous.

1. Imagine that you are an animal breeder. You want to focus on breeding animals for people in your area. The data you collect will allow you to concentrate on the animal breeds that most people in your area want, and the care that these people would expect you to provide for the animals.
2. Imagine that you are a software games engineer. You want to focus on designing a game for 16–18 year olds that encourages them to engage in their studies. The data that you collect will allow you to design a game that works for people in your college, with the possibility of the game becoming national and perhaps even international.
3. Imagine that you are in charge of swimming at the local swimming pool. You want to focus on providing swimming lessons for all who use the baths. The data you collect will allow you to draw up a timetable of when the lessons should take place and the ages and stages to which these lessons would apply.

Exercise 1A

1 Look at the human variables that are listed in i–vi below.

a. List the variables that are qualitative and those that are quantitative.

b Identify which of the following quantitative data are discrete or continuous.

i Hair colour

ii Body temperature

iii Number of teeth

iv Religious faith

v Confidence that you will get this question correct

vi Time it takes to travel from home to college

2 **a** Write down two quantitative variables about your favourite band.

b Identify the variables as discrete data or continuous data.

3 Write down two qualitative variables about the government.

4 You are asked to complete an evaluation form to assess a school lesson, by rating it. Choose from: 1 (excellent), 2 (very good), 3 (average), 4 (poor) or 5 (very poor).

Is this data quantitative or qualitative? Why?

5 A botanist finds a skeleton of a newt, as shown.

a Write down three variables that the botanist might record.

b Say whether the variables are qualitative or quantitative.

c Identify the quantitative data as discrete or continuous.

Primary and secondary data

- **Primary data** is data that is collected by a researcher or research company through direct surveys, observations, or interviews with the subjects of the data. Primary data is expensive and time-consuming to obtain, but you know how it was collected – it has been observed or collected directly from first-hand experience.
- **Secondary data** is data that already exists, for example, data from the Office for National Statistics, or from other sources. It can also be data that has been processed in some way, for example, by grouping. Secondary data is cheap and easy to obtain but it might be outdated and from an unknown source. Therefore this type of data could be biased.

Primary data: data that comes directly from first-hand experience.

Secondary data: data that already exists or has been processed.

Mathematics in the real world

Portakabin is a UK-based company that was founded in 1961. It sells and hires out portable and modular buildings. The buildings are used, for example, as offices, nurseries, schools, hospitals, call centres and laboratories. *Portakabin* uses both primary and secondary data to keep up with the way customers' needs for accommodation may change.

Portakabin collects primary data from its customers and staff. The sales team collects data on a regular basis, by talking to customers. The team uses focus groups (groups of people selected to discuss a product before it is launched or to provide feedback on issues) in the working environment to find out what affects workers' performance and productivity. These results are used to develop new products and services. Surveys are used to extract quantitative data.

To improve working conditions, *Portakabin* uses secondary data. They found out from a Gallup survey that 66% of British workers consider that the quality of the working environment is important. It was also reported that noise disturbs 33% of workers, with the result that these employees are four times more likely to become disengaged from working.

Discussion

Choose one of the following scenarios and discuss it with your peers. In each scenario, identify the data you would collect, explaining why and whether this data would be primary, secondary, or both.

1. Imagine that you are an animal breeder and you wish to design and market a new food product for an animal of your choice. Decide if you will concentrate on what you think the person looking after the animal wants in terms of cost, essential food, or luxury food, or if you will concentrate on the needs of the animal and whether the animal will eat the food product.
2. Imagine that you are a software games engineer and you wish to design a new game based on grumpy guinea pigs. Decide if you will concentrate on a particular age range, the quality of the graphics, or the cost of the game to the consumer.
3. Imagine that you are in charge of the local swimming pool and you wish to design a new water feature. Decide if you will concentrate on a particular age range, a certain level of swimming ability, or the cost that will be charged to use the water feature.

Exercise 1B

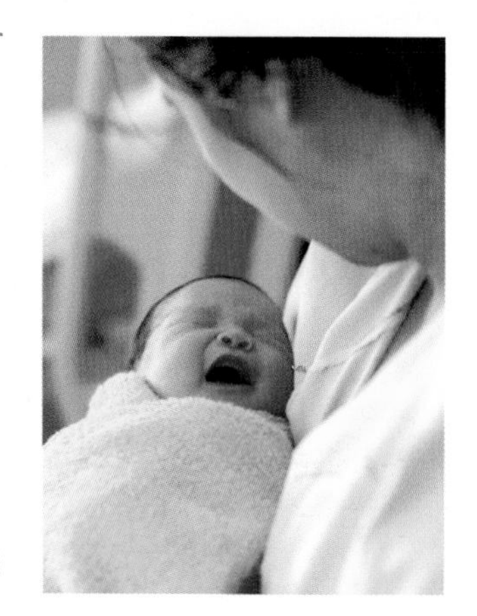

1 The local maternity clinic is due for a revamp and you are involved.

Decide if you would focus on what patients want, or on what staff would like.

Assume that money is not an option, so you want the best available facilities and equipment.

a What secondary data might you use?

b What primary data might you collect?

2 A householder has a smart meter in her home, which she uses to record her use of gas and electricity.

The smart meter is linked to the energy company, which records and prepares the bills.

The householder records the amount of gas and electricity she uses each month.

The energy company sends a bill and a summary of the monthly usage during the year.

a Is the householder collecting primary or secondary data?

b Is the householder receiving primary or secondary data?

3 The police report that car crime has been reduced in a certain area.

Explain how the police could have used primary and secondary data to reach this conclusion.

4 A campaigning group wants to limit the bonuses paid to bankers.

They need to find out how much is being paid by various banks to the different categories of bankers.

Should they use secondary data or collect primary data, or both?

Think about the advantages, disadvantages, and possible bias of each.

5 Sam, a fruit wholesaler, wants to sell gooseberries.

a What primary data might Sam collect?

b What secondary data might Sam use?

Collecting data

Data can be collected by direct observation, interviews, surveys, experiments and testing. Scientific data in Physics and Chemistry is often collected by doing experiments. In Psychology, data is often collected by testing, observation and interviews, though not exclusively. In experiments there are generally two variables: the variable that you can control, which is called the **independent** (or **explanatory**) variable, and the variable that results, which is called the **dependent** (or **response**) variable. In short, the dependent variable is dependent on the independent variable.

For example, if a farmer wanted to find out the effect of various amounts of fertiliser on his crops, then the amount of fertiliser the farmer puts on the field will be the independent variable, because the farmer can control that. The crops that result will be the dependent variable, as this would depend on the amount of fertiliser the farmer applies.

Mathematics in the real world

In 2014, the Office for National Statistics published a report to show that 'the extraction of oil and gas continues to decline'. The chart in the report shows that over the period 2000–2012, there has been a decline in the quantities of oil and gas extraction.

However, if you are interested in the whole picture of oil and gas supply in the UK over time, the chart does not show the whole picture. It does not show estimates that there are up to 1.3 billion tonnes of undiscovered oil resources, and up to 1010 billion cubic metres of undiscovered gas resources available in the UK.

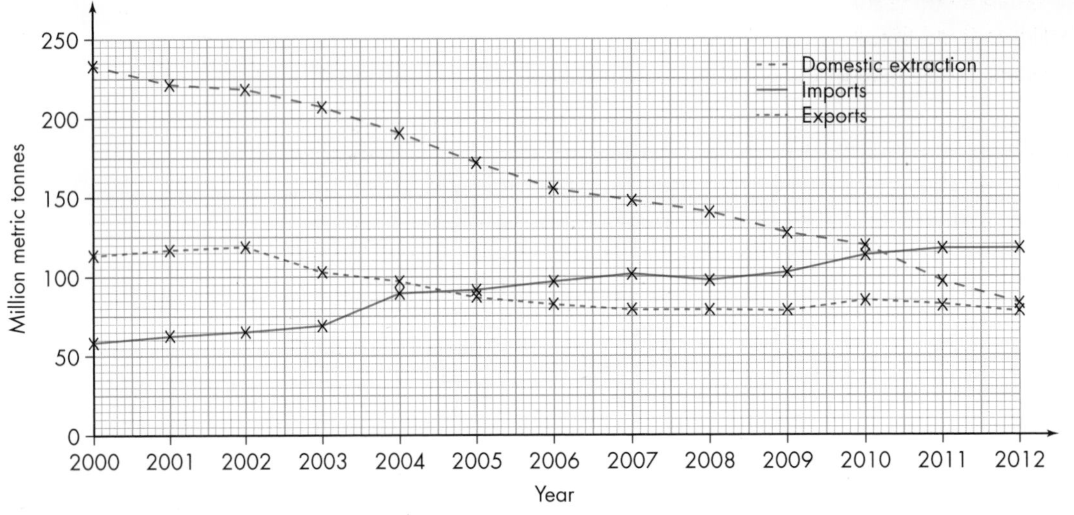

Discussion

Choose one of the following scenarios and discuss it with your peers. In each scenario, talk about how and why you would collect the data you require and what would be the independent and dependent variables.

1. Imagine that you are an animal breeder and you want to design and market a new food product for an animal of your choice.
2. Imagine that you are a software games engineer and you wish to design a new game based on grumpy guinea pigs.
3. Imagine that you are in charge of swimming at the local swimming pool and you wish to design a new water feature.

Exercise 1C

1 A global software company decides to introduce a new operating system.

They need to research whether to sell the operating system for a one-off charge, charge a fixed amount each month, or charge an amount each month based on how much the system is used.

Advise the company on using primary or secondary data and *how* and *why* they should collect each type of data.

2 The intensive care ward at the local maternity clinic thinks that premature babies develop faster if they spend more time with their mothers.

a What data should the intensive care ward collect?

b Why should they collect it?

c How should they collect the data?

d What are the independent and dependent variables?

3 A high street coffee chain decides to move from using spoons to wooden stirrers.

Before they make the change they decide to do some research.

a What data should they collect?

b Why should they collect the data?

c How should they collect the data?

d What are the independent and dependent variables?

4 Think about your possible career (or interests) and how you will use data to examine something that interests you. Decide what information you would collect and whether you would use primary and secondary data (or both). Also, why should you collect the data? Use what you have learnt so far to write a short description of this (about 150 words). Justify the decisions you make.

D2: Collecting and sampling data

Learning objectives

You will learn how to

- Deduce properties of populations from a sample, whilst realising the limitations.
- Appreciate the advantages and disadvantages of various sampling methods.

Introduction

Sampling is used by researchers to collect data from some of the **population**, in order to make conclusions about all of the population.

A population is generally connected with a complete set of people. However, in statistics, a population can also be everything involved in the study, for example: it might be all the microprocessors made by a firm, or to a diamond merchant, all the diamonds from a particular mine. As a rough guide, if there are n items in the population then the sample size should be $\sqrt{n}$. There are various ways to sample data and this chapter will look at the advantages of **random**, **cluster**, **stratified** and **quota** sampling methods taking into account any **bias**.

Sampling: used to collect data from part of a population.

Population: a complete set of items that share a common property.

Mathematics in the real world

In the 2015 General Election a pre-election poll organised by YouGov surveyed 20 000 people to give an indication of the likely outcome of the election. On Election Day, an exit poll was taken at a sample of polling stations (about 100 out of 40 000), with a sample of about 100–200 voters at each polling station. The problem with exit polls is that the voters or polling stations selected might not be typical of the UK population as a whole (and the people asked might not tell the truth). However, exit polls are generally more accurate than a pre-election poll because the people involved have actually voted. In a pre-election poll, the people asked might not be the people who vote. The exit poll will thus produce a very different outcome to those of the pre-election polls.

The table shows the differences between the pre-election poll, the exit poll and the actual results. At the time, the result predicted by the exit poll was so different to all the pre-election polls that many people did not take it seriously, but in the end, the exit poll proved much closer to the actual result than the pre-election poll.

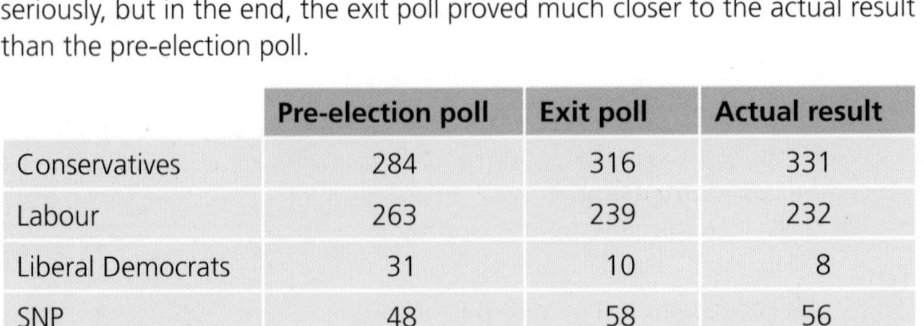

	Pre-election poll	Exit poll	Actual result
Conservatives	284	316	331
Labour	263	239	232
Liberal Democrats	31	10	8
SNP	48	58	56

Bias

In statistics, **bias** means that results are distorted. This can happen in many ways, for example, picking a non-representative sample such as asking all the students who study Mathematics in college where the College Prom should be held. Other ways in which bias could affect the results is by people being untruthful, errors in processing or recording results, poor survey design, and getting no response to a survey.

Bias: a distortion of results.

Discussion

Choose one of the following scenarios and discuss it with your peers. In each scenario, explain how you test the item mentioned – with a census or a sample – and why you chose that method.

1. Imagine that you are an animal breeder. You have been told about a new food product that can help to improve your animals' health.
2. Imagine that you are a software games engineer. You have designed a new game for older teenagers.
3. Imagine that you are in charge of swimming lessons at the local swimming pool. You want to know parents' reactions to changing the times of the lessons for 3-year-olds.

Exercise 1D

1 An estate agency wants to know the following information. Should the agency use a census or a sample?

- **a** The number of rooms in a bungalow in England
- **b** The number of rooms in a bungalow in an adjacent street
- **c** The number of people who offer less than the asking price of a house
- **d** How long an advert should appear on a website

2 The Government wants to close the bars in the House of Commons.
What population will be affected?

3 A community college wants to close all its facilities on a Sunday.
What population will be affected?

4 The group leader of the village toddler group (ages 2–3) wants to plan an end-of-year excursion.

- **a** Should the group leader use a census or a sample to decide where they should go?
- **b** What population will be affected?

Sampling methods

- **Random sampling** is used to give every item in the population (the sampling frame) the same chance of being chosen. Each item is given a number, then numbers are picked at random, using either random number tables or a computer to identify the items to be measured or surveyed.
- **Cluster sampling** is often used in market research. The population is divided into clusters (groups), then a sample of clusters is selected using random sampling, and all items in those clusters are surveyed.

Random sample: used to collect data from part of a population without bias.

Cluster sample: used to collect data from all members of a randomly selected cluster (or group).

- **Stratified sampling** is used to ensure that each stratum (or layer) is properly represented in the sample. The population is divided into strata (layers) then a random sample from each stratum is selected using random sampling and all selected items are surveyed.
  For example, in a college of 150 males and 250 females, a stratified sample of 80 students is to be selected to participate in a survey about transport. Since male students represent 150 of the college population of 400, then $\frac{150}{400}$ of the sample of 80 should be male. $\frac{150}{400} \times 80 = 30$ So, 30 males must be selected randomly. Therefore 50 female students will also be selected randomly.
- **Quota sampling** is also used in market research. The person doing the surveying is told the quota (amount) of items from each section to be surveyed, and is then free to select those items as she or he wishes.

Stratified sample: used to collect random data from a stratum (or layer) where the number of items selected is proportional to the size of the stratum in the population.

Quota sample: used to collect used to collect data chosen by the sampler from a stratum (or layer) where the number of items selected is proportional to the size of the stratum in the population.

Census or sample?

A **census** is used when every member of a population provides data. The UK National Census is taken every 10 years. It is used to collect data from every household so that plans can be made about many things such as the number of schools, colleges and teachers needed, or how much the NHS might have to spend on the care of elderly people. A large census like this is very expensive, so a **sample** is often used to save both time and money. Conclusions can be made from the sample about the whole population from which it came. The larger the sample the more accurate it is likely to be, but increasing the sample size costs extra time and money.

Census: used when every member of the population provides data.

Sample: part of a population.

Mathematics in the real world

- **Random sampling** is used in the National Lottery when six balls are selected at random from 49.
- **Cluster sampling** has become a popular method to perform immunisation coverage surveys. This method has proved to be very suitable in dispersed rural populations in Kenya.

  The names of children aged 5–14, who attended school were known, and 30 clusters were needed.

  The total school population was divided by 30 to determine the sampling interval, a. Then a random number, n, was chosen and the school of the nth child in alphabetical order was the first cluster to be surveyed. The school of the $(n + a)$th child in alphabetical order was the second cluster to be surveyed. The school of the $(n + 2a)$th child in alphabetical order was the next cluster to be surveyed, and so on, until 30 clusters (in this case, primary schools) were surveyed. If the same school is selected twice, then the school of the next child (the $(n + ra + 1)$ th child) on the list is substituted in its place.
- **Stratified sampling** is used for a political survey. If the survey needs to reflect the diversity of the population, the researcher would want to include participants of various groups such as race or religion, based on their proportionality to the total population. A stratified survey could therefore claim to be more representative of the population than a survey done using other sampling methods.
- **Quota sampling** is something you may have experienced when people who are doing a market survey in the street stop you and ask if you would answer some questions about a certain product. The marketing

people will have been told that they need to ask a certain number of people aged, for example, 16–20, 21–30, 31–40. Thus, they are stratifying the population, but rather than selecting the people in each group randomly, they are free to choose who they will question.

Advantages and disadvantages of various types of sampling

Type	Conditions of use	Advantages	Disadvantages
Random	Population members are similar to each other	Free from bias	Tedious and time-consuming
Cluster	Population consists of units rather than individuals (for example, the types of trees in parks in different areas of the UK)	Cheaper than random sampling, as it can reduce travel and admin costs; can show regional variation	Not a genuine random sample; can be subject to bias if only a few clusters are used
Stratified	Population members are similar to each other but contain several easily identifiable groups (such as gender and religion)	More accurate than simple random sampling, as a fair proportion of responses from each stratum is obtained; can show different tendencies in each stratum	Tedious and time- consuming
Quota	When there are several easily identifiable groups (such as gender and religion) and stratified sampling is not possible	Simple to take	Not genuinely random, since the surveyor picks the items and so is likely to be biased

Discussion

Choose one of the following scenarios and discuss it with your peers.

1. Imagine that you are an animal breeder and you wish to design and market a new food for an animal of your choice. What kind of sampling method would you use to find out what food the animal owners would buy? Why? Would you use the same sampling method to find what food the animals would eat? Why? In each case say what the sampling frame is.

2. Imagine that you are a software games engineer and you wish to design a new game based on grumpy guinea pigs. What kind of sampling method would you use to find out what features the gamers would buy? Why? What sampling frame would you use? Why?

3. Imagine that you are in charge of swimming at the local pool and you wish to design a new water feature. What kind of sampling method would you use to find out what features the bathers would use? Why? What sampling frame would you use? Why?

Time and money

There is always a fine balance to be found between removing bias and increasing sample size. If a sample is too small then it could easily be biased, as it may not be representative of the population. However, if the sample is large, this increases the cost of the survey, which may then become costly to conduct. Therefore, there is a need to balance the accuracy of the survey with the cost. At the start of this section it was mentioned that if there are n items in the population then the sample size should be $\sqrt{n}$.

Exercise 1E

In each of the following questions, estimate the size of the population and then suggest a suitable sample size. At this point in the course the estimate of the population size is not important. You will learn more about this kind of estimation in Chapter 3. What matters is that the sample size should be suitable for the population estimated.

1 The local maternity clinic is due for a revamp.

The registrar decides to give a survey to all the people who enter the clinic on Friday to find out their opinion on what catering facilities should be provided.

Explain why the sample is biased and suggest a better way of sampling.

2 A householder has a smart meter to record the use of gas and electricity.

This is linked to the energy company so that the bills can be prepared.

The householder wishes to estimate the use of his gas and electricity over a year to help with budgeting.

The householder records the amount of gas and electricity used in March and uses this to work out an estimate of the gas and electricity used throughout the year.

Explain why the sample is biased and suggest a better way of sampling.

3 The police want to know if car crime has been reduced in County Durham.

a What is the population they should use?

b What would be a suitable method of sampling?

c Why would that be suitable?

4 A campaigning group wants to limit the bonuses paid to bankers.

They decide that they need to know the amounts that bankers are paid, what non-bankers think of certain amounts of bonuses, and if it would be better to pay bankers a higher salary to avoid confrontation in future.

a What would be a suitable method of sampling to find out the information they want?

b Why would the method be suitable?

5 A biologist has an interest in oak galls (a species of small wasps).

Oak galls are in small spheres formed on oak trees; when the grubs inside turn into tiny wasps they burrow out by making small holes.

To investigate the number of oak galls that come from oak trees, the biologist collects all the galls that have fallen into his garden from a nearby oak tree.

a What is the population?

b Why might the sample be biased?

c Explain a better sampling method.

6 Look at your answer to Exercise 1C question **4**, about your possible career (or interests) and how you will be using data to examine something that interests you. Think about how you would collect the data you described. What possible sampling methods might you use? How could you avoid bias? Write a short description of about 150 words, justifying the decisions you make.

D3: Representing data numerically

Learning objectives

You will learn how to:

- Calculate measures of location (mean, median, mode, quartiles and percentiles) and spread (range, interquartile range and standard deviation).
- Interpret these measures and use them to make conclusions.

Introduction

Most of these measures (**mean, median, mode** and **quartiles, range** and **interquartile range**) you will have met before in your GCSE course. This section will focus on **percentiles** and **standard deviation,** although there will be some questions to remind you of what you have met before.

Here is a summary of the measures you should have met before.

Measure	Advantages	Disadvantages	Example
Mean	Probably the most used average; uses every item of data	Might not be representative if there is an extreme value	1, 50, 52, 56, 56, 56, 58 Mean = $\frac{1+50+52+56+56+56+58}{7} = 47$ This is a lower value than most of the data.
Median	Only looks at the middle value, so not affected by extreme values	Can be misleading because it does not consider all the values	1, 1, 1, 4, 56, 56, 58 Median is the 4th value; median = 4 Here the numbers below the median are close together, but the others and a long way above the median.
Mode	Easy to find: no calculation needed; the only average that can be used with qualitative data	Can be misleading; inappropriate if the data contains few duplicates	1, 1, 1, 4, 56, 56, 58 Mode is 1 This is not representative of the data as a whole.
Quartiles	They divide the data into four equal groups	Not easy to calculate if the data set is small	Often referred to as the **lower quartile**, **median** and **upper quartile** 0, 3, 7, 10, 12, 17 so the median lies between 7 and 10, that is, 8.5. The lower quartile (LQ) divides the data below the median (0, 3, 7) into two groups, and is therefore 3. The upper quartile (UQ) divides the data below the median (10, 12, 17) into two groups and is therefore 12.
Range	It measures the spread of the data	It only uses the two extreme values, which may not be representative of how the data is spread	1, 51, 51, 54, 56, 56, 98 The range is 98 – 1 = 97, but without the end values the range would be only 5.

Interquartile range (IQR)	Measures the range of the middle 50% of the data, so is not influenced by extreme values	Half the data plays no part in this measure of spread	0, 3, 7, 10, 12, 17, 20 LQ is the $\frac{7+1}{4}$ = 2nd value: LQ = 3 UQ is the $\frac{3(7+1)}{4}$ = 6th value: LQ = 17 IQR = UQ – LQ = 17 – 3 = 14

Measures of location are sometimes referred to as **measures of position** and **measures of spread** as **measures of dispersion**.

- **Percentiles** divide data into 100 equal parts, just as quartiles divide data into four equal parts (quarters). Therefore the lower quartile Q_1 is also the 25th percentile P_{25}, the median or Q_2 is the 50th percentile P_{50} and the upper quartile Q_3 is the 75th percentile, P_{75}. Sometimes the range between the 10th and 90th percentile is used, $P_{90} - P_{10}$, as the measure of spread because it uses more data and still avoids extreme values.

Percentile: a value below which a given percentage of the observations fall.

Mathematics in the real world

The World Health Organization publishes charts so that parents can see how their child is growing compared to other children. This also allows health workers to judge whether a child is progressing well or needs help.

Here is part of the chart for girls aged 0–24 months.

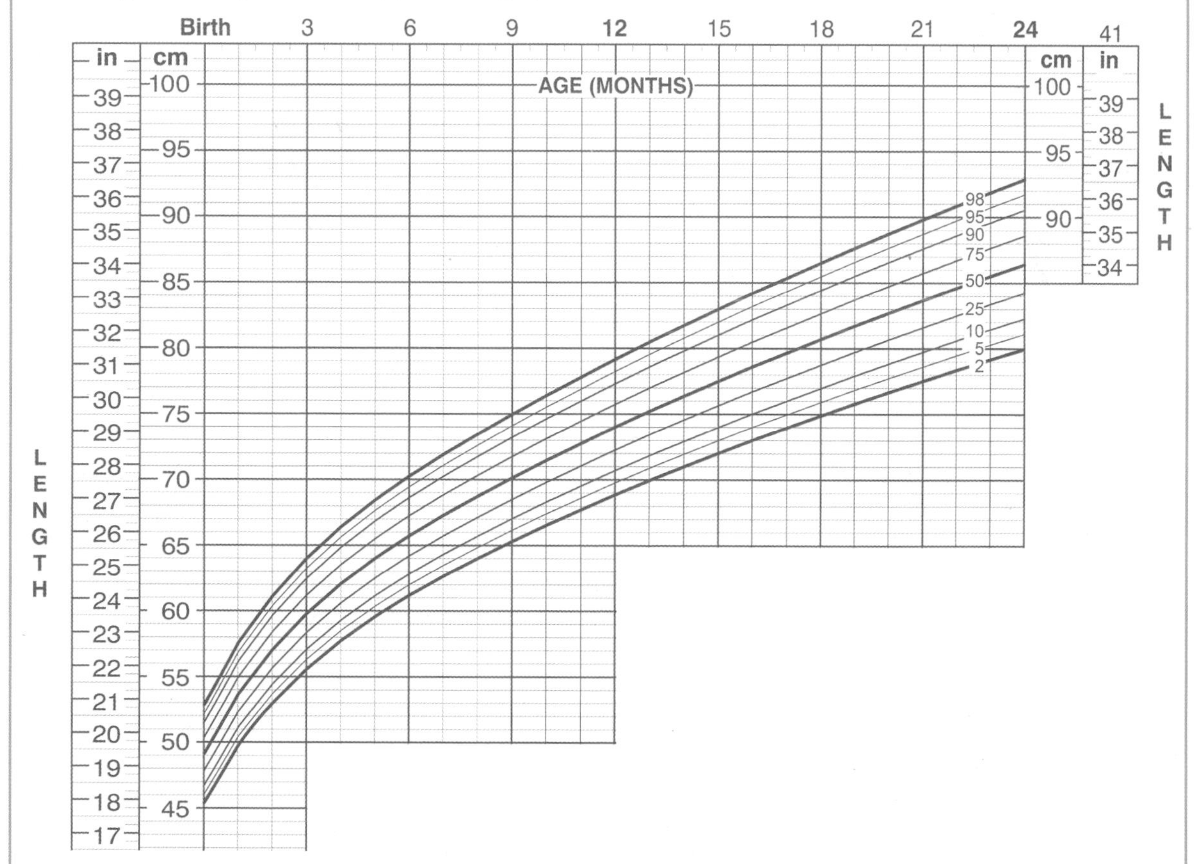

The numbers on the curves (98, 95, 90, 75, 50, 25, 10, 5, and 2) refer to P_{98}, P_{95}, P_{90}, P_{75}, P_{50}, P_{25}, P_{10}, P_5 and P_2. These are strategic percentiles, as they correspond to strategic points in a population that will become more apparent when dealing with standard deviation.

Discussion

Choose some of the following questions and discuss them with your peers.

1. Look at the chart. What information do you think it shows?
2. In which month is the length of the baby increasing most? How do you know without looking at the numbers on the scales?
3. Would you be worried if a 12-month-old girl was 71 cm long?

Exercise 1F

Use the chart above to help you to answer these questions.

1 Find the median length in centimetres of the following.

a A 6-month-old girl

b A 19-month-old girl

2 Work out the interquartile range in inches of a 15-month-old girl.

3 Work out P_{90} for a 17-month-old girl.

4 What percentage of 12-month-old girls would you expect to be longer than 79 cm?

5 What percentage of 17-month-old girls would you expect to be shorter than 77 cm?

6 The length of a baby girl is 70 cm. Between what ages might she be?

7 Work out $P_{90} - P_{10}$ for a two-year-old girl.

8 Seventy-five per cent of what age of baby girls is above 72 cm?

Standard deviation

Standard deviation: a measure of spread that uses all the data.

- **Standard deviation** is a measure of the spread of data using every item of data. It is especially useful when comparing values from different sets of data and when analysing the position of an item of data in a **normal** population. An advantage of standard deviation is that it uses all the data. A disadvantage is that it takes longer to calculate than the range or interquartile range if you do not use a calculator.

The following graph shows an example of standard deviation being used to compare two tests, one in Mathematics and one in Physics. The mean mark is the same in each case: 50%. The standard deviation in the Mathematics test is 5% and the standard deviation in the Physics test is 10%. This means that the marks in the Physics test are more spread out than the marks in the Mathematics test.

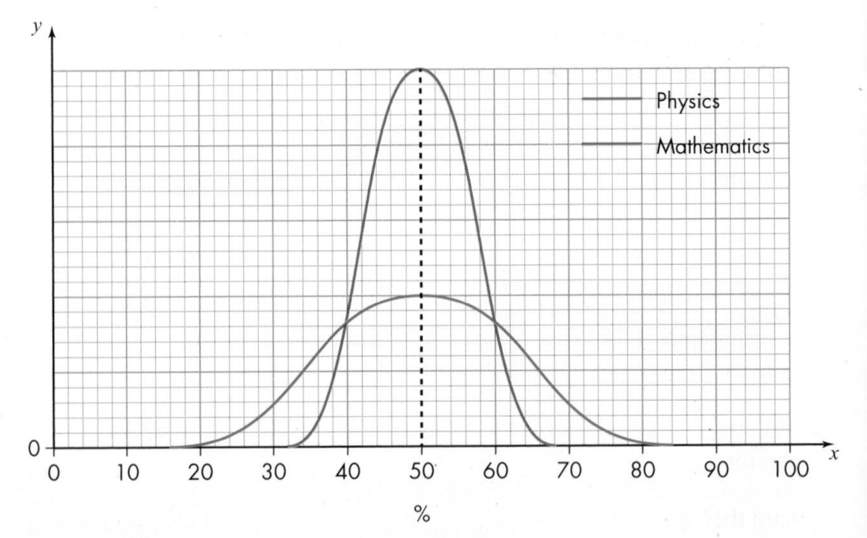

Calculating standard deviation

The standard deviation is sometimes called the 'root mean square deviation from the mean' because that is how it is calculated!

The reason why you square, then square root, is because if you added up the differences from the mean, that part would come to zero as some would be negative and some positive. Squaring a negative number makes it positive, so you square the differences from the mean before you add them up and then divide by the number of items. After you have done that you take the square root because squaring and square rooting are inverses of each other.

For a set of data which is expressed as $x_1, x_2, \ldots x_n$ with a mean $\bar{x}$ the standard deviation σ is

$$\sqrt{\frac{\sum(x-\bar{x})^2}{n}}$$

Beyond the spec

This looks like a very complicated formula, so here is a step by step method to show how it works with a set of data {2, 3, 6, 9, 15}.

Step	Algebraic notation	Example
1 Find the mean.	$\bar{x}$	$\frac{2+3+6+9+15}{5} = 7$
2 Find the deviations from the mean.	$x-\bar{x}$	2 – 7, 3 – 7, 6 – 7, 9 – 7, 15 – 7 that is –5, –4, –1, 2, 8
3 Square these deviations.	$(x-\bar{x})^2$	25, 16, 1, 4, 64
4 Find the mean of these squared deviations.	$\frac{\sum(x-\bar{x})^2}{n}$	$\frac{25+16+1+4+64}{5} = \frac{110}{5} = 22$
5 Find the square root of this result.	$\sqrt{\frac{\sum(x-\bar{x})^2}{n}}$	$\sqrt{22} = 4.69\ldots$

There is an easier formula to use which involve doing fewer subtractions.

$$\sigma = \sqrt{\frac{\sum x^2}{n} - \bar{x}^2}$$

Again, this looks like a very complicated formula, so here is a step-by-step method to show how it works with a set of data {2, 3, 6, 9, 15}.

Step	Algebraic notation	Example
1 Find the mean.	$\bar{x}$	$\frac{2+3+6+9+15}{5} = 7$
2 Square each item of data and sum them.	$\sum x^2$	4 + 9 + 36 + 81 + 225 = 355
3 Find the mean of the sum of the squares.	$\frac{\sum x^2}{n}$	$\frac{355}{5} = 71$
4 Subtract the square of the mean.	$\frac{\sum x^2}{n} - \bar{x}^2$	71 – 49 = 22
5 Find the square root of this result.	$\sqrt{\frac{\sum x^2}{n} - \bar{x}^2}$	$\sqrt{22} = 4.69\ldots$

For discrete frequency distributions the formulae are:

$$\sigma = \sqrt{\frac{\sum f(x-\bar{x})^2}{\sum f}} \text{ or } \sigma = \sqrt{\frac{\sum fx^2}{\sum f} - \bar{x}^2}.$$

$\sum f$, the sum of the frequencies, is used instead of n.

For grouped frequency distributions you use the mid-point of the group for x.

In the real world, no one works out standard deviation by hand. Everyone uses a calculator or computer. If you use a calculator, read the manual carefully to see how to enter the data efficiently. There is no difference in the marks awarded in the examination if you use the statistical buttons on a calculator or work it out step by step.

Mathematics in the real world

Intelligence quotient (IQ) scores are based around a normal distribution of intelligence where the mean score is 100 (this is what a person with average intelligence would expect to achieve - it is just a convenient number on a scale) and the standard deviation is 15 (again, just a convenient number with which to work).

About 68% of the population has an IQ within one standard deviation, either side of the mean, that is, between 85 and 115; around 95% have an IQ within two standard deviations (70 to 130) and about 99.7% within three standard deviations (55 to 145). If IQ had been set up with a mean score of 10 and a standard deviation of 3, then about 68% of the population would still have an IQ within one standard deviation, either side of the mean, that is, between 7 and 13; around 95% have an IQ within two standard deviations (4 to 16) and about 99.7% within three standard deviations (1 to 19).

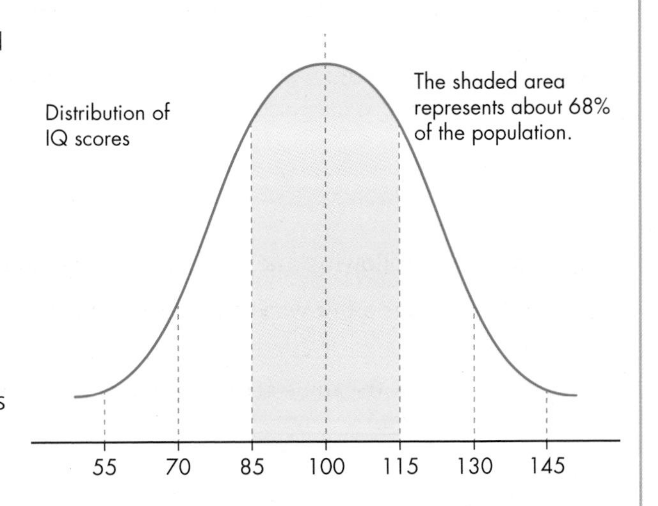

Beyond the spec

Use of standard deviation in standardising scores

Standard scores (or **z-scores**) are often used to compare data from different samples where the means and standard deviations are known.

Here is some real data showing one year's results from the Biomedical Admissions Test given to applicants who want to study medicine and veterinary courses at Cambridge University.

Gender	Mean	Standard deviation
Female	5.03	0.60
Male	5.23	0.75

Imagine that as an admissions tutor you have only one place and two candidates: a male who scored 6.43 on the test and a female who scored 6.23 on the test. Which of these represents the better performance?

To compare the scores, calculate how many standard deviations each score is from the mean.

The woman's score is $\frac{6.23-5.03}{0.60}=\frac{1.2}{0.60}=2$ standard deviations above the mean.

The man's score is $\frac{6.43-5.23}{0.75}=\frac{1.2}{0.75}=1.6$ standard deviations above the mean.

Therefore the woman's performance is better than the man's performance because it is more standard deviations away from the mean.

This means that among the women, the score of 6.25 represented a better performance in that group, than a score of 6.43 by the man in his group.

This sort of standardising can be used in situations where it is known that one group is known to generally score higher than another group. Say, for example, that a university or college wants to have similar numbers of boys and girls on a particular course. They know that girls generally achieve higher scores in a particular test, therefore they can standardise the scores to achieve the required result.

Discussion

Choose some of the following questions and discuss them with your peers.

1. Do you think this is a fair way of comparing performances? If so, why? If not, why not?
2. What do you think the male score should be to be comparable with the female score, above?
3. What do you think the female score should be to be comparable with the male score, above?
4. What can you say about a standardised score if it is above average?

Exercise 1G

In all these questions you can use the statistical buttons on a calculator (no extra credit is given for using a step-by-step method). If your calculator uses buttons with σ_n and σ_{n-1}, the preferred use is the σ_{n-1} button, but there are no penalties for using the σ_n button.

1 Calculate the mean and standard deviation of the following.

Check your results using a calculator.

a 3, 5, 6, 9, 11, 12, 17

b 30, 50, 60, 90, 110, 120, 170

c 9, 11, 12, 15, 17, 18, 23

2 Here are the lengths of the stirrers from Exercise 1C, question 3.

Length in millimetres	11.0	14.0	17.8	19.0
Frequency	2	52	14	13

a Calculate the mean and standard deviation length of stirrer.

b What can you say about the mean length compared with the median and modal length?

3 Based on the information in this bar chart of retweet lengths, calculate the mean and standard deviation of a length of retweet.

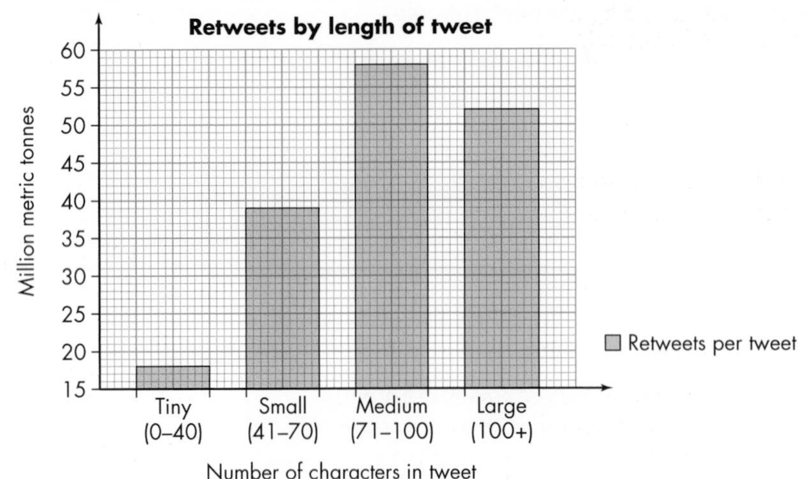

Data source: Retweets by Length of Tweet, Track Social.

Beyond the spec

4 Use the formulae to calculate the mean and standard deviation for the following data.

a $\frac{\sum x^2}{n} = 69, \bar{x}^2 = 44$

b $\frac{\sum (x-\bar{x})^2}{n} = 16, n = 8, \sum x = 72$

c $n = 10, \sum x^2 = 612.5, \sum x = 35$

5 Find the better result in each of the following. Use standardised scores to decide.

a 56 where $\bar{x} = 50$ and $\sigma = 3$ or 45 where $\bar{x} = 50$ and $\sigma = 3$.

b 23 where $\bar{x} = 30$ and $\sigma = 4$ or 57 where $\bar{x} = 65$ and $\sigma = 3.7$.

c 4.5 where $\bar{x} = 3.9$ and $\sigma = 0.45$ or 3.9 where $\bar{x} = 2.1$ and $\sigma = 1.2$.

---

6 Look at your answers to Exercise 1C, question **4** and Exercise 1E, question **6** about your possible career (or interests) and how you will use and collect data to examine something that interests you. Think about how you would analyse the data you described. What possible statistical measures might you calculate? Why would you use them? Write a short description of about 150 words, justifying the decisions you make.

D4: Representing data diagrammatically

Learning objective

You will learn how to:

- Construct and interpret diagrams for grouped discrete data and continuous data, know their appropriate use and reach conclusions based on these diagrams.

Introduction

There are many ways to represent data in diagrams and you will have had considerable experience in drawing and interpreting them in your GCSE course. The diagrams you will concentrate on here (**histograms**, **cumulative frequency graphs**, **box-and-whisker plots** and **stem-and-leaf diagrams**) will develop those skills.

Here is a resume of the diagrams with which you will be familiar, and their distinguishing features.

Diagram	Advantages	Disadvantages	Example
Pictogram	Data is easily counted and represented using symbols; often used with qualitative data	Difficult to draw and hard to represent fractions of symbols	Number of countries participating in the International Mathematical Olympiads 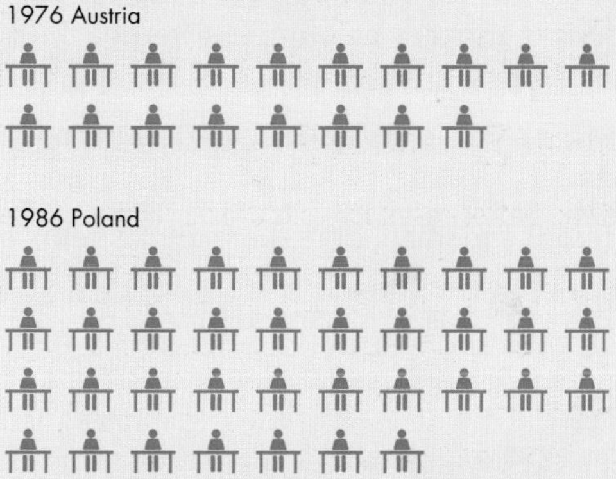 1976 Austria 1986 Poland
Bar chart (including multiple and composite charts)	Good for comparing data in different categories; easily understood	Shows the number of items, but not their actual values	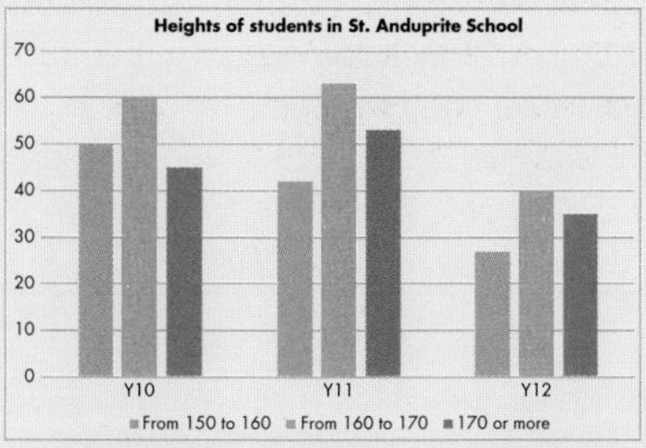 Heights of students in St. Anduprite School Y10 Y11 Y12 From 150 to 160, From 160 to 170, 170 or more
Pie charts (including proportional charts)	Good visual representation of the proportions; often used with qualitative data	Individual data is lost; not really suitable if there are many categories	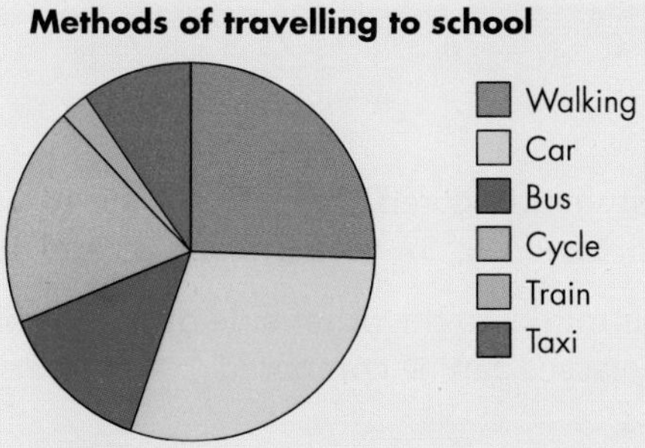 Methods of travelling to school Walking, Car, Bus, Cycle, Train, Taxi

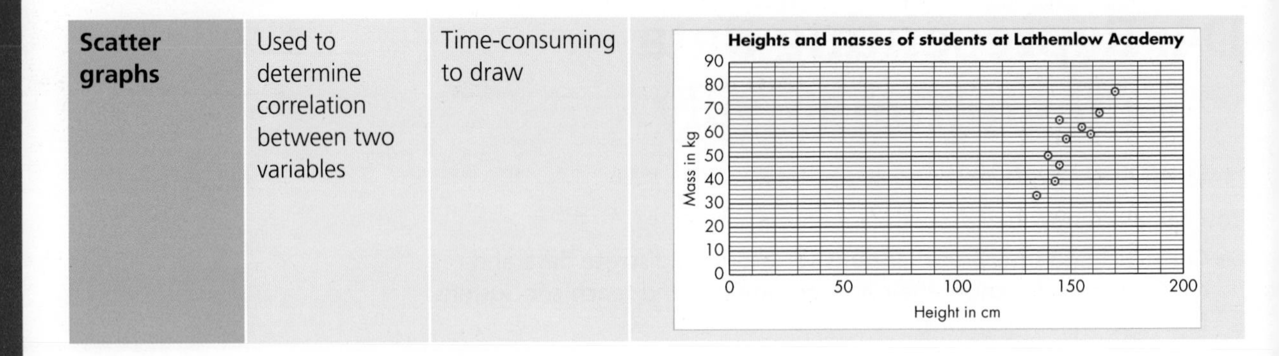

Scatter graphs	Used to determine correlation between two variables	Time-consuming to draw	

Stem-and-leaf diagrams

These are simple diagrams that are good for comparing small amounts of data. The data is not lost.

They are not really appropriate if there is a lot of data, as the appearance can be off-putting.

Example 1

Two classes sit the same test that is marked out of 50.

Class A students score 42, 36, 27, 27, 13, 41, 48, 17, 29, 45, 17, 10, 30, 22, 8, 38, 41, 34, 28, 26, 46, 18, 32, 17, 24, 38, 16, 21 and 12 marks.

Put the scores into a stem-and-leaf diagram by writing the stems (the digits in the tens column) in a vertical line, then adding the leaves (the units digits) in order alongside the appropriate stem.

Notice that a key is always included, showing how the chart represents the data.

Then find the median: here there are 29 items of data, so the median is the 15th item. Counting from the largest item downwards (or the smallest item upwards) this is 27 (circled in red).

The median divides the data into two groups of 14 items each, so the quartiles will divide each of these groups equally and be found halfway between the seventh and eighth values, from the top, and halfway between the seventh and eighth values from the bottom, that is, the lower quartile is 17 (halfway between two 17s) and the upper quartile is 38 (halfway between the two 38s).

Class A

Key 3 | 6 = 36 marks

4	1	1	2	5	6	8		
3	0	2	4	6	8	8		
2	1	2	4	6	7	(7)	8	9
1	0	2	3	6	7	7	7	8
0	8							

Class B students scored 21, 36, 45, 38, 15, 46, 29, 23, 35, 35, 19, 13, 40, 15, 24, 27, 41, 35, 14, 45, 37, 37, 28, 16, 42, 26, 33 and 46.

Add their marks to the other side of the stem-and-leaf diagram, so that the two classes can be compared.

Marks in a test

Class B		Class A
6 6 5 5 2 1 0	4	1 1 2 5 6 8
8 7 7 6 5 5 5 3	3	0 2 4 6 8 8
9 8 7 6 4 3 1	2	1 2 4 6 7 7 8 9
9 6 5 5 4 3	1	0 2 3 6 7 7 7 8
	0	8

Key 5 | 3 | 6 means 35 marks in Class B and 36 marks in Class A

Box-and-whisker plots

Box-and-whisker plots make it fairly easy to find quartiles and interquartile ranges. They are good for comparing two data sets and summarising data. However, some work must be done on the data before drawing the plot diagram, so drawing a stem-and-leaf diagram first will help a lot.

In box-and-whisker plot diagrams, the individual data is lost and the mean and mode are not identifiable. They are probably best drawn on squared paper.

Example 2

Use the same data as we used in the stem-and-leaf diagram above.

From the stem-and-leaf diagram, the data is summarised below:

	Minimum mark	Lower quartile	Median	Upper quartile	Maximum mark
Class A	8	17	27	38	48
Class B	13	21.5	34	39	46

The 'box' is a rectangle that is drawn from the lower to the upper quartile; the line inside it represents the median.

Draw the 'whiskers' at the minimum and maximum values, joined to the box, as shown.

The two box-and-whisker plot diagrams are shown above each other, so that you can compare the two classes easily. The plot diagrams can be drawn horizontally like this, or turned through 90°.

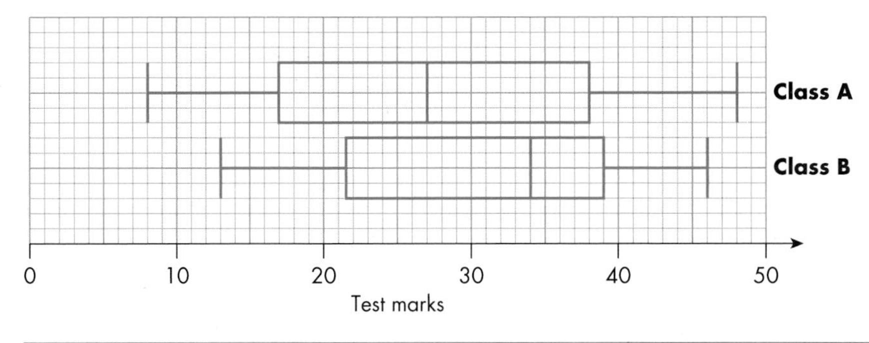

Discussion

Compare the two classes. Think about comparing the measures of spread and the measures of location. Which class do you think did better? Why?

Cumulative frequency graphs

These are used when grouped data is given, so the individual data have been lost.

It is easy to find percentiles and quartiles using cumulative frequency diagrams.

Example 3

Here is a grouped frequency table showing the time spent on mobile phone calls.

It shows, for example, that 28 calls lasted between 10 and 20 minutes.

Time, t, in minutes	$0 < t \leq 5$	$5 < t \leq 10$	$10 < t \leq 20$	$20 < t \leq 30$	$30 < t \leq 60$
Frequency	8	15	28	20	9

Now work out the cumulative frequencies by adding up the frequencies as you go along.

Time, t, in minutes	$0 < t \leq 5$	$0 < t \leq 10$	$0 < t \leq 20$	$0 < t \leq 30$	$0 < t \leq 60$
Cumulative frequency	8	23	51	71	80

Notice that the time intervals have also changed: 51 calls lasted less than or equal to 20 minutes.

Now plot these cumulative frequencies on graph paper, making sure that you plot the cumulative frequencies at the end of the interval, that is, at (5, 8), (10, 23), (20, 51), (30, 71) and (60, 80).

You can also plot the point (0, 0), as no calls lasted 0 minutes or less.

Join the points with a smooth curve (not straight line segments) to create the graph shown here.

Find the median by drawing a line from 40 (half of 80) on the cumulative frequency axis to the graph, then down to the time axis: here you can see that the median is 16 minutes.

Read the upper quartile (24 minutes) and lower quartile (9 minutes) in a similar manner.

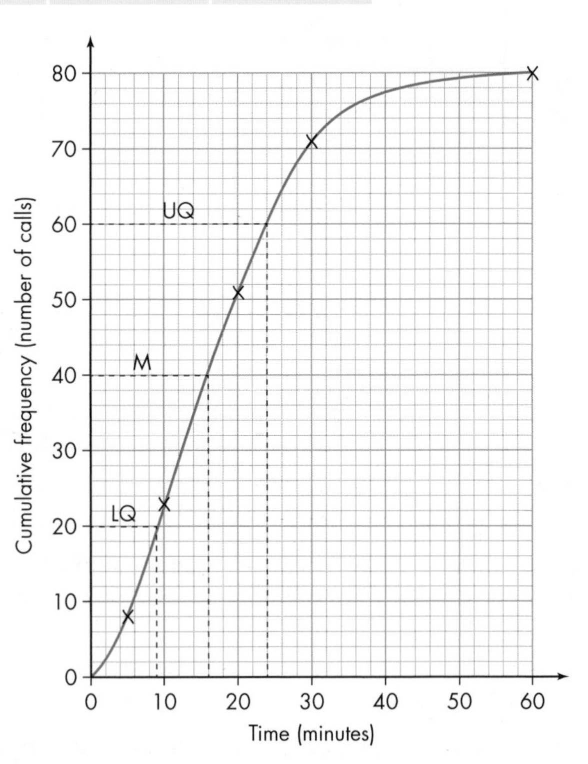

Histograms

Histograms are used especially when grouped continuous data is given, more so when the group intervals are not all the same. The difference between bar charts and histograms is that with bar charts, the <u>height</u> of the column is proportional to the frequency of that interval; with histograms, it is the <u>area</u> of the column that is proportional to the frequency. This means that the vertical axis should always be labelled with <u>frequency density</u>, not just frequency.

Frequency density: the number of items in a given unit.

Example 4

Here is a grouped frequency table showing the time spent on mobile phone calls that you used for the cumulative frequency graph.

Time, t, in minutes	$0 < t \leqslant 5$	$5 < t \leqslant 10$	$10 < t \leqslant 20$	$20 < t \leqslant 30$	$30 < t \leqslant 60$
Frequency	8	15	28	20	9

Work out the frequency densities (in this case the number of calls per minute) by dividing each frequency by the width of its interval.

Time, t, in minutes	$0 < t \leqslant 5$	$5 < t \leqslant 10$	$10 < t \leqslant 20$	$20 < t \leqslant 30$	$30 < t \leqslant 60$
Frequency	8	15	28	20	9
Frequency density	$8 \div 5 = 1.6$	$15 \div 5 = 3$	$28 \div 10 = 2.8$	$20 \div 10 = 2$	$9 \div 30 = 0.3$

Finally, put this information into the histogram, as shown.

It is useful (but not essential) to shade a square stating the frequency it represents.

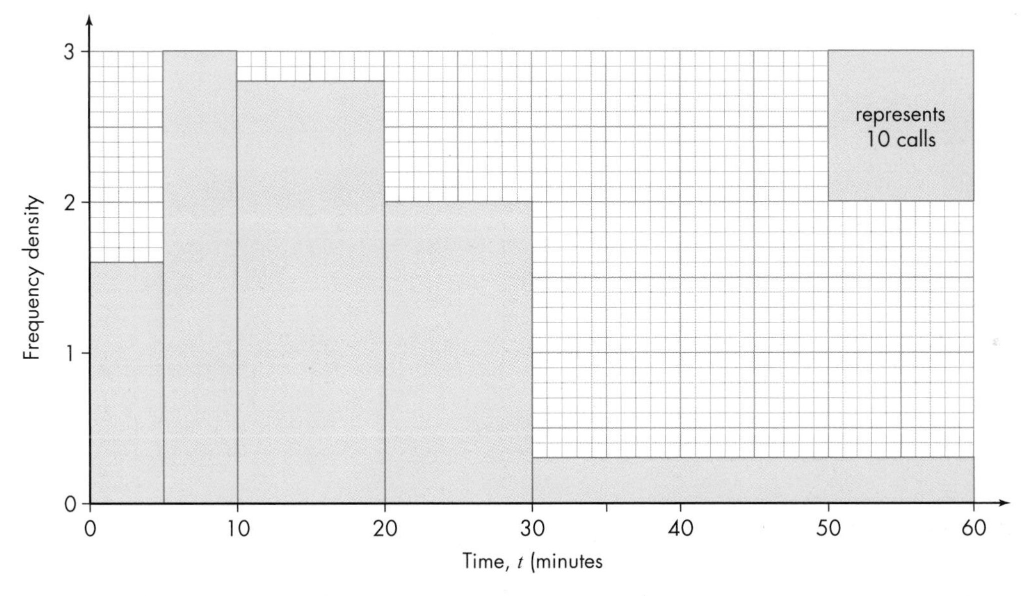

Notice that if you multiply the height of the column by its width, you obtain the number of calls (the frequency) of that interval.

Histograms are used because they do not distort the visual impression given by a bar chart when there are unequal class intervals.

Exercise 1H

1 Here are the scores of Sam and Ella after throwing darts at a dartboard.

Sam scored 27, 54, 16, 1, 39, 5, 60, 25, 8, 40 and 20.

Ella scored 26, 12, 51, 20, 50, 19, 48, 57, 30, 24, 21 and 15.

a Draw a back to back stem-and-leaf diagram showing Sam and Ella's scores.

b Draw box-and-whisker plot diagrams showing their scores.

c Write a short report to compare their scores, using your diagrams to help you.

2 Elsa and Christof find a snowdrift and decide to make snowballs for Olaf.

Here is a grouped frequency table showing the mass in grams of each snowball.

Mass, m, in grams	$40 < m \leqslant 50$	$50 < m \leqslant 60$	$60 < m \leqslant 80$	$80 < m \leqslant 100$	$100 < m \leqslant 105$
Frequency	9	30	38	32	11

a Draw a cumulative frequency graph and use it to estimate:

i the median mass of a snowball

ii the upper and lower quartiles masses of the snowballs.

b Construct a frequency density table and use it to draw a histogram displaying the masses.

3 Here is a scatter graph of the actual age and reading age of a group of sixth formers. Describe the distribution, then use the scatter graph to answer the following questions.

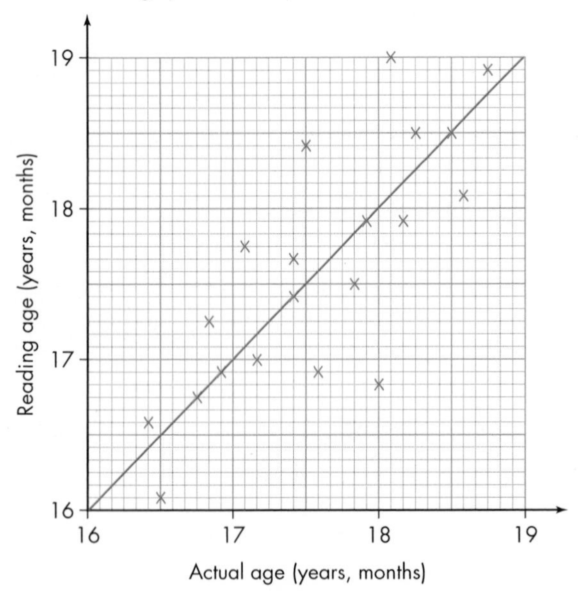

a What is the reading age of the student who is 17 years and 10 months?

b What is the range of reading ages for the students?

c What is the greatest difference between the reading age and actual age of a student?

d What is the median reading age?

e What percentage of students has the same actual and reading age?

4 This diagram shows the percentage test marks of two student groups for the same test.

Use it to determine which of the following statements are true.

a The median is the same for both classes.

b Q_1 for Class Y is 20 marks less than Q_1 for Class X.

c The interquartile range for Class Y is twice the interquartile range of Class X.

d $P_{75} - P_{50}$ is the same for both classes.

e $P_{25} - P_0$ is the same for both classes.

f The standard deviation for Class Y is less than the standard deviation for Class X.

g The mean mark is the same for both groups.

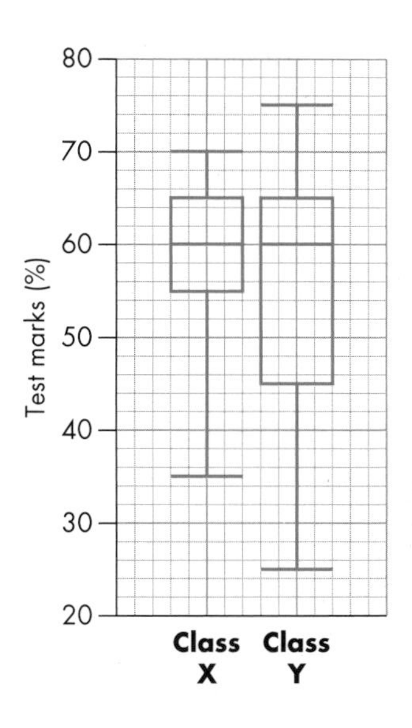

5 This diagram shows the time spent on 80 mobile phone calls.

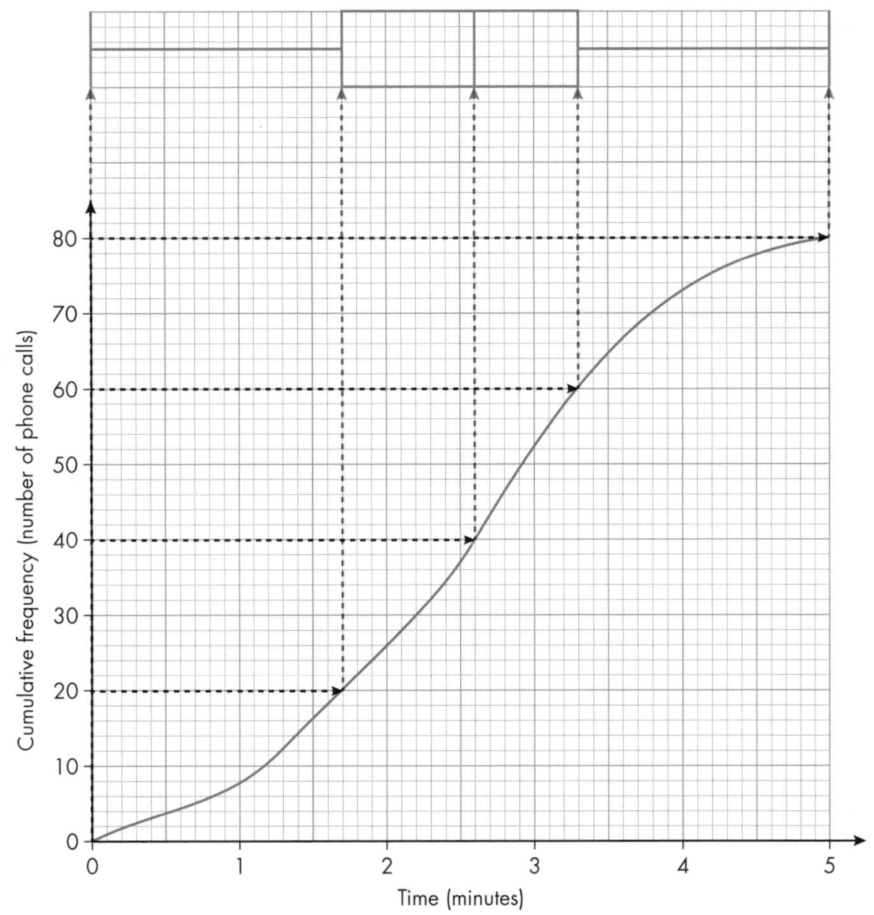

Use the diagram to estimate the following statistics.

a The median time spent on mobile phone calls.

b The range of time spent on mobile phone calls.

c $Q_3 - Q_1$.

d P_{40}.

e How many calls lasted longer than 2 minutes 12 seconds?

6 Donald kept a record of the time he spent playing *Incandescent Ducks*.

He put the data into a table and a histogram, but Daisy distracted him before he finished.

Copy and complete the table and histogram.

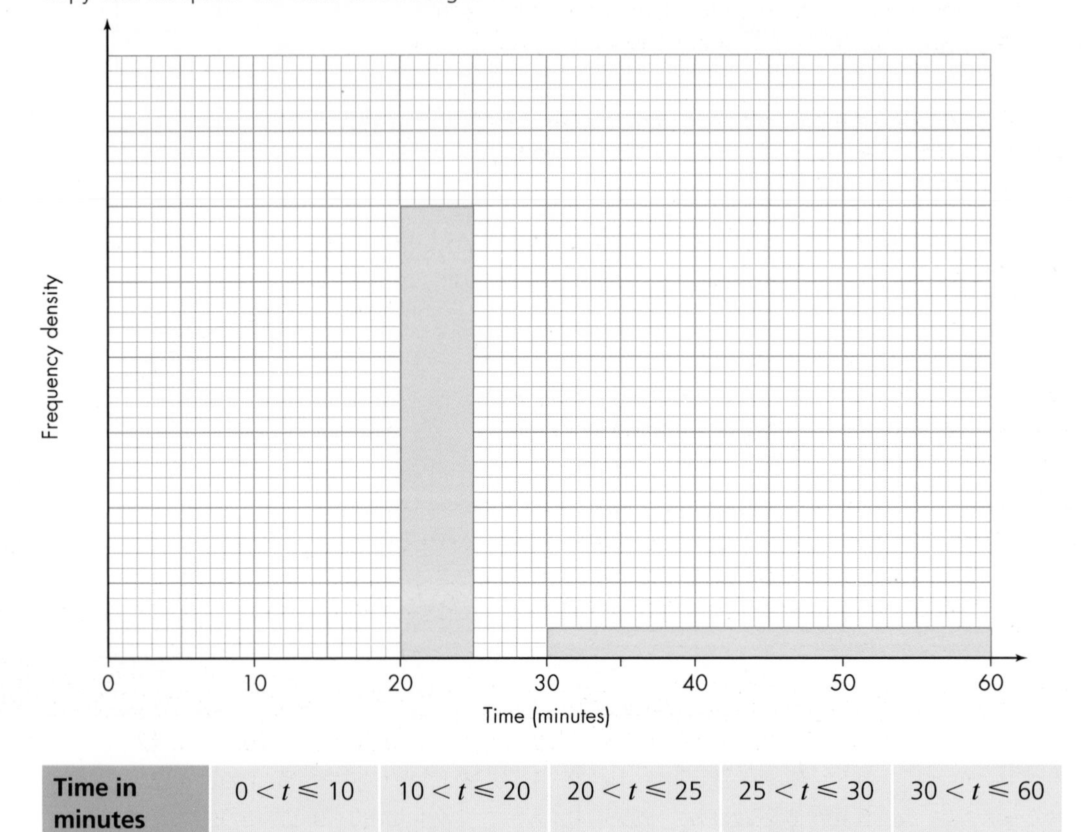

Time in minutes	$0 < t \leqslant 10$	$10 < t \leqslant 20$	$20 < t \leqslant 25$	$25 < t \leqslant 30$	$30 < t \leqslant 60$
Frequency	15	25	15	25	

7 Here are five box plots showing the results of five experiments to determine the speed of light, each experiment consisting of 20 runs.

a Which experiment would you consider to be the most reliable?

b Which would you consider to be the least reliable?

Support your decisions with statistical calculations derived from the box plots.

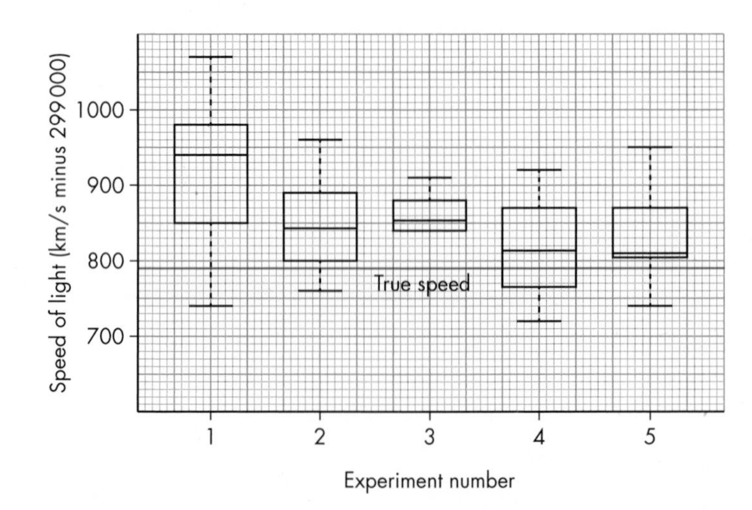

Analysis of data

8 Look at your answers to Exercise 1C, question **4** and Exercise 1E, question **6** about your possible career (or interests) and how you will be using and collecting data to examine something that interests you. Now think about how you would display the data you described. What possible diagrams might you use? Would you decide to go for clarity or novelty in your diagrams? Write a short description of about 150 words, justifying the decisions you make.

9 A market researcher was asked to find what games were played by college students. The researcher conducted a survey of 100 students at two colleges and obtained the following results.

Game	Spy Mouse	Angry Birds	Candy Crush	Cut the Rope	Tiny Death Star
Simmonds	15	25	38	13	9
Thorntons	8	27	17	29	19

Construct an appropriate diagram to show these results, giving reasons why you chose that type of diagram and why you rejected at least one of the other types.

10 A college statistician wanted to compare the reaction times between students in Year 12 and Year 13. The students used Sheep Dash on the Internet, which gave them their average reaction time. The following table shows the results. The average reaction times have been rounded to the nearest hundredth of a second.

Year	12	13
Times in seconds	0.36 0.87 0.94 0.91 0.15 0.51 0.41 0.73 0.44 0.34 0.99 0.20 0.80 0.95 0.22 0.37 0.94 0.11 0.55 0.86	0.98 0.28 0.28 0.72 0.42 0.31 0.97 0.84 0.62 0.21 0.73 0.30 0.25 0.95 0.70 0.63 0.31 0.20 0.39 0.24

Construct an appropriate diagram to show these results, giving reasons why you chose that type of diagram and why you rejected at least one of the other types.

Case study

Claire has type 1 diabetes. She checks her blood sugar levels before meals so that she can inject the right amount of insulin that her body needs to convert the carbohydrate she eats into energy that her body can use. She also checks her blood sugar levels when she feels it is getting low (hypoglycaemia). Her target is to try and keep her blood sugar level between 3.9 and 7.8 mmol/L (millimoles per litre). If it is below 4 she drinks something that has a high sugar content to bring her level up. If it is above 10 she increases the amount of insulin she takes the next time, or she exercises to help lower the blood sugar level.

Here is the raw data over three days in May. Claire treats this as primary data and recognises it as continuous quantitative data.

Logbook

May 06, 2015 - Jun 02, 2015 (28 days)

	12a	1a	2a	3a	4a	5a	6a	7a	8a	9a	10a	11a	12p	1p	2p	3p	4p	5p	6p	7p	8p	9p	10p	11p	Daily Totals	
Wednesday, May 6, 2015																										
Glucose (mmol/L)							14.0					3.8			8.1	10.7		6.9						4.3	Average (6)	8.0
Notes																										
Thursday, May 7, 2015																										
Glucose (mmol/L)								7.0					12.3				4.2							4.9	Average (4)	7.1
Notes																										
Friday, May 8, 2015																										
Glucose (mmol/L)								6.3					7.2		4.3				8.8			9.3			Average (5)	7.2
Notes																										

Claire attends a diabetic clinic once or twice a year to talk to a nurse who specialises in diabetes.

Claire takes along a summary of her blood test results. The nurse receives Claire's logbook and treats that as secondary data. The nurse sees that Claire is doing regular checks. Results that are above Claire's target are lightly-shaded; those below target are shaded dark.

Claire and her nurse discuss the data. This time they decide that no major change is needed in Claire's dosage because she runs to college on a Wednesday which explains why her level was low before lunch. She has also had a blood sample taken, which shows that her control over the last six months has been good. Since the blood sample was taken and subjected to a full analysis, the nurse can compare it with Claire's previous results. This use of data over both short and long term means that Claire and her nurse are both confident that she is in very good health and can take care of herself with minimum fuss.

Both the nurse and Claire have become skilled at using statistics in this way to control Claire's diabetes. In the long run it helps Claire to lead a full life with few or no complications, and so cuts down on the cost to the NHS.

Here are Claire's summary results.

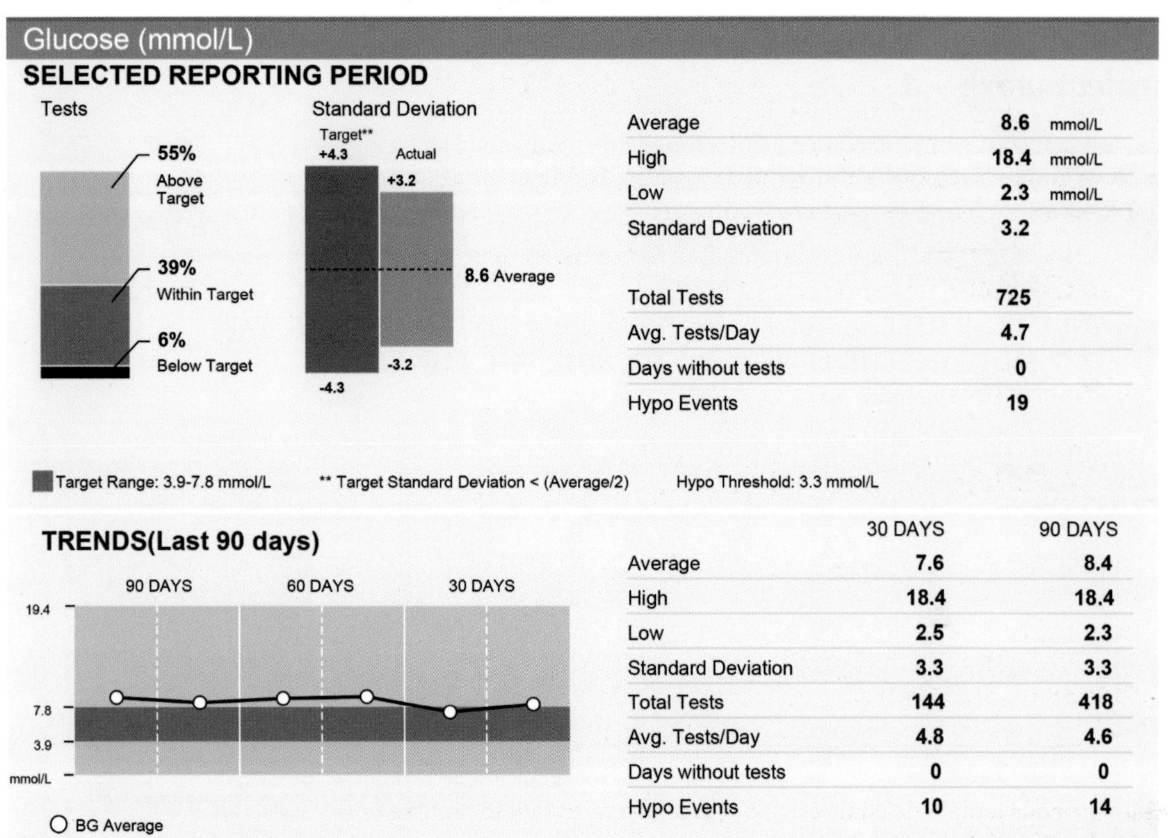
Snapshot

Jan 01, 2015 - Jun 02, 2015 (153 days)

Glucose (mmol/L)

SELECTED REPORTING PERIOD

Average	8.6	mmol/L
High	18.4	mmol/L
Low	2.3	mmol/L
Standard Deviation	3.2	
Total Tests	725	
Avg. Tests/Day	4.7	
Days without tests	0	
Hypo Events	19	

Target Range: 3.9-7.8 mmol/L ** Target Standard Deviation < (Average/2) Hypo Threshold: 3.3 mmol/L

TRENDS(Last 90 days)

	30 DAYS	90 DAYS
Average	7.6	8.4
High	18.4	18.4
Low	2.5	2.3
Standard Deviation	3.3	3.3
Total Tests	144	418
Avg. Tests/Day	4.8	4.6
Days without tests	0	0
Hypo Events	10	14

By presenting the data in this format, the results are shown numerically as well as in graphs and charts. Claire is pleased that the trend in her blood sugar results shows that she is getting very close to the target zone. She is also very pleased that her standard deviation is well within the target zone for that measure: a guide for this is that it should be less than half the average result.

Since Claire has a reasonable knowledge of statistics she can interpret these results intelligently, helping her to understand her condition. As a result, Claire is able to lead a life that is relatively free of complications.

Project work

To help you bring together all the techniques you have developed from this chapter, choose one of the following projects. They are all concerned with the battle of Trafalgar (21 October 1805).

Project work – 1

The following information comes from a broadsheet published after the Battle of Trafalgar. It shows statistical data of the two fleets that took part in the battle.

THE ENGLISH FLEET consisted of
27 SHIPS OF THE LINE

	GUNS	MEN
VICTORY	110	837
ROYAL SOVEREIGN	110	837
BRITANNIA	100	800
TEMERAIRE	98	738
PRINCE	98	738
TONNANT	84	650
BELLEISLE	74	590
REVENGE	74	590
MARS	74	590
NEPTUNE	98	738
SPARTIATE	74	590
DEFIANCE	74	590
CONQUERER	74	590
DEFENCE	74	590
COLLOSUS	74	590
LEVIATHAN	74	590
ACHILLE	74	590
BELLEROPHON	74	590
MINOTAUR	74	590
ORION	74	590
SWIFTSURE	74	590
POLYPHEMUS	64	500
AFRICA	64	500
AGAMEMNON	64	500
DREADNOUGHT	98	738
AJAX	80	850
THUNDERER	74	590
Total	2,178	17,076

THE Combined FLEETS OF FRANCE & SPAIN
33 SHIPS OF THE LINE

		GUNS	MEN	
SANTISUMA TRINIDAD	(S)	140	1200	Taken and Destroyed
BUCENTAURE	(F)	84	800	Taken and Destroyed
RAYO	(S)	100	1000	Taken and Destroyed
PRINCIPE de ASTURIAS	(S)	112	1100	Escaped
INDOMTABLE	(F)	84	800	Destroyed
FOUGOUEX	(F)	74	700	Taken and Destroyed
ACHILLE	(F)	74	700	Blown-up
SANTA ANA	(S)	112	1100	Taken, but got away
MONTANES	(S)	74	700	Escaped
HEROS	(F)	74	700	Escaped
SAN LEANDRO	(S)	64	600	Dismasted, but Escaped
SAN JUSTO	(S)	74	700	Dismasted, but Escaped
SAN ILDEFONSO	(S)	74	700	Taken
LE SWIFTSURE	(F)	74	700	Taken, formerly English
ALGESIRAS	(F)	74	700	Taken, but got away
PLUTON	(F)	74	700	Escaped, much damaged
NEPTUNE	(F)	84	800	Escaped
BAHAMA	(S)	74	700	Taken
SAN NEPOMUCENO	(S)	74	700	Taken
MONARCA	(S)	74	700	Destroyed
SAN FRANCISCO de ASIS	(S)	74	700	Destroyed
ARGONAUTE	(F)	74	700	On shore at Cadiz
LE BERWICK	(F)	74	700	Taken and Destroyed, formerly English
L'AIGLE	(F)	74	700	Taken and Destroyed
INTREPIDE	(F)	74	700	Taken and Burnt
SAN AGUSTIN	(S)	74	700	Taken and Burnt
REDOUBTABLE	(F)	74	700	Taken and Destroyed
ARGONAUTA	(S)	80	800	Taken and Destroyed
FORMIDABLE	(F)	80	800	Escaped } but later captured
MONT-BLANC	(F)	74	700	Escaped } but later captured
SCIPION	(F)	74	700	Escaped } but later captured
DUGUAY-TROUIN	(F)	74	790	Escaped } but later captured
NEPTUNO	(S)	80	800	Destroyed
Total		2,652	25,200	

The Combined Enemy Fleet superior in Guns 474 in Men 8,124

(F) Denotes French ships (S) Denotes Spanish ships

Four ADMIRALS taken
One killed, and Three wounded.

Imagine that you are a statistician in 1805. Write a report for King George III, comparing the fleets. Your report should contain both numerical and graphical representations.

Project work – 2

What happened to the sailors on board the British ships at Trafalgar?

The following table lists the details of the numbers killed and wounded in each ship.

The fleet was split into two columns, the Weather column, which was north of the Lee column, both sailing eastwards in a line. The ships are listed with the top ship at the front of the column.

Weather column	Killed	Wounded
Victory	57	102
Temeraire	47	76
Neptune	10	34
Leviathan	4	22
Britannia	10	42
Conqueror	3	9
Agamemnon	2	8
Ajax	2	9
Orion	1	23
Minotaur	3	22
Spartiate	3	20

On its own		
Africa	8*	44

Lee column	Killed	Wounded
Royal Sovereign	47	94
Belleisle	33	94*
Mars	29	71*
Tonnant	26	50
Bellerophon	27	127*
Colossus	40	160
Achille	13	59
Dreadnought	7	26
Polyphemus	2	4
Revenge	28	51
Swiftsure	9	8
Defiance	17	53
Thunderer	4	12
Defence	7	29
Prince	0	0

* Indicates that not all sources agree on this figure

Write a report for King George III (who was on the throne in 1805). Use charts and diagrams to represent the casualties on board each ship.

You might like to rank the ships in order of safety and use appropriate averages.

What were the probabilities of being wounded or killed on each ship?

Project work – 3

What happened to the sailors on board the Combined Fleet of France and Spain?

In 1906, Edward Fraser estimated the numbers shown in this table.

Nationality	Killed	Wounded	Prisoners
French	3 373	1155	>4 000
Spanish	1022	1383	3 000–4 000

Write a report, comparing the number killed and wounded with those on the British fleet.

Check your progress

How confident are you feeling in your level of knowledge? What do you need to practise more?

Spec reference	Learning objective			
D1.1	Appreciate the difference between qualitative and quantitative data			
D1.2	Appreciate the difference between primary and secondary data			
D1.3	Collect quantitative and qualitative primary and secondary data			
D2.1	Infer properties of populations or distributions from a sample, whilst knowing the limitations of sampling			
D2.2	Appreciate the strengths and limitations of random, cluster, stratified and quota sampling methods and applying this understanding when designing sampling strategies			
D3.1	Calculate/identify mean, median, mode, quartiles, percentiles, range, interquartile range and standard deviation			
D3.2	Interpret these numerical measures and reach conclusions based on these measures.			
D4.1	Construct and interpret diagrams for grouped discrete data and continuous data, know their appropriate use and reach conclusions based on these diagrams			

2 Maths for personal finance

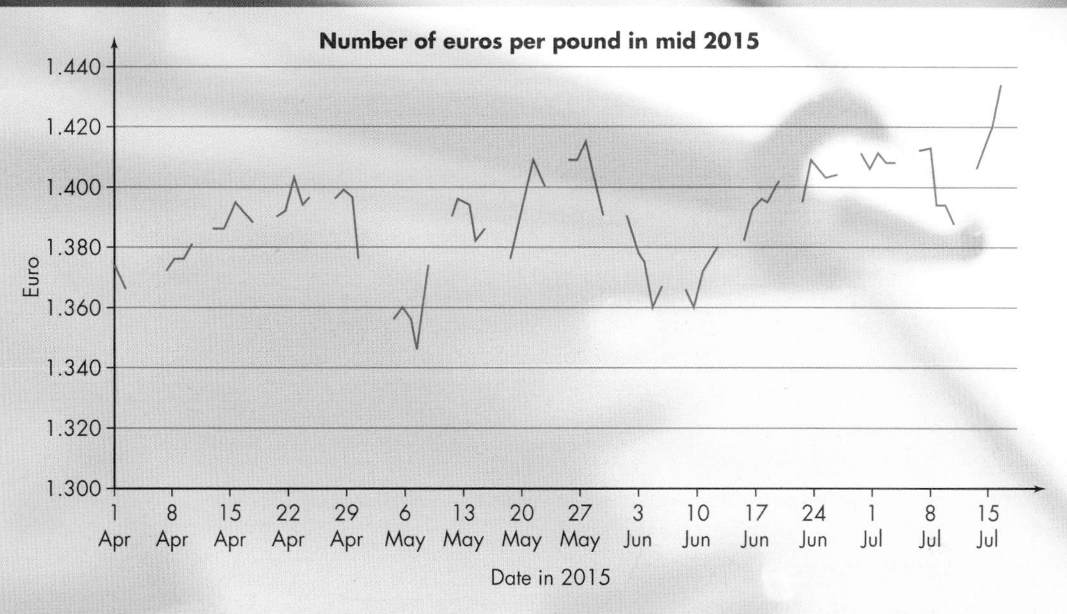

In 2015 the euro (€) was losing value against the pound (£). This meant that if you were going on holiday to a country that used the euro, you got a better rate of exchange. The graph shows how the rate changed. You can see there were some days when it was far better to buy your euros than others.

For example, if you had decided to take £200 spending money with you and you changed this on 10 June at a rate of €1.36 for £1 you would have received €272.

However, if you had waited until 15 July and exchanged at a rate of €1.42 for £1 you would have obtained €284 – a difference of €12.

So, what can you deduce from this graph? Here are some questions to get you started.

1 What is the significance of the intervals on the horizontal axis?

2 Why are there some gaps in the graph? Why are some gaps greater than others?

3 How can you estimate the average rate over the period of the graph?

The graph here was produced on a spreadsheet from the data entered into it. If you have time, explore how to make charts and graphs using any software available to you.

F1: Numerical calculations

Learning objectives

You will learn how to:

- Substitute numerical values into various expressions evaluating them correctly.
- Apply and interpret limits of accuracy and find approximate solutions to financial problems.

Introduction

Personal finance is all about managing your own monetary affairs, such as income from your job, savings, loans (including student loans), credit cards and budgeting. You also need to know about tax and National Insurance.

Using a spreadsheet can help you manage your finances. Spreadsheets help you avoid making careless mistakes with calculations, but remember to check the answers carefully to make sure you have entered all the data and formulae correctly.

Interest is the amount of money earned on savings or the amount of money added to a loan.

The spreadsheet shown here shows the monthly costs of using a credit card to buy a new phone. The monthly costs have to take account of the amount repaid and the amount of **interest** charged.

	A	B	C	D
1	AMEZERCARD		Purchase	£560.00
2			iPhone 7	
3				
4	month	total outstanding	monthly repayment	interest added
5	1	£560.00	£11.20	£5.60
6	2	£554.40	£11.09	£5.54
7	3	£548.86	£10.98	£5.49
8	4	£543.37	£10.87	£5.43
9	5	£537.93	£10.76	£5.38
10	6	£532.55	£10.65	£5.33
11	7	£527.23	£10.54	£5.27
12	8	£521.96	£10.44	£5.22

This is the amount you want to borrow.

The interest of 1% is calculated on the amount at the start of the month before the repayment is made.

How do you think these amounts are calculated?

The minimum monthly payment is 2%.

Most of the cells in the spreadsheet use calculated values, using the formulae shown in the second spreadsheet. You can use the copy-and-paste function to copy formulae, to avoid making mistakes.

	A	B	C	D
1	AMEZERCARD		Purchase	560
2			iPhone 7	
3				
4	month	total outstanding	monthly repayment	interest added
5	1	=D1	=0.02*B5	=B5*0.01
6	2	=B5-C5+D5	=0.02*B6	=B6*0.01
7	3	=B6-C6+D6	=0.02*B7	=B7*0.01
8	4	=B7-C7+D7	=0.02*B8	[illegible]
9	5	=B8-C8+D8	=0.02*B9	[illegible]
10	6	=B9-C9+D9	=0.02*B10	[illegible]
11	7	=B10-C10+D10	=0.02*B11	[illegible]

This is the amount you want to borrow.

The interest of 1% is calculated on the amount at the start of the month before the repayment is made.

The minimum monthly payment is 2%.

Exercise 2A

1 Why is the formula in cell B5 expressed as =D1?

2 What is the significance of the 0.02 in the cells in Column C?

3 Look at the formula in cell D5. Why is cell B5 multiplied by 0.01?

4 Compare the formulae in Columns C and D.
Why does it not matter whether the cell number comes first or last in the calculation?

5 Why is the formula in cell B6 expressed as =B5-C5+D5?

6 If the person buying this iPhone7 only ever pays the minimum payment, how long will it take to pay off the credit card?
Why do you think credit cards ask for a minimum payment of 2% or £5, whichever is more?

Priority of operations

In complicated calculations, the mathematical operations must be carried out in the correct order. You can remember the order using the mnemonics **BIDMAS** or **BODMAS**:

- **B**rackets – carry out any calculations inside brackets first
- **I**ndices or **O**thers – then work out the indices (such as squares, cubes, square roots) or others (such as trig functions)
- **D**ivision – then do the divisions
- **M**ultiplication – and the multiplications
- **A**ddition – next do the additions
- **S**ubtraction – and finish with any subtractions.

Example 1

$(7 - 4) \times 5$

$= 3 \times 5 = 15$ The **b**rackets are worked out first.

Example 2

$3 + 17 \times 4 \div 2$

$= 3 + 17 \times 2$ Work out the **d**ivision $4 \div 2 = 2$ first

$= 3 + 34$ and then the **m**ultiplication $17 \times 2 = 34$

$= 37$ and then the **a**ddition $3 + 34 = 37$.

Example 3

$2^4 + 5 \times 3$

$= 16 + 5 \times 3$ Work out the **i**ndices first $2^4 = 16$

$= 16 + 15$ and then the **m**ultiplication $5 \times 3 = 15$

$= 31$ and finally the **a**ddition $16 + 15 = 31$.

Example 4

$5 + \sin (18 \times 2 - 6)°$

$= 5 + \sin (36 - 6)°$ Work out the brackets first. In this case, the brackets have a multiplication and a subtraction – do the multiplication first.

$= 5 + \sin 30°$

$= 5 + 0.5$ Then do the trig function, sin 30°

$= 5.5$ and finish with the addition.

Cash flow

Cash flow is a measure of cash coming in and cash going out. It can be applied to businesses or to your own personal finances, and can be used to show whether you are living within your means.

Exercise 2B

1 **Cash flow (C)** is calculated as (C) = Income (I) minus Expenditure (E):

$C = I - E$

a Calculate C if I = £12 260 and E = £9820.

b What can you say if $C < 0$?

c Open a spreadsheet like the one shown and insert a formula into cell C3 so that it calculates C directly. Format the columns so that you don't have to enter the £ sign every time.

	A	B	C	D
1	Income, *I*	Expenditure, *E*	Cash Flow, C	
2				
3	£13,200	£8,640	£4,560	
4				

d Copy the formula down the column and use the spreadsheet to work out the cash flows with these values for income and expenditure.

i Income £68 265, Expenditure £53 198

ii Income £18 652, Expenditure £12 465

iii Income £47 315, Expenditure £51 770

iv Income £8281, Expenditure £5915

2 **The Leverage ratio** compares debt with income.

Your debts and liabilities consist of loans, phone contracts, gym memberships and amounts owed.

Your income is all the money you receive, from wages and any allowance you get from your parents.

$$\text{Leverage ratio} = \frac{\text{Total liabilities + Total debts}}{\text{Total income}}.$$

a Is a leverage ratio less than 1 good or bad? Why?

b Calculate the leverage ratio if you have total liabilities of £159, total debts of £26 and total income of £210.

c Calculate the monthly leverage ratio if in one year you have total liabilities of £600, your monthly total debts are £18 and your weekly total income is £20. Use a spreadsheet like the one shown.

	A	B	C	D
1	Total monthly income	Total monthly liabilities	Total monthly debts	Leverage ratio
2				
3	£80.00	£68.00	£15.00	1.04
4				
5				

d Copy your worksheet and change it to calculate the annual leverage ratio if the income is given as a total **annual** income. How would you change the formula if you were paid weekly?

Limits of accuracy

The results of financial calculations may give exact whole numbers, but often end in many places of decimals. To make things simpler you can **round** or **truncate** amounts to leave you with something that is appropriate and easily understood.

When you give a measurement you should state its accuracy, such as 67 mm to the nearest mm. In mathematical terms this means that the length, l, lies within the **interval** $66.5 \leq l < 67.5$.

Here the **error** is ±0.5 mm. By giving the measurement as 67 mm to the nearest mm the true measurement will not be more than 0.5 mm different from the stated measurement.

The range is expressed in terms of the **lower** and **upper bounds**.

- The **lower bound** is 67 – the error = 67 – 0.5 = 66.5 mm.
- The **upper bound** is 67 + the error = 67 + 0.5 = 67.5mm.

Although the length could not be 67.5 mm (or else it would be rounded to 68 mm), that is what you give as the upper bound – the length can get as close as you like to it.

Rounding has an error of plus or minus half the unit to which it is rounded.

If you **truncate** an amount then the error is slightly different. Truncation removes the end part of the amount without rounding up. For example if I want to buy 2.4 m of curtain material, the retailer will measure off a greater length, such as 2.47 m.

In mathematical terms this means that the length, l, lies within the interval $2.4 \leq l < 2.5$.

In the case of truncation to the nearest 0.1 m, the error is 0.1 m. The true measurement cannot be more than 0.1 m more than the stated measurement.

Truncation therefore has an error equal to the unit to which it is rounded.

Giving your age is an example of truncation. You could be aged 16 years and 11 months but you are still aged 16 despite being nearer to 17 years of age. The error here is 1 year.

With regards to personal finance, money is generally rounded to the nearest penny when interest is calculated. For example, if you want to exchange £100 into euros, you would get €131.20 at a rate of €1.312 to the pound. However exchange bureaux usually only deal with notes, and not coins, so they have to give you a whole number of euros. They would ask you to pay for €130 or €135. Dividing these amounts by 1.312 gives a cost of £99.08536… or £102.8963…, and these amounts would then be rounded to £99.09 and £102.90.

In this case the error is 0.5p – the difference between the actual value and the requested amount will not be more than half a penny.

The **error** is the maximum value an amount can differ from a calculated value.

The **lower bound** is the minimum value of a rounded amount.

The **upper bound** is the maximum value of a rounded amount.

Truncate means to chop off without rounding.

Mathematics in the real world

Her Majesty's Revenue and Customs (HMRC) truncate and round up when doing tax calculations. When you declare the interest on savings, HMRC truncate the amount you submit, so you do not pay tax on a very small amount of your savings. If you declare savings interest of £56.78, HMRC truncate this to £56, so you are not paying tax on the 78p. However if you have paid tax, HMRC always round that up to the next pound, so if you have paid tax of £432.10 they credit you with having paid £433.

Discussion

Discuss with your peers examples of times when you have rounded or truncated amounts and why it has been sensible to do this.

Exercise 2C

1 For each of these, write an interval using inequality signs with the lower and upper bounds. Specify the size of the error for each measurement.

- **a** Cost of a phone: £350 rounded to the nearest £10
- **b** Volume of a drinks can: 330 ml truncated to the nearest 10 ml
- **c** Broadband download speed: 1.6 Mb/s rounded to the nearest 0.1 Mb/s
- **d** Broadband upload speed: 0.55 Mb/s rounded to the nearest 0.01 Mb/s
- **e** Mass of a £2 coin: 12.00 g rounded to two decimal places
- **f** Diameter of a £2 coin: 28.40 mm rounded to two decimal places

2 Using the information about tax calculations above, what figures would HMRC use in the following situations?

- **a** You declare savings interest of £37.86.
- **b** You declare paid tax of £67.45.

3 Using the information about tax calculations above, what is the error if HMRC credit you with having paid the following amounts?

- **a** They use a figure for savings interest of £43.
- **b** They credit you with paying £39 in tax.

4 Here are some of the specifications for a Honda CB250 motorbike.

Type		Accuracy
Capacity	234 cc	To the nearest cc
Compression ratio	9.2 : 1	To 1 decimal place
Maximum power	14.6 kW @ 9000 rpm	To the nearest 0.1 kW and nearest 500 rpm
Dry weight	132 kg	To the nearest kg
Fuel capacity	16 litres	To the nearest litre

1 litre = 0.220 gallons to 3 significant figures.

- **a** Write each of the specifications as an interval using inequality signs with lower and upper bounds. State the error in each case.

Seventeen bikers kept a record of how many miles to the gallon (mpg) they got from their Honda CB250 bikes. Their results, recorded to the nearest whole mpg, are shown in the stem and leaf diagram.

- **b** Use this data to calculate the mean mpg and state the error.

Stem	Leaf
7	1 2 3 6
6	0 4 5 7 7 7 7
5	6
4	4 5 8
3	4 7

Key 3 | 4 represents 34 mpg

- **c** Use your answers from parts **a** and **b** to calculate the upper and lower bounds of the distance covered on a full tank of fuel. (Take care with units!)

5 The average distance of the moon from Earth is 384 403 km but this can vary by 43 592 km.

Write down the upper and lower bounds for the distance of the moon from Earth and then explain why it is not possible to give the distance to one significant figure.

6 A fast food company publishes its nutritional data online.

The energy of its products is given per portion in kcal to the nearest 5 kcal.

Its trademark product contains 490 kcal.

The company states that this is 25% of the GDA (General Daily Allowance) for an average woman in her 20s. The percentages are given to the nearest whole number.

a Using this data work out the upper and lower bounds for the following.

i The energy content of the product

ii The percentage of the GDA

iii The GDA for the woman described

b What would be a sensible single figure to give for this GDA?

Approximate solutions

It is often useful to consider approximate solutions to financial calculations. Approximate calculations can be done quickly and will help tell you whether a decision or plan is sensible or achievable, and will help confirm whether exact calculations are correct.

Example 5

Emi has a part-time job and last month her weekly earnings were:

£24.45 £27.80 £34.10 £26.69.

She thinks she can buy a reasonable car for about £1200.

If she starts to save all her earnings, how long will it take her before she has enough to buy a car?

Round each amount to the nearest £10 and add the rounded amounts:

£20 + £30 + £30 + £30 = £110.

£1200 ÷ £110 ≈ 11 so it will take her about 11 months, assuming she earns roughly the same amount each month.

You could have added all the amounts exactly and found the average and done an exact division, but approximating the amounts gives a reasonable solution because Emi's earnings vary and she might also wish to spend some of the money on other things during the year.

Do the exact calculation. What's the difference?

Example 6

Deepak's father wants to decorate the rooms in his house.

He estimates the area of the interior walls in one room is 25 m^2.

There are six rooms that size and one room twice that area.

One litre of paint covers 13 m^2.

A 2.5-litre tin costs £16.99.

A DIY store has an offer of two 2.5-litre tins for £25.

Approximately how much will Deepak's father have to pay for paint?

Total area to cover $= 6 \times 25 + 50 = 200$ m^2

A 2.5-litre tin of paint covers $2.5 \times 13 \approx 32$ m^2

Number of tins needed $= \frac{200}{32} \approx 7$

The best buy will be to buy three lots of two 2.5-litre tins and one 2.5-litre tin of paint.

Total cost $= 3 \times £25 + £16.99 \approx £92$

Exercise 2D

1 A weekly bus ticket costs Sam £10.

She has to attend college for 195 days each year.

Estimate how much she will save if she buys a bike for £95 and uses that all the time except in December and January. College is closed for two weeks over Christmas and New Year.

2 A college lunch costs £1.85. To save money, Nelson decides to make his own cheese sandwiches. Each week, he buys a large loaf (£1.49) and 600 grams of cheese.

Assuming he doesn't get fed up with cheese sandwiches, estimate how much he saves in a college year of 195 days.

3 Joe has a part-time job cleaning at a local hotel at weekends. He is paid £6.80 per hour on Saturdays and time and a quarter on Sundays. He works 5 hours on Saturdays and 3 hours on Sundays.

Estimate how much he earns in a year.

4 Alex and Sally organise a running race. They record the income from entrance fees.

On the day of the race, there were 289 adults each paying £16, 58 under-18s each paying half price and 164 concessions paying £12 each.

Organising the race costs three-quarters of the entrance fees.

Estimate the profit Alex and Sally make.

Discussion

Discuss some of these scenarios with your peers.

1. How much do you expect your income to be this year? What about your expenditure?
2. What major items do you expect to buy in the next year? How will you pay for them?
3. Did your last holiday cost more or less than you expected? Why?

F2: Percentages

Learning objectives

You will learn how to:

- Interpret and work with percentages efficiently using multiplying factors.
- Solve problems involving percentage change including finding the original value.

Introduction

Percentages, fractions and decimals are all ways of representing numbers. For example $50\% = \frac{50}{100} = \frac{1}{2} = 0.5$. Note that $50\% = \frac{50}{100}$ is written first because 'per cent' means 'out of hundred'.

If you want to find 67% of a number then you need to find 67 hundredths of it. The easiest way to do this is to multiply by 0.67 which is the decimal representation of $\frac{67}{100}$.

Discussion and group work

Discuss with your peers what the word 'percentage' means to you. Work in a small group and pool your knowledge about percentages. Think about where you have met percentages outside education. On an A3 sheet of paper make a poster outlining the methods you know for calculating percentages and give examples of how you can use that knowledge to work out problems you might meet.

Mathematics in the real world

The discussion groups will have come up with many examples of applications of percentage. Spend a few minutes looking at the posters produced by the other groups.

Example 7

Yolanda pays off 67% of an electricity bill of £84.

How much does she pay?

$84 \times 0.67 = 56.28$

Answer £56.28

Note that both the calculation and answer are written down.

Writing the calculation and the answer helps identify where any errors might have been made. When multiplying by a decimal less than 1, the answer must always be smaller than the number you started with.

Example 8

Will pays off £93 from a debt of £136.

What percentage has he paid?

$\frac{93}{136} = 93 \div 136 = 0.683\,824\ldots$

Answer 68%

It is not necessary to write the fraction. It is here to show you the connection between fractions, decimals and percentages.

Percentages are generally given correct to the nearest whole number.

Example 9

Jack and Beth have sales targets in their shop.

One month, Jack made sales of £4.4k against his £5k target.

Beth made sales of £5.45k against her £6k target.

Who did better?

$4.4 \div 5 = 0.88 = 88\%$

$5.45 \div 6 = 0.9083\ldots = 91\%$

Answer: Beth did better.

Exercise 2E

Do not use the % key on your calculator.

1 Write these percentages as decimals.

a 91% **b** 147% **c** 3% **d** 0.7%

2 Write these decimals as percentages.

a 0.38 **b** 0.04 **c** 0.008 **d** 2.34

3 Calculate the following amounts using the method shown in Example 7.

a 39% of £81 **b** 48% of £53 **c** 22% of £810

d 3% of £257 **e** 87% of £49 **f** 3.7% of £44

4 The nutritional values of a cup of porridge made with milk are given as follows.

	Food group	Amount per serving (g)	% of an average adult's daily recommended intake
a	Total fat	5.3	8%
b	of which saturates	2.4	12%
c	Sugars	9.2	10%
d	Salt	0.20	3%

Work out the average adult's daily recommended intake of each food group in grams.

Specify the error for each of your answers.

(Assume that the amount per serving has been given correct to two significant figures and the percentage is exact.)

5 Work out these percentages.

a 36p as a percentage of £25

b £14 as a percentage of £369

c £265 as a percentage of £3000

d £4.50 as a percentage of £14 000

6 The nutritional values of various types of pizza are given below.

	Type	Serving size (g)	Calories	Calories from fat
a	Fire Alarm	236	620	240
b	Bacon & Cheese	238	730	310
c	Lovers	225	740	320
d	Buffylow	225	630	220
e	Cheese	213	600	220

For each type of pizza calculate the percentage of the calories that come from fat.

Which type of pizza contains the most calories per gram?

7 Which of these coffee drinks has the highest percentage of caffeine?

	Type	Volume (fluid ounces)	Caffeine (mg)
a	Frappuccino	9.5	90
b	Iced	11	115
c	Decaf	16	25
d	Mocha	50.7	539
e	Double shot	6.5	125
f	Americano	16	225
g	Cappuccino	16	150
h	Ready brew	8	135

Percentages over 100%

Percentage increases are often expressed using percentages over 100%. The number line shows the equivalences between decimals and percentages.

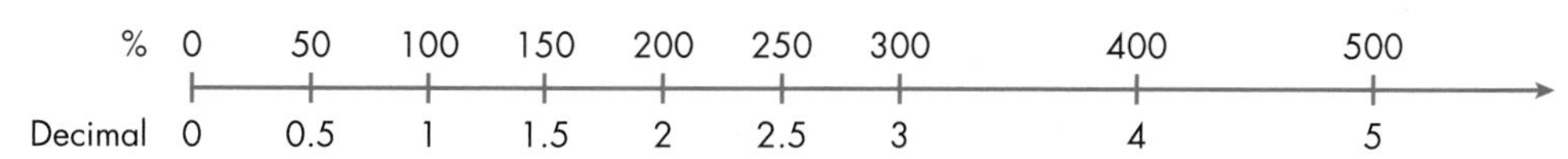

So for example, $150\% = \frac{150}{100} = 1.5$.

You can deduce other results, such as 349% = 3.49.

Percentage changes

If you increase an amount by 14%, you will have 100 + 14 = 114% of the original amount. To work out the new amount, you multiply the original amount by 1.14.

If you decrease an amount by 14%, you will have 100 – 14 = 86% of the original amount. To work out the new amount, you multiply the original amount by 0.86.

The numbers 1.14 and 0.86 in these calculations are the **multipliers**.

Using multipliers simplifies your calculations.

Example 10

Increase £84 by 67%

$1.67 \times 84 = 140.28$

Answer £140.28

Example 11

Decrease £84 by 67%

Since the amount is reduced by 67%, only 33% remains, so you need to find 33% of £84.

$0.33 \times 84 = 27.72$

Answer £27.72

Example 12

A price is increased by 15%.

The new price is £79.

What was the original price?

Let the original price be £x.

Then $x \times 1.15 = 79$

So $x = \frac{79}{1.15} = 68.6956...$

Answer £68.70

Example 13

A mass is decreased by 8%.

The new mass is 5.3 kg.

What was the original mass?

Let the original mass be x kg.

Then $x \times 0.92 = 5.3$.

So $x = \frac{5.3}{0.92} = 5.7608...$

Answer 5.76 kg

Exercise 2F

1 Copy and complete the table.

	Fraction	Decimal	Percentage
a			275%
b		4.67	
c	$6\frac{1}{4}$		
d		10.09	
e			607%

2 Calculate the new values.

a Increase 78 m by 35%

b Decrease £29 by 17%

c Increase 658 kg by 135%

d Decrease 2.9 cl by 6%

e Increase 468 tonnes by 304%

3 You inherit investments worth £1000.

Calculate the value after two years if the following rates apply.

a An increase of 10% followed by a decrease of 10%

b A decrease of 10% followed by an increase of 10%

c An increase of 5% followed by a decrease of 5%

d A decrease of 20% followed by an increase of 20%

e An increase of 1% followed by a decrease of 1%

What do you notice about your results?

4 What single percentage change is the same as the following?

a An increase of 12% followed by an increase of 12%

b A decrease of 12% followed by a decrease of 12%

c An increase of 5% followed by a decrease of 9%

d A decrease of 28% followed by an increase of 65%

e An increase of 25% followed by a decrease of 20%

5 Calculate the original values of the following.

a After an increase of 14% a can of fizzy drink costs 39 pence.

b After a decrease of 8% a pair of designer jeans costs £286.12.

c After two increases of 11% the monthly rent of a flat is £510.

d After two decreases of 3% the mass of a snack bar is 21.3 g.

6 In September 2015 the bank base rate was 0.5%.

It had remained at that level since March 2009, when it became the lowest rate in over 300 years.

Date	10/4/08	8/10/08	6/11/08	4/12/08	8/1/09	5/2/09	5/3/09
Rate	5%	4.5%	3%	2%	1.5%	1%	0.5%

a Who benefits when rates drop? Those with mortgages and loans or those with savings?

b Explain, with calculations, the effect the change in the rate had if you had

i savings of £300 **ii** a loan of £2000.

F3: Interest rates

Learning objectives

You will learn how to:

- Work with simple and compound interest.
- Appreciate the use of percentages in savings and investments.

Introduction

When you pay money into a building society or bank savings account, you will receive an **interest payment**. The amount of the interest payment is based on:

- the **original amount** you pay in
- the **interest rate**
- the **period of time** on which the interest is calculated, known as the **term**.

Simple interest is paid only on the original amount invested (called the **principal**) or the original amount borrowed.

Compound interest is the interest that is earned not only on the principal but also on all the interest earned previously. Compound interest is paid to you when you save money and is paid by you when you borrow it.

If you deposit £100 in a savings account with a interest rate of 2% **per annum** (each year), you will receive an interest payment of £100 × 2% = £2 after 1 year.

If you leave the £2 in the savings account, you will have a total of £100 + £2 = £102. At the end of the next year, you will receive an interest payment of £102 × 2% = £2.04, so you have a total of £102 + £2.04 = £104.04.

The interest has been **compounded**.

Simple interest is the amount paid only on the original amount invested or borrowed.

Compound interest is the amount paid on the total amount when the interest is calculated.

Calculating simple interest

The formula for calculating the simple interest is

$$I = Prt,$$

where:

- I is the interest
- P is the principal (the original amount)
- r is the interest rate
- t is the time, usually measured in years.

Notice that r is a decimal, not a percentage, so if the interest rate is given as 5% then use $r = 0.05$.

Example 14

What is the simple interest when £1200 is invested for 5 years at 6%?

$P = 1200$, $r = 0.06$ and $t = 5$.

So $I = 1200 \times 0.06 \times 5 = 360$.

Answer £360

Example 15

An amount is invested for 3 years at 2%. Interest of £78 is earned.

Work out the principal amount.

$I = 78$, $r = 0.02$ and $t = 3$

Substitute these values into the formula to get $78 = P \times 0.02 \times 3$

Rearranging gives $P = \frac{78}{0.02 \times 3} = 1300$

Answer £1300

Calculating compound interest

The formula for calculating the amount accumulated using compound interest is

$$A = P(1 + r)^t,$$

where:

- A is the amount accumulated
- P is the principal
- r is the interest rate
- t is the time, usually measured in years.

Remember r is the decimal equivalent of the percentage rate.

Example 16

What is the final amount when compound interest is paid on £1200 for 5 years with an annual interest rate of 6%?

$P = £1200$, $r = 0.06$, $t = 5$

So $A = 1200(1 + 0.06)^5 = 1200 \times 1.06^5 = 1605.8706...$

Answer £1605.87

The total interest paid is £1605.87 – £1200 = £405.87.

Compare the compound interest of £405.87 with the simple interest calculated in Example 14 of just £360.

Example 17

£800 is invested for 3 years and earns £99.89 in compound interest.

Work out the interest rate.

Add the original amount and the interest to find the total amount accumulated.

$A = £800 + £99.89 = £899.89$

Let $x = 1 + r$. This makes things simpler to write and solve.

Substitute these values into the formula to get $899.89 = 800x^3$

Rearranging gives $x^3 = \frac{899.89}{800}$

$x = \sqrt[3]{\frac{899.89}{800}} = \sqrt[3]{1.124\,86} = 1.0399...$

Remember you set $x = 1 + r$, so $r = x - 1$:

$r = 1.04 - 1 = 0.04 = 4\%$

Answer 4%

The Rule of 72

The Rule of 72 gives an estimate of how long it will take (T) in years to double an investment that pays r% compound interest per annum (each year). It states

$$T \approx \frac{72}{r}.$$

It is used in financial situations to give people an idea of how long it will take to double their money if they invest a lump sum.

Example 18

Use the Rule of 72 to find approximately how long it would take an investment of £100 to double with a compound interest rate of 3%.

Then use your answer to see exactly what the investment would be worth after that many years.

With a compound interest rate of 3%, it would take approximately $\frac{72}{3} = 24$ years to double.

After 24 years an investment of £100 at 3% compound interest would be worth:

$$£100(1+0.03)^{24} = 100 \times 1.03^{24} = £203.2794\ldots,$$

a bit more than double the initial investment, but after 23 years the investment would be worth £197.3586... – it would not yet have doubled.

Exercise 2G

1 In 2015 a building society offered different annual interest rates depending on the amount of money in the account.

- balances of £1000 to £2000 1%
- balances from £2000 to £3000 2%
- balances from £3000 to £20 000 3%

All rates are gross (before tax of 20% is deducted).

a How much interest would a balance of £17 500 receive before tax if it was left in the account for 2 years?

b What would the saver receive after the tax was deducted?

2 Open a spreadsheet like the one shown and insert a formula into cell D3 so that it calculates simple interest, I directly.

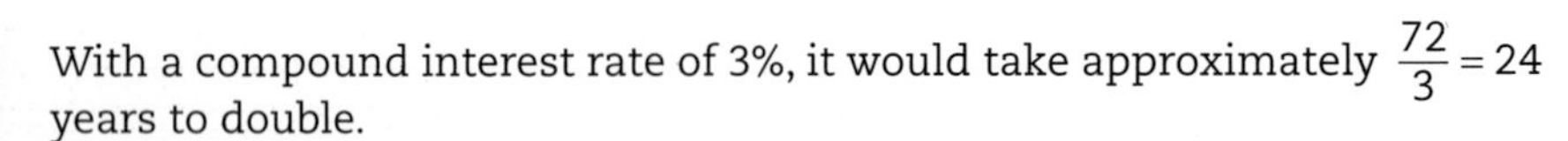

	A	B	C	D	E	F
1	Principal, *P*, in pounds	Annual rate, *r*%	Time, *T*, in years	Interest, *I*, in pounds	Tax on interest at 20%	Amount received
2						
3	£600.00	4	5	£120.00	£24.00	£96.00
4						

Copy the formula down the column. Enter various amounts for P, r and T to test your formula works.

Extend your spreadsheet to calculate the tax and amount received after tax.

Add a column that calculates the final amount in the account at the end of the period.

3 Use the Rule of 72: $T \approx \frac{72}{r}$ in this question.

a Calculate T if $r = 4$.

b If you want to double an investment in 10 years, what value of r would you need?

c Open a spreadsheet like the one shown and insert a formula into cell B3 so that it calculates T directly. Copy the formula down the column. Enter various amounts for r and T to test your formula.

	A	B
1	**Annual rate, *r*%**	**Time, *T*, in years**
2		
3	3.6	20
4		

d What formula would you put in cell A3 to calculate the rate if you knew the time in which your investment needed to double?

4 Building societies operate with compound interest.

The spreadsheet shows how an investment of £1000 grows under simple and compound interest over a period of 10 years at 8%.

	A	B	C	D	E
1	**Principal, *P*, in pounds**	£1,000.00		**Annual rate, *r*%**	8
2					
3		**Simple interest**		**Compound interest**	
4	End of year	Interest	Amount accumulated	Interest	Amount accumulated
5		Simple interest		Compound interest	
6	1	£80.00	£1,080.00	£80.00	£1,080.00
7	2	£80.00	£1,160.00	£86.40	£1,166.40
8	3	£80.00	£1,240.00	£93.31	£1,259.71
9	4	£80.00	£1,320.00	£100.78	£1,360.49
10	5	£80.00	£1,400.00	£108.84	£1,469.33
11	6	£80.00	£1,480.00	£117.55	£1,586.87
12	7	£80.00	£1,560.00	£126.95	£1,713.82
13	8	£80.00	£1,640.00	£137.11	£1,850.93
14	9	£80.00	£1,720.00	£148.07	£1,999.00
15	10	£80.00	£1,800.00	£159.92	£2,158.92

Here are the formulae used in the spreadsheet above.

	A	B	C	D	E
1	**Principal, *P*, in pounds**	1000		**Annual rate, *r*%**	8
2					
3		**Simple interest**		**Compound interest**	
4	End of year	Interest	Amount accumulated	Interest	Amount accumulated
5	1	=B$1*E$1/100	=B1+B5	=B1*E$1/100	=B1+D5
6	2	=B$1*E$1/100	=C5+B6	=E5*E$1/100	=E5+D6
7	3	=B$1*E$1/100	=C6+B7	=E6*E$1/100	=E6+D7

a Why do the formulae have a string sign ($) between E and 1?

b What is the significance of the green triangle in the cells in Column D?

c Construct your own spreadsheet to work out simple and compound interest.

d Use the Rule of 72 to calculate an estimate of the time taken to double the amount. How accurate is this estimate?

Annual Equivalent Rate

The **Annual Equivalent Rate (AER)** is used to compare the annual interest rates between savings where the interest is calculated over different periods. Some savings accounts calculate the interest daily, while others calculate weekly, monthly or annually. To make the comparison you use a standard period of **one year**.

AER is the annual interest rate on savings.

If a savings account has a compound interest rate of 1% monthly, the amount in the account will increase as follows:

- after 1 month $A = P \times 1.01$
- after 2 months $A = P \times 1.01 \times 1.01 = P \times 1.01^2$
- after a year (12 months) $A = P \times 1.01^{12} = 1.126\,825\ldots$

So a compound interest rate of 1% monthly corresponds to an AER of 12.68%.

Notice that the AER is more than the simple interest rate, which would be $12 \times 1\% = 12\%$.

When the interest rate is only 1%, the difference is not very big, but the spreadsheet shows what happens when the monthly percentage rate increases.

	A	B	C	D	E	F	G	H	I	J	K
1	Monthly rate	1%	2%	3%	4%	5%	6%	7%	8%	9%	10%
2	AER	12.68%	26.82%	42.58%	60.10%	79.59%	101.22%	125.22%	151.82%	181.27%	213.84%
3	Equivalent simple interest rate	12.00%	24.00%	36.00%	48.00%	60.00%	72.00%	84.00%	96.00%	108.00%	120.00%

An AER of 12.68% is very high, and most savings accounts in 2015 had AERs of between 0.5% and 2%.

The AER, r, is given by

$$r = \left(1 + \frac{i}{n}\right)^n - 1$$

where i is the nominal interest rate *as a decimal* and n is the number of compounding periods per year. You do not have to remember this formula: it is on the formula sheet you will be given for an examination.

If you check this with the table above for a nominal rate of 7% paid monthly, which would nominally be $7\% \times 12 = 84\%$ annually, you have

$$r = \left(1 + \frac{0.84}{12}\right)^{12} - 1 = 1.07^{12} - 1 = 2.252\,191\ldots - 1 = 1.252\,191\ldots$$

and this as a percentage is 125.22%.

Discussion

Discuss the following with your peers.

How often is interest applied to your savings accounts? Yearly or half yearly? How much difference does it make?

Look for some AERs in advertisements in newspapers or online and compare them with the nominal rates given.

Exercise 2H

1 Work out the simple interest and the amount accumulated for the following.

a £500 is invested for 4 years at 3%.

b £200 is invested for 7 years at 2%.

c £456 is invested for 8 years at 9%.

2 Work out the missing quantity.

a The principal invested at 4% compounded annually over 2 years if the interest is £53.

b The time if the interest on a principal of £327 invested at 11% compounded annually is £63.27.

c The rate if the interest on a principal of £932 invested for 5 years is £43.

3 Work out the amount accumulated and the total interest received for the following.

a £600 is invested for 5 years at 4% compound interest.

b £532 is invested for 6 years at 2.5% compound interest.

c £398 is invested for 4 years at 1.6% compound interest.

4 Work out the annual compound interest rate for the following.

a Interest of £187.68 is received when a principal of £2300 is invested for 2 years.

b Interest of £1375.32 is received when a principal of £7200 is invested for 3 years.

c Interest of £48.36 is received when a principal of £3412 is invested for 4 years.

5 Work out the AER for the following nominal rates.

a 3.5% per quarter (every 3 months)

b 15% per quarter

c 0.6% per week (take a year as 52 weeks)

d 0.6% per day (take a year to be 365 days)

6 **a** What monthly rate would yield an AER of 30%?

b What weekly rate would yield an AER of 30%?

F4: Repayments and the cost of credit

Learning objectives

You will learn how to:

- Work with student loans and mortgages.

Introduction

Student loans and mortgages are probably the biggest financial commitments you will make in your life. Both types of loan can involve large sums of money and will usually take many years to repay, so it is important that you understand what you are dealing with.

Student loans are used to help pay for further and higher education courses. Student finance can have three different elements:

- tuition fee loan
- maintenance loan – for full-time students only
- maintenance grant or Special Support Grant – for full-time students only.

Tuition fee loans pay for your course. Maintenance loans and grants help with living costs (such as accommodation, books and bills, etc.). The difference between a loan and a grant is that you have to pay back loans but you do not have to pay back grants.

The conditions on student finance vary from time to time. Up-to-date information can be found on the internet: search for *student loans*.

You apply for student finance every year. You do not need a confirmed place at university or college to apply. This section is based on the student finance package from 1 September 2016.

A **mortgage** is a loan taken out to buy property (usually a house) or land. The loan is secured against the value of your home until the mortgage is paid off. If you cannot keep up the payments on your mortgage, the lender can repossess (take back) your home.

A **mortgage** is a loan taken out to buy property.

There are various types of mortgages.

- A **repayment mortgage** pays back the initial loan and the interest.
- An **interest-only mortgage** only pays the interest. It does not reduce the amount of the initial loan borrowed. At the end of the mortgage period the borrower still has to pay off the amount borrowed. Repayments each month are cheaper, but there is a substantial sum to find at the end of the mortgage.

Discussion

Discuss the following with your peers.

1. What do you think are the advantages and disadvantages of repayment and interest-only mortgages?
2. What do you know about student loans? Look up more details about them and consider the pros and cons of taking out loans.
3. Why might you take out a student loan even if you have enough savings not to need one?

After doing your research on student loans, give a short presentation (about 10 minutes) in groups of three or four, about what you have found.

Mathematics in the real world

The Student Finance website has a useful repayment calculator which calculates repayments based on the size of the loan, the length of the course and the starting salary. In 2015, you only started to repay your student loan when you were earning £21 000 or more.

The graph shows the repayments for a tuition loan of £9000 and maintenance loan of £8000 taken out each year over three years starting in September 2016, so the total amount borrowed is £51 000. A starting salary of £25 000 has been entered.

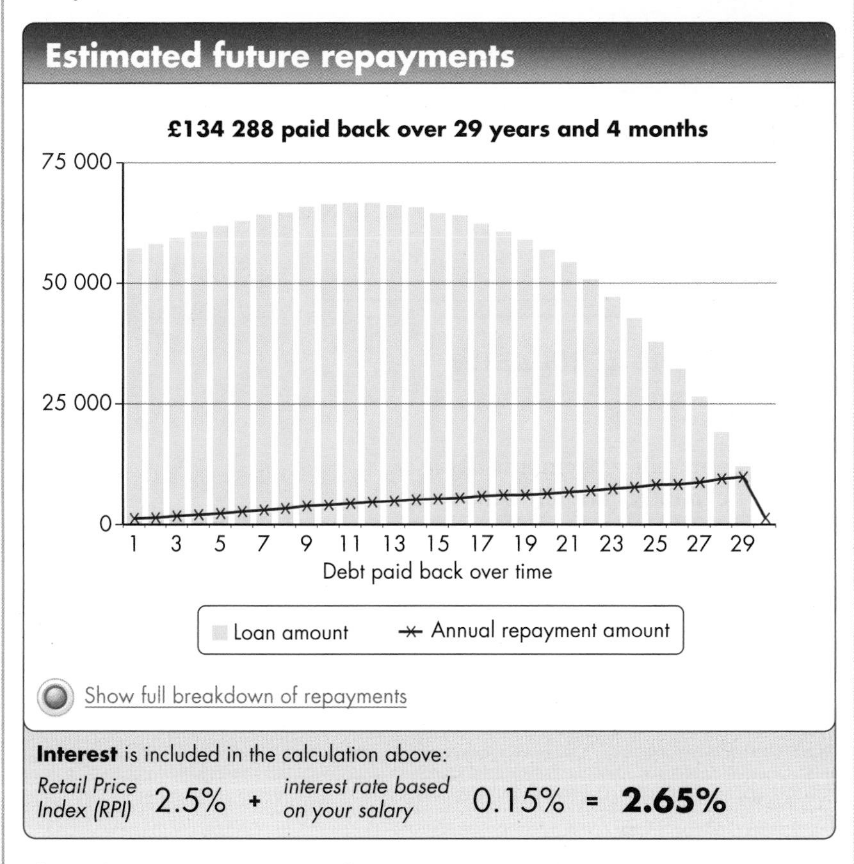

The website tells you that the first repayment would be £8 in April 2019.

It would take 29 years and 4 months to repay the loan and the total paid would be £134 288.

Discussion

Discuss with your peers what happens to the loan and annual repayments as your income goes up or down?

Exercise 2I

Use the repayment calculator on the Student Finance website to answer these questions.

1 For a three-year course, an annual tuition fee of £7500 and an expected starting salary of £22 000, find:

a the total amount borrowed

b the total amount repaid

c the time to repay.

2 For a four-year course, an annual tuition fee of £9000 and an expected starting salary of £22 000, find:

a the total amount borrowed

b the total amount repaid

c the time to repay.

If a maintenance loan of £4000 is also taken, find:

d the extra amount repaid

e the extra time to repay.

3 For a three-year course, an annual tuition fee of £6000, a maintenance loan of £5000 per year and an expected starting salary of £22 000, find:

a the total amount borrowed

b the total amount repaid

c the time to repay.

If the maintenance loan is not taken, find:

d how much less is repaid

e how much more quickly the loan is repaid.

Calculating student loans without website assistance

This section deals with calculating student loan repayments using a calculator and a spreadsheet. You will need to be familiar with doing these calculations.

Interest rates vary from year to year, as they are based on the RPI (Retail Price Index) and also on a sliding scale if your income is between £21 000 and £41 000. However, for the purpose of this book, use the following information.

- Annual repayments are 9% of everything earned in excess of £21 000.
- Simple interest of 2% is added to the outstanding loan for income between £21 000 and £41 000.
- Simple interest of 4% is added to the outstanding loan for income of £41 000 or more.

The government website offers the following advice on how to work out your monthly repayment.

1. Take away £21 000 from your annual salary before tax.
2. Work out 9% of the remainder.
3. Divide by 12.
4. Round the answer down to the nearest pound.

Example 19

Devin's starting salary is £28 000 a year.

How much will his monthly repayments be?

Using the steps above you have:

1. £28 000 − £21 000 = £7000
2. £7000 × 0.09 = £630
3. £630 ÷ 12 = £52.50
4. Monthly repayment is £52.

Example 20

Devin's starting salary is £28 000 a year.

This increases by £2000 each year.

Calculate his monthly repayments for years 1, 2 and 3.

Now it makes sense to use a spreadsheet as you need multiple calculations.

Here is the spreadsheet of results and the formulae used to obtain them.

	A	B	C	D
1	Start of year	Salary (£)	Annual repayment (£)	Monthly repayment (£)
2	1	28000	630	52
3	2	30000	810	67
4	3	32000	990	82

	A	B	C	D
1	Start of year	Salary (£)	Annual repayment (£)	Monthly repayment (£)
2	1	28000	=0.09*(B2-21000)	=INT(C2/12)
3	=A2+1	=B2+2000	=0.09*(B3-21000)	=INT(C3/12)
4	=A3+1	=B3+2000	=0.09*(B4-21000)	=INT(C4/12)

Remember that multiplying by 0.09 is how to calculate 9% of an amount. The use of =INT() in the spreadsheet rounds the value in the brackets down to the nearest whole number.

You also need to be able to work out how much any outstanding loan would be and to do that you need to add the interest to the amount of the loan at the start of each year then subtract the amount repaid over the year.

Example 21

Devin had a student loan of £24 000.

His starting salary is £28 000 a year.

This increases by £2000 each year.

Calculate how much he owes at the end of years 1, 2 and 3.

When will he have reduced the loan to less than £20 000?

Here is the spreadsheet of results.

	A	B	C	D	E	F
1	Start of year	Salary (£)	Annual repayment (£)	Monthly repayment (£)	Interest (£)	Amount of outstanding at end of year (£)
2	1	28000	630	52	480.00	23856.00
3	2	30000	810	67	477.12	23529.12
4	3	32000	990	82	470.58	23015.70
5	4	34000	1170	97	460.31	22312.02
6	5	36000	1350	112	446.24	21414.26
7	6	38000	1530	127	428.29	20318.54
8	7	40000	1710	142	406.37	19020.91

At the end of year 1 Devin owes £23 856, at the end of year 2 he owes £23 529.12 and at the end of year 3 he owes £23 015.70.

His loan will be less than £20 000 after 7 years.

The formulae used in the spreadsheet are shown below.

	A	B	C	D	E	F
1	Start of year	Salary (£)	Annual repayment (£)	Monthly repayment (£)	Interest (£)	Amount of outstanding at end of year (£)
2	1	28000	=0.09*(B2-21000)	=INT(C2/12)	=0.02*24000	=24000+E2-12*D2
3	=A2+1	=B2+2000	=0.09*(B3-21000)	=INT(C3/12)	=0.02*F2	=F2+E3-12*D3
4	=A3+1	=B3+2000	=0.09*(B4-21000)	=INT(C4/12)	=0.02*F3	=F3+E4-12*D4
5	=A4+1	=B4+2000	=0.09*(B5-21000)	=INT(C5/12)	=0.02*F4	=F4+E5-12*D5
6	=A5+1	=B5+2000	=0.09*(B6-21000)	=INT(C6/12)	=0.02*F5	=F5+E6-12*D6
7	=A6+1	=B6+2000	=0.09*(B7-21000)	=INT(C7/12)	=0.02*F6	=F6+E7-12*D7
8	=A7+1	=B7+2000	=0.09*(B8-21000)	=INT(C8/12)	=0.02*F7	=F7+E8-12*D8

The amount outstanding at the end of the year is the amount owing at the start of the year plus the interest on the loan minus the amount paid during the year.

The amount paid during the year is calculated by multiplying the monthly payment by 12. Note that this will not necessarily be equal to the annual repayment because the monthly repayment is rounded down to the nearest pound.

Exercise 2J

For this exercise you will need to use the following information.

- Annual repayments are 9% of everything earned in excess of £21 000.
- Simple interest of 2% is added to the outstanding loan for income between £21 000 and £41 000.
- Simple interest of 4% is added to the outstanding loan for income of £41 000 or more.

Set out your work neatly and feel free to use a spreadsheet to reduce the time spent calculating multiple results.

1 Eva's starting salary is £32 000 a year.

How much will her monthly payments be in her first year?

2 Frank's starting salary is £25 000 a year.

How much will his monthly payments be in his first year?

3 Gaia's starting salary is £35 000. This increases by £1800 each year.

Calculate her monthly repayments for years 1, 2 and 3.

4 Hamid's earnings in his first year are £29 000. These increase by £1500 each year.

Calculate his monthly repayments for years 1, 2 and 3.

5 Gaia's outstanding loan when she starts working is £27 000.

Using the information in question **3** about her salary, calculate how much she owes at the end of years 1, 2 and 3.

6 Hamid's outstanding loan when he starts earning is £30 000.

Using the information in question **4** about his earnings, calculate how much he owes at the end of years 1, 2 and 3.

Mortgage costs

If you want to buy a house, you will probably have to take out a mortgage. When you have a mortgage, the lender (the bank or building society that gives you the mortgage) will send you a statement every year or every six months showing what you have paid and the outstanding debt.

The example shows a statement for an interest-only mortgage.

TRANSACTION STATEMENT OF RESIDENTIAL LOAN		FOR PERIOD ENDED 31/08/15	
DATE	TRANSACTION DETAILS	DEBITS	CREDITS
01/01/15	BROUGHT FORWARD BALANCE	110,615.25	
	INTEREST CHARGED TO 31/08/15	3,317.66	
27/01/15	DIRECT DEBIT RECEIPT		321.71
25/02/15	DIRECT DEBIT RECEIPT		321.71
25/03/15	DIRECT DEBIT RECEIPT		321.71
25/04/15	DIRECT DEBIT RECEIPT		321.71
27/05/15	DIRECT DEBIT RECEIPT		511.60
25/06/15	DIRECT DEBIT RECEIPT		511.60
18/07/15	LEGAL/DEED/ PRODUCTION FEE	70.00	
25/07/15	DIRECT DEBIT RECEIPT		511.60
26/08/15	DIRECT DEBIT RECEIPT		497.77
	TOTAL DEBITS	114,002.91	
	−TOTAL CREDITS		3,319.41
31/08/15	=TRANSFERRED BALANCE−INCLUDING INTEREST TO 31/08/15	110,683.50	
	THIS IS NOT A REDEMPTION FIGURE		

The interest rate varied throughout the year.

- In January, the interest rate was 3.49%.
- In May, the interest rate increased to 5.55% and the repayments went up to £511.60.
- In August, the interest rate decreased to 5.40% and the repayments went down to £497.77.

Annual Percentage Rate

The Annual Percentage Rate (**APR**) is similar to the AER but generally AER is used with savings accounts and APR in relation to borrowing money.

APR is the annual interest rate on loans.

The APR includes any costs of borrowing the money, such as fees paid to advisers at the start (front-end fees) or annual charges (such as credit card membership fees).

APRs for very short-term loans (sometimes called **pay-day loans**) can be very high.

Discussion

Discuss the following with your peers.

Are all credit cards the same? Find out the APR of various credit cards online and discuss the merits of each. Why do you think there is a difference in rates?

Calculating the APR

The formula used in relation to APR is given on the formula sheet. It is:

$$C = \sum_{k=1}^{m} \left(\frac{A_k}{(1+i)^{t_k}} \right)$$

where

- £C is the amount of the loan
- m is the number of payments
- i is the APR expressed as a decimal

- $£A_k$ is the amount of the kth repayment
- t_k is the interval in years between the start of the loan and the kth repayment.

It assumes there are no additional fees.

This complicated formula is a shorthand way of writing the **sum** of a number of terms.

Σ is the Greek capital letter sigma and is used to denote the sum of a number of terms.

The letters at the bottom and top of Σ indicate that you calculate the sum of all the terms from the first to the mth.

Most of the time the repayments will be equal so $A_1, A_2, A_3, ..., A_m$ will all be equal.

Example 22

You borrow £400 from a money lender to be paid back in five equal monthly instalments at an APR of 1509%.

How much is the monthly payment, A?

Here $C = 400$, $m = 5$, $i = 15.09$, $A = A_1 = A_2 = A_3 = A_4 = A_5$ and t_k is a fraction with a denominator of 12 because there are 12 months in a year.

$$\text{So } 400 = \sum_{k=1}^{5}\left(\frac{A_k}{(1+15.09)^{t_k}}\right) = \frac{A_1}{16.09^{1/12}} + \frac{A_2}{16.09^{2/12}} + \frac{A_3}{16.09^{3/12}} + \frac{A_4}{16.09^{4/12}} + \frac{A_5}{16.09^{5/12}}$$

$$400 = \frac{A}{16.09^{1/12}} + \frac{A}{16.09^{2/12}} + \frac{A}{16.09^{3/12}} + \frac{A}{16.09^{4/12}} + \frac{A}{16.09^{5/12}}$$

$$400 = A\left(\frac{1}{16.09^{1/12}} + \frac{1}{16.09^{2/12}} + \frac{1}{16.09^{3/12}} + \frac{1}{16.09^{4/12}} + \frac{1}{16.09^{5/12}}\right)$$

$$400 = A(0.793\,329\ldots + 0.629\,371\ldots + 0.499\,299\ldots + 0.396\,108\ldots + 0.314\,244\ldots)$$

$$400 = A \times 2.632\,35$$

$$A = 151.9554$$

So the monthly payment is £151.96 and you will have paid a total of £759.80.

Borrowing money can be very expensive.

Example 23

Find the amount of the loan if there are three annual payments of £500 at an APR of 25%.

Here $m = 3$, $i = 0.25$, $A_1 = A_2 = A_3 = 500$ and $t_1 = 1$, $t_2 = 2$ and $t_3 = 3$ because the payments are annual.

$$C = \sum_{k=1}^{3}\left(\frac{A_k}{(1+0.25)^{t_k}}\right)$$

$$= \frac{A_1}{1.25^1} + \frac{A_2}{1.25^2} + \frac{A_3}{1.25^3}$$

$$= \frac{500}{1.25^1} + \frac{500}{1.25^2} + \frac{500}{1.25^3}$$

$$= 976$$

The loan is therefore £976. (Notice that you had to pay out a total of £1500 for this.)

Exercise 2K

1 You borrow £500 to be paid back in four equal annual payments at an APR of 32%.

How much is each instalment?

2 You borrow £700 to be paid back in three equal monthly payments at an APR of 744%.

How much is each instalment?

3 Look online at the short-term cost of borrowing money either from a bank or money lender. Check that the APR is correct for the interest rate stated.

(Remember that there may be some hidden charges, so your answer may not be exact.)

4 Work out the amount of the loan and how much is repaid in the following cases.

- **a** Three equal annual payments of £600 with an APR of 27%.
- **b** Four equal annual payments of £100 with an APR of 23%.
- **c** Four equal monthly payments of £100 with an APR of 730%.

5 If you are thinking about going on to higher education and will be applying for student finance, use a student loan repayment calculator to investigate the amounts you will be paying back. See what the difference is for different salaries.

F5: Graphical representation

Learning objectives

You will learn how to:

- Create graphs and interpret results from graphs in financial contexts.

Using spreadsheets to draw graphs

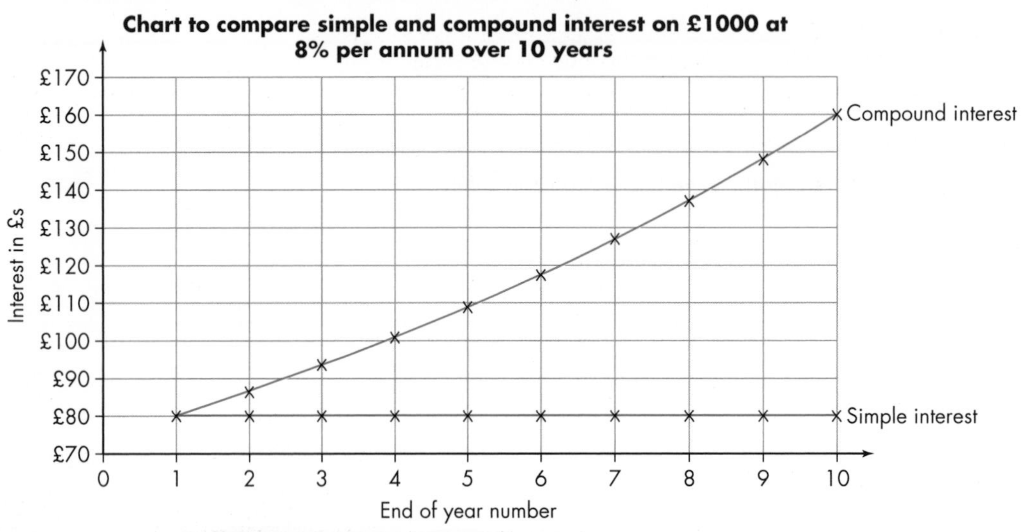

Data source: Shanghai Stock Exchange Composite Index, SHCOMP:IND, Bloomberg L.P.

Spreadsheets have graph-plotting functions which can help visualise and explain financial calculations.

The graph here shows the simple and compound interest earned on £1000 at 8% per annum over 10 years that was displayed numerically earlier in this chapter. Be careful when interpreting graphs. Notice that the vertical scale here starts at £70, so it appears that the compound interest is nine times as much as the simple interest at the end of 10 years when it is really just about twice the amount.

Discussion

Discuss the following with your peers.

What are the important features of this chart?

There are lines as well as points on the chart. Why?

What are the advantages of using the graph-plotting function on a spreadsheet?

Exercise 2L

1 The table shows the amounts to be repaid on a student loan of £45 000 assuming a starting salary of £60 000.

Year	0	1	2	3	4	5	6	7	8
Amount owing (£)	50 134	49 561	47 859	44 915	40 765	35 430	28 686	20 554	11 027
Amount paid (£)	3249	4320	5441	6459	7392	8481	9473	10 398	11 347

Show this data on an appropriate graph and work out the total amount repaid.

2 In June and July 2015 China's stock market suffered a large crash.

Here are the figures of the Shanghai Composite Index between 12 June and 8 July 2015.

Date	12 June	15 June	18 June	23 June	24 June	29 June	30 June	3 July	6 July	8 July
Index	5166	5063	4785	4576	4690	4053	4277	3687	3776	3507

Show this data on an appropriate graph and work out the percentage drop over the period.

3 This chart shows the amount spent on the NHS in the UK as a percentage of the UK's Gross Domestic Product (GDP). The chart shows periods of recession, when the GDP goes down. In other years, the GDP goes up.

Write down three things that you can deduce from this chart.

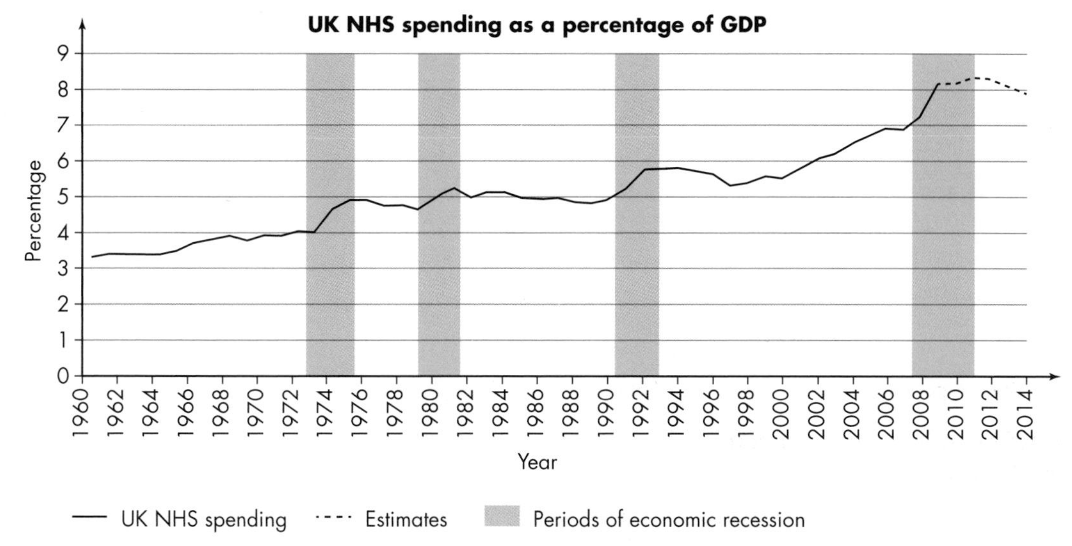

Data source: Spending on health and social care over the next 50 years, John Appleby/The King's Fund/Organisation for Economic Co-operation and Development (2012).

4 The chart shows sales information about iPhones.

a Write down three things that you can deduce from this chart.

b Suggest why there are significant peaks in the chart?

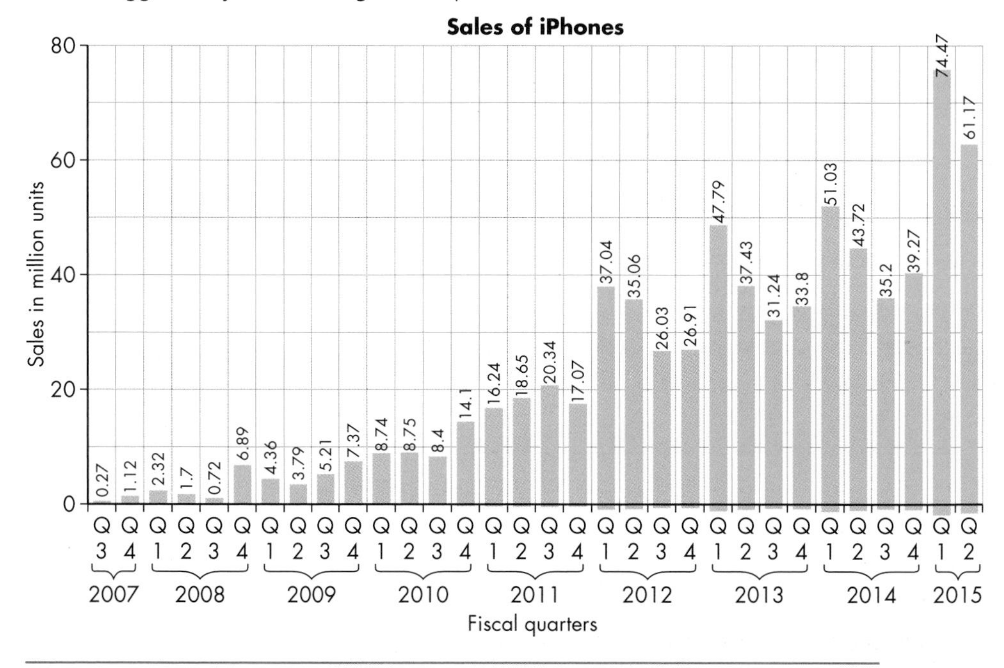

F6: Taxation

Learning objectives

You will learn how to:

- Work with income tax, National Insurance and Value Added Tax (VAT).

Introduction

When you have a job, you will usually be paid weekly or monthly. You should receive a payslip which shows your pay and pay rates, and your **deductions**. The deductions are items such as:

- income tax
- National Insurance (NI)
- pension contributions.

Income tax is a tax you pay on the money you earn.

As well as paying income tax on your pay, you also have to pay income tax on some savings.

You get a tax-free personal allowance, which is the amount you can earn before you start to pay tax. In 2015, the standard personal allowance was £10 600. It was reduced by £1 for every £2 you earned above £100 000.

Tax bands

The table shows the tax applied to different levels of pay for the tax year 2015–16.

Taxable income	£0 to £31 785	£31 786 to £150 000	Over £150 000
Rate	Basic rate of 20% People with the standard personal allowance start paying this on income over £10 600.	Higher rate of 40% People with the standard personal allowance start paying this on income over £42 385.	Additional rate of 45%

Example 24

Calculate the amount of tax a person on an annual salary of £50 000 must pay.

On an annual salary of £50 000, your tax is made up of:

40% on £7615 (the amount over £42 385) = £3046

20% on £31 785 (£42 385 – £10 600) = £6357

so the total tax deducted annually is

£3046 + £6357 = £9403

or $\frac{£9403}{12} = £783.58$ each month.

National Insurance

You pay National Insurance (NI) contributions to qualify for certain benefits including the State Pension and Maternity Allowance.

National Insurance is a contribution you pay for certain state benefits.

In 2015–16 you paid National Insurance if you were:

- 16 or over
- an employee earning more than £155 a week (£672 a month or £8060 a year)
- self-employed and making a profit of £5965 or more a year.

You should be sent a National Insurance number automatically just before your 16th birthday. You need your NI number when you:

- start a new job
- claim benefits
- apply for a student loan.

Most people with a job pay Class 1 NI rates. The rates for 2015–16 are shown in the table.

Your pay	Class 1 National Insurance rate
Less than £672 a month	0%
£672 to £3532 a month	12%
Over £3532 a month	2% (on the excess above £3532)

In 2015–16 you did not pay National Insurance on the first £8060 of your annual earnings.

Example 25

Calculate the monthly NI payment a person on an annual salary of £50 000 must pay.

On an annual salary of £50 000, the monthly income subject to NI is:

$$\frac{£50\,000 - £8060}{12} = £3495$$

so the monthly NI payment is

$12\% \times £3495 = £419.40$

In Example 24, the tax deducted on an annual salary of £50 000 was calculated to be £783.58 per month. The total monthly tax and NI deductions are:

$£419.40 + £783.58 = £1202.98$

Value Added Tax (VAT)

VAT is a tax on goods and services. In 2015, the standard rate of VAT was 20%, but there was a reduced rate of 5% on some goods such as domestic fuel. VAT is not applied to most foods and children's clothing.

VAT (Value Added Tax) is a tax on most goods and services.

Discussion

Discuss the following with your peers.

1. Why is income tax necessary?
2. Why is National Insurance necessary?

Exercise 2M

1 Work out how much tax is paid by these people.

a A teacher who earns £37 496 per annum

b A company director who earns £90 000 per annum

c A shop assistant who earns £16 365 per annum

2 How much Class 1 National Insurance is paid on these monthly incomes?

a £750 **b** £1500 **c** £3875

3 How much VAT is added to these goods before they are put on sale?

a A watch with a pre-VAT price of £47

b A quarterly gas bill with a pre-VAT cost of £345

4 How much do you expect to earn in the first year of your career?

Use that amount to work out how much tax and National Insurance you would pay.

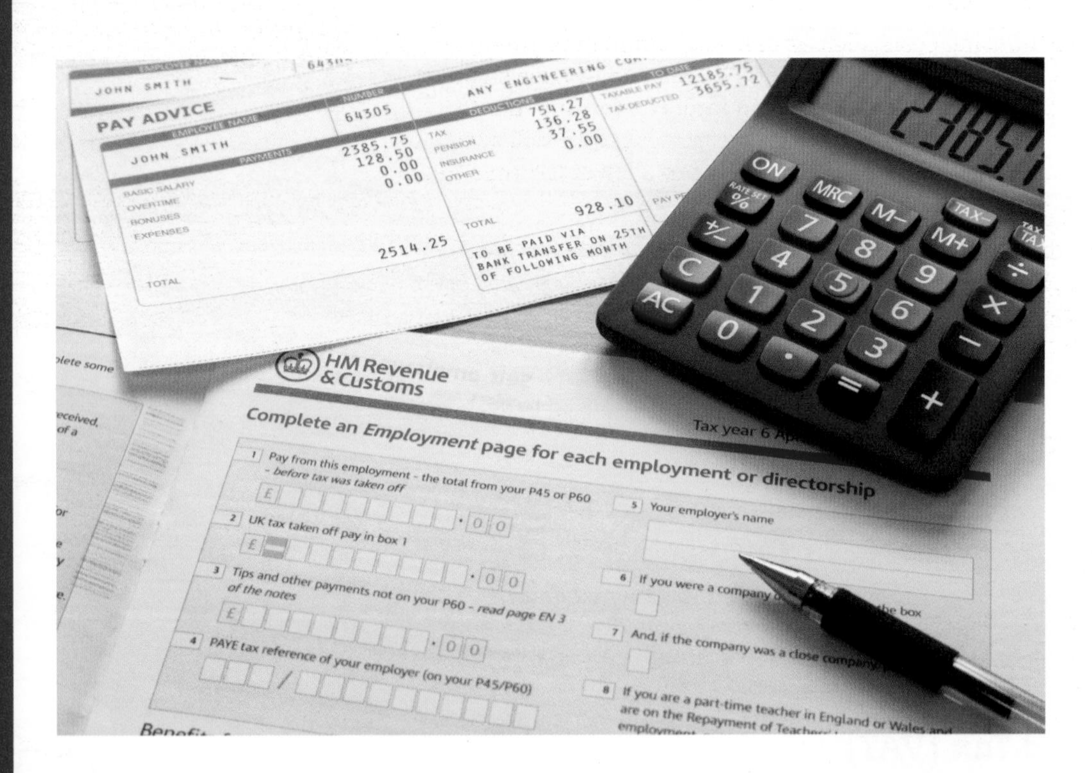

F7: Solution to financial problems

Learning objectives

You will learn how to:

- Deal with inflation, including the Retail Price Index (RPI) and Consumer Price Index (CPI).
- Find solutions to complex financial problems and work with currency exchange rates.

Introduction

Inflation is the rate of increase in prices for goods and services.

There are a number of different measures of inflation in use. The ones used most often are the **Retail Price Index** (**RPI**) and the **Consumer Price Index** (**CPI**). These look at the prices of many goods and services, such as milk, cinema tickets and bread, and track the changes in the prices of these over time.

Inflation rates are expressed as percentages. If the CPI is 3%, this means that on average, the prices of products and services are 3% higher now than they were a year ago.

The RPI includes housing costs such as mortgage interest payments and council tax, whereas the CPI does not. The RPI and CPI use slightly different formulae to calculate the inflation rate. The formula the CPI uses takes into account that when prices rise, some people will switch to products that have gone up by less. This results in a lower CPI than RPI in nearly all cases.

The data from the CPI and RPI rates are used by the government and businesses, and play an important role in setting economic policy. State benefits and many occupational pensions rise in line with the CPI. Government index-linked savings products and some train ticket prices rise in line with the RPI.

Inflation is the rate of increase in prices for goods and services.

The **Retail Price Index** (**RPI**) is used to track changes in prices of goods and commodities.

The **Consumer Price Index** (**CPI**) is also used to track changes in prices of goods and commodities.

How is inflation calculated?

Every month the Office for National Statistics (ONS) collects the price of more than 100 000 goods and services from a wide range of retailers across the country, including online retailers.

All these prices are combined using information on average household spending patterns to produce an overall prices index. It also takes into account how much people spend on different items. Therefore items are weighted so that the more people spend on them, the more importance they are given in the index.

For example, people spend more on fuel than on postage stamps, so a rise in the price of petrol and diesel would affect the overall rate of inflation more than a rise in the cost of postage stamps.

Mathematics in the real world

Hyperinflation

Interest rates and inflation are usually quite low (less than 6%) but in cases of **hyperinflation**, inflation rates can rise well above 100% and prices go up very quickly. After World War I, the German currency, the mark, lost its value and Germany suffered from hyperinflation.

The German government issued higher and higher denomination notes to cope with the demand for large amounts of money. The highest denomination bank note was for 100 trillion marks. Bank notes became worthless so quickly that they were sometimes used as wallpaper. To combat the rapidly rising inflation some towns issued their own currency, known as notgeld.

Zimbabwe suffered from hyperinflation in the early 21st century. In 1980, 1 Zimbabwe dollar (ZWD) was worth about US$1.25. By 2004, hyperinflation had reached 624% and then 1730% in 2006.

On 1 August 2006, the Reserve bank of Zimbabwe revalued the currency so that 1000 ZWD became 1 ZWN. By the end of June 2007, the yearly inflation rate had reached 11 000% and by 16 July, inflation had risen to 2 200 000%. A further revaluation was made on 1 August 2008 when 10^{10} ZWN became 1 ZWR. On 1 August 2008 it was estimated that inflation was over 11 250 000% for June that year and the next month Zimbabwe's annual inflation was 231 000 000%. This meant that prices doubled every 17.3 days.

After that there were no official figures, but it has been estimated that by mid November 2008 inflation peaked at an annual rate of 89.7×10^{21} per cent. At that rate, prices were doubling every 24.7 hours.

Solving problems

In August 2008, Zimbabwe's annual inflation was 231 000 000%. This means that prices doubled every 17.3 days. How do you know that prices doubled that quickly?

You can check it by calculating how many periods of 17.3 days there are in a year.

$365 \div 17.3 = 21.0983$ (to 6 s.f.)

So you work out $2^{21.0983} = 2\ 245\ 000$ (to 5 s.f.)

This is expressed as a percentage as 224 500 000% (to 5 s.f.).

This is close to the given figure of 231 000 000%.

Exchange rates

If you go abroad, you pay for things using the local currency. In most European countries, this will be the euro.

The table shows some exchange rates for common holiday destinations.

Pound, £	Euro, €	US dollar, $	Turkish Lira, TL	
1.00	1.3572	1.5175	3.9282	Sell
1.00	1.5879	1.7755	4.5960	Buy

You often get a better rate if you buy a larger amount.

Some exchange bureaux will buy back at the same rate, but to do that you often have to pay an extra charge. Most bureaux buy back at a different rate. It is one way that they make a profit on currency exchanges.

Example 26

George wants to change £750 into euros.

How many euros will he get?

He comes back from holiday with €80.

How much does he get when he converts his euros back into pounds?

The rate in the above table is €1.3572 to the pound, so for £750 he would expect to get:

$1.3572 \times 750 =$ €1017.90.

When he converts €80 back into pounds he gets:

$80 \div 1.5879 =$ £50.38.

Exercise 2N

1 Kate's annual pay rise is calculated by multiplying her pay by the inflation rate at the end of December.

Her salary is £15 600 and the inflation rate is 1.2%.

a What is her new salary?

b How much more is she paid each month?

Kate's monthly train season ticket goes up by the inflation rate + 1%.

Her monthly season ticket costs £236.

c What is the cost of the new monthly season ticket?

d How much better or worse off will she be after her salary goes up and her season ticket goes up?

2 Use the table of exchange rates given earlier in the chapter.

a On one chart draw the conversion graphs for buying and selling pounds and euros.

Use your chart to convert

i £1200 to euros

ii €700 to pounds.

b On another chart draw the conversion graphs for buying and selling pounds and Turkish Lira.

Use your chart to convert

i £500 to Turkish lira

ii 800 Turkish lira to pounds.

Check you have drawn your conversion charts correctly by doing the exchange rate calculations directly.

3 Where would you like to go on holiday?

Find out the exchange rate and do some research into how much a fast-food burger costs in that country and compare it with the UK.

Budgeting

Budgeting is about creating a plan for how you spend your money. The spending plan is called a **budget**. On a personal level, budgeting helps you work out whether you will have enough money to do the things you need to do or would like to do.

Budgeting is about balancing your expenses with your income to avoid running up debts. Budgeting will help you set priorities for your spending so that you can afford to spend on the things that are the most important to you.

Budgeting is also essential for all business activities. Businesses need to ensure they can afford to carry out their activities and projects without losing money. Tracking budgets as a project progresses is a vital part of running any kind of business.

Example 27

Danielle is organising a conference for 20 teachers. She needs 4 speakers. She wants to pay the main speaker £200. The other three speakers will be paid £150 each. They will not have to pay for their lunch and refreshments.

The venue costs are shown in the table

Room hire	£500
Audio visuals	£100
Buffet lunch	£12 per person
Mid-morning and afternoon refreshments	£4 per person per refreshment
Refreshments on arrival	£3 per person

Danielle has to pay 20% VAT on all these items.

She wants to spend £150 on advertising and make a profit of at least £2000 from the event. The profit will be her earnings from the event.

How much should she charge per person?

Catering per person	$3 + 4 + 12 + 4$	23
Number of people	20 teachers + 4 speakers + Danielle	25
Total cost of catering	$25 \times$ £23	£575
Venue costs	Room hire	£500
	Audio visuals	£100

So the total cost of catering and venue costs, including VAT, will be:

(Total cost of catering + venue costs) $\times$ 1.2

= £(575 + 500 + 100) $\times$ 1.2 = £1410

The cost of the speakers and advertising is:

£200 (main speaker) + 3 $\times$ £150 (other speakers) + £150 (advertising) = £800

In addition, Danielle wants to make a profit of £2000.

So the total cost of organising the event is:

£1410 + £800 + £2000 = £4210.

The amount she has to receive from the 20 teachers is therefore

£4210 $\div$ 20 = £210.50

As an experienced conference organiser, Danielle knows it's always a good idea to add a **contingency**, so she rounds the conference fee to £220 to cover any unforeseen expenses.

Exercise 20

You can use a spreadsheet for these questions

1 A college canteen offers set meals and drinks at the following prices:

Breakfast	£1.95	Coffee	60p
Lunch	£3.30	Tea	50p
Dinner	£4.10	Drink cans	70p

Guy is at college for 200 days of the year. He eats three meals in the canteen every day (including weekends) and has coffee with breakfast and also mid-morning, tea with lunch and also mid-afternoon and a can of drink with his dinner.

He pays rent of £345 per month and has to pay that every month for a year whether or not he is in residence.

How much does he have left from his maintenance grant of £8000 for transport and other essentials?

2 Lottie and her three friends share an apartment while at college. The landlady charges a total of £1350 rent per month (on a yearly lease) for the apartment that sleeps 4. Bills (gas, electricity and wifi) are not included in the rent and must be paid by Lottie and her friends.

Lottie estimates that the quarterly bills will be

gas	£200
electricity	£300
wifi	£120.

Since Lottie is to be responsible for the direct debits paying all these bills, her friends decide that they will each pay a quarter of the utility bills and give Lottie their share of the bills, but Lottie should only pay one half of what each of the others pay in rent.

- **a** What is the total annual cost of the rent plus utilities?
- **b** What is Lottie's share of the rent?
- **c** How much does Lottie pay in total each year?

3 Jitesh runs a football club for under-10s at the local primary school. The school gives him a budget of £500 to cover equipment and other costs.

Goals are £23.99 each and footballs are £8.99 each.

Shirts, socks and shorts are £10 per kit (minimum order, 6 kits).

He has to pay the caretaker £15 for each session.

Jitesh wants to run the club for one hour each week for 20 under-10s, providing each one with kit. The children are expected to bring their own football boots.

- **a** Decide how many goals and footballs Jitesh should buy.
- **b** On the basis of your answer to part **a**, how many weeks can the club run for before the budget runs out?
- **c** If Jitesh wants to run the club for all 39 weeks of a school year, how much should he charge the parents of each child?

Case study

Adam keeps a record of the miles he drives so that he can estimate his business expenses on his tax return. He keeps track of this in a spreadsheet, part of which is shown here. The first column shows the miles that are not work related. The other columns refer to his three main work categories and other works. Some of the data is hidden to save space.

He uses formulae in the spreadsheet to add up the miles in each column and then to calculate the percentage used on business.

He keeps a separate record of his tax, insurance and garage bills so that he knows how much his business use of the car costs him.

Adam also keeps a separate spreadsheet on his taxable income and allowances. His savings are taxed at source at 20%. On his tax form he has to say how much interest he has received and how much tax has been deducted already. This will show if his total earnings take him into the 40% tax band.

He receives some income from benefits and pensions. He also has some part-time work and does some freelance work. His benefits and pensions are taxed and the part-time work is taxed at source by his employer.

As well as his personal allowance, Adam can claim tax relief on his professional subscriptions as these are essential to the work he does.

Remember that **income** is **rounded down** and **allowances** are **rounded up** to the nearest pound.

You see that his total taxable income is £26 917 – £400 = £26 517. This falls within the 20% tax band.

His personal allowance is £10 600 so that is deducted to mean he has to pay 20% tax on £15 917. So he must pay £15 917 × 0.2 = £3183.40 in tax.

He has already paid a total of £2315.40, however, so he just has to pay an extra £3183.40 – £2315.40 = £868 this year.

G31 | fx =100*G29/(G28+G29)

	A	B	C	D	E	F	G
1	Home use	Work 1	Work 2	Work 3	Other works		
27	86.7				301.3		
28	10.2				151.6	home mileage	2418
29	22.4				10.3	business mileage	10978.2
30	48.7				233.9		
31	33.3				9.1	% business use =	82
32	2.9				346.7		
33	5.2				277.6		
34	31.9				141.9		
35	10				151.7		
36	10.1				8.4		
37	66.9				9.3		
38	5.2				9.1		
39	27.1				8.4		
40	31				341.2		
41	2.2				281.8		
42	5.1				9.1		
43	143.7				10.9		
44	12.6						
45	49.6						
46	5.1						
47	106.2						
48	5.1						
49	8.1						
50	12.8	782.6	3747.6	1452.3	4995.7		
51	5.1						

	A	B	C
1			tax paid
2	Interest on savings (after tax)	£229.09	£57.26
3			
4	Earnings	taxable pay	tax paid
5	as employee	£7,724.91	£1,544.80
6	self employed source 1	£1,650.00	
7	self employed source 2	£193.74	
8	self employed source 3	£200.89	
9	self employed source 4	£1,975.00	
10	self employed source 5	£1,430.00	
11	Pension 1	£5,136.90	£745.40
12	Pension 2	£8,606.04	£25.20
13			
14	Totals	£26,917.48	£2,315.40
15			
16	Allowances	non-taxable	
17	Professional subscription 1	£52.43	
18	Professional subscription 2	£90.00	
19	Professional subscription 3	£65.00	
20	Professional subscription 4	£35.00	
21	Professional subscription 5	£60.00	
22	Professional subscription 6	£97.00	
23			
24	Total	£399.43	

Project work

Project work – 1

Find out more about the CPI, the goods that are used in its production and the weightings of the goods and services. You will find the Office for National Statistics on the internet very useful.

Project work – 2

Choose one of the following and cost it out.

1 You are planning your 18th birthday party. Work out a budget for the party. You'll need to think about:
   - number of guests
   - cost of a venue
   - cost of invitations
   - cost of hiring a DJ or a band
   - cost of food and drink.
2 You want to spend your holidays travelling around Europe for a period of your choice. Decide on an itinerary and the total cost of the holiday. You have to fund this yourself. How will you pay for it?
3 You are planning to go to university in the future. Set out a budget for your first year based on either living in college or sharing a flat with a friend.
4 You want to redecorate your bedroom. Plan what needs to be done and cost it out as accurately as possible. Your parents insist that you pay half the cost so budget carefully.
5 A rich uncle is going to give you a second-hand car for your 18th birthday. Estimate the running costs for a year. Use online research to find the costs of road tax and car insurance. Can you afford to run a car? Would it be cheaper to use public transport?

Check your progress

How confident are you feeling in your level of knowledge? What do you need to practise more?

Spec reference	Learning objective			
F1.1	Substitute numerical values into formulae, spreadsheets and financial expressions			
F1.2	Use conventional notation for priority of operations, including brackets, powers, roots and reciprocals			
F1.3	Apply and interpret limits of accuracy, specify simple error intervals due to truncation or rounding			
F1.4	Find approximate solutions to problems in financial contexts			
F2.1	Interpret percentages and percentage changes as a fraction or a decimal and interpret these multiplicatively			
F2.2	Express one quantity as a percentage of another			
F2.3	Compare two quantities using percentages			
F2.4	Work with percentages over 100%			
F2.5	Solve problems involving percentage change, including finding the original value			
F3.1	Calculate simple and compound interest, including the Annual Equivalent Rate (AER)			
F3.2	Work with savings and investments			
F4.1	Work with student loans and mortgages, including the Annual Percentage Rate (APR)			
F5.1	Plot points to create graphs and interpret results from graphs in financial contexts			
F6.1	Work with income tax, National Insurance and Value Added Tax (VAT)			
F7.1	Deal with the effect of inflation, including the Retail Price Index (RPI) and Consumer Price Index (CPI)			
F7.2	Set up, solve and interpret the solutions to financial problems, including those that involve compound interest using iterative methods			
F7.3	Work with currency exchange rates including commission			
F7.4	Manage a budget			

3 Estimation

There is an anecdote about how, scientists at a university once solved a problem about increasing milk production at a dairy farm by collecting data at the farm then going away for a couple of weeks to analyse that data. When they returned to the farmer they said "We have the solution, but it only works for a spherical cow in a vacuum."

Clearly a cow is a very complex shape! To make the situation easier the scientists had looked for a simple shape that could resemble a cow and had chosen a sphere. It would also depend on many other factors, so to keep things simple the scientists decided to ignore the many other factors.

Modelling is mainly applied mathematics and gives a means of constructing mathematical situations in which problems can be solved. Most of the mathematics you will have met so far at GCSE Mathematics has been clinically presented and might have seemed at times to have little to do with the real world. That is because mathematics in the real world does not come as neat little questions, but as larger challenges that are solved by making appropriate assumptions.

For example, Newton's mechanics is a model that approximates what happens in the real world. It is quite sufficient for everyday life situations as long as you deal with speeds well below the speed of light and with macro-particles. In this chapter you will be developing confidence and competence to tackle complex problems that have no single correct answer. Your work will depend on the assumptions that you make and so it is very important that you write these down so that anyone who looks at your work will know what you are assuming right at the start. Provided that your assumptions are reasonable, you will come to good sensible answers. The use of spreadsheets and

calculators (including graphing calculators) will be assumed so you are encouraged to use them.

So, to get things started, how would you model a cow? What shape might be better than a sphere?

What about a snake? Would you model that in the same way as a cow?

How might geographers model a volcano?

How might car manufacturers model a car?

Discussion

Discuss with your peers instances when you have used mathematics in the real world to come up with a solution to a challenge that you have faced. These may vary from estimating whether you can cross a road safely to budgeting for Christmas. How did you solve these problems? What technology (if any) did you use? Why did you want to solve those problems? Were you successful? Why? / Why not?

E1: The modelling cycle

Learning objectives

You will learn how to:

- Represent a situation mathematically, make assumptions and simplifications.
- Select and use appropriate mathematical techniques for problems and situations.
- Interpret results in the context of a given problem.
- Evaluate methods and solutions including how they may be affected by assumptions made.

Introduction

The complete modelling cycle is shown here:

Modelling is the process of applying mathematics to a real-world problem to solve it.

Validating is checking that the solution found is realistic.

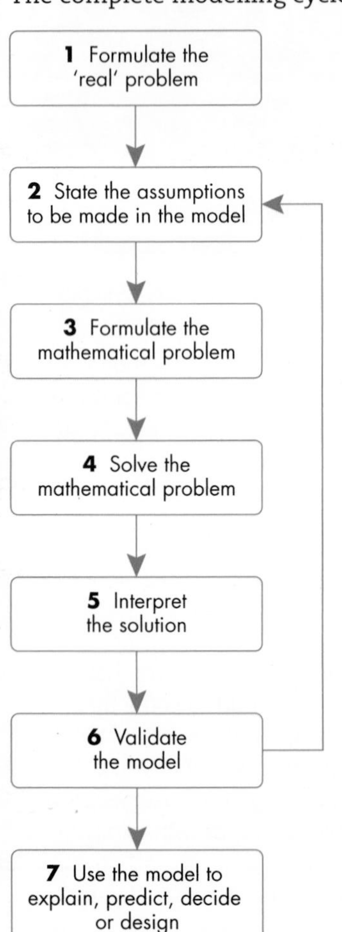

It the model cannot be validated (for example if the solution in the mathematical world does not match what happens in the real world) then the assumptions made need to be changed and the cycle repeated until the model is validated.

A simplified version of the model is shown here:

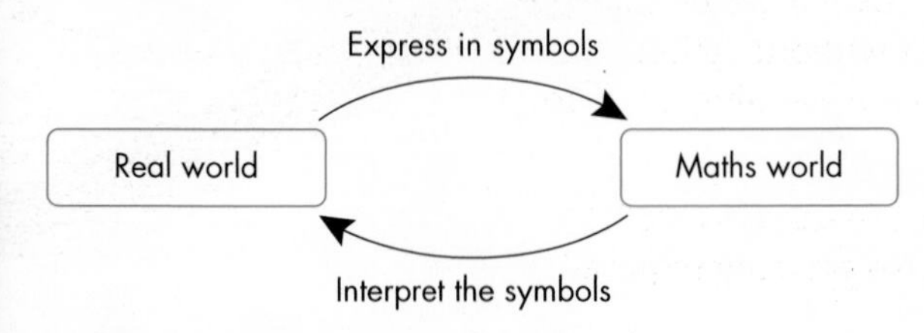

Working through the modelling cycle: Fizzy drinks

What is the best shape for fizzy drinks cans?

Step 1: The problem

To figure out why fizzy drinks cans are the shape they are.

Step 2: The assumptions

The firm that makes the cans wants them to:

a. contain a suitable amount of drink (330 ml);

b. be as cheap as possible, so needs to use as little material as possible;

c. come out of dispensing machines easily;

d. be easily packable into boxes so as little space as possible is wasted.

To model the situation it is simplified by ignoring the smaller details such as making any necessary welding to join the lid to the can, the problems with incorporating a ring-pull in the lid, any hollows in the base, etc. Once these have been put to one side the best shape can be found and then the details can be worked on separately.

Discussion

Discuss these questions in groups of three or four.

1. What shapes could the cans be?
   (Ignore the criteria above for the time being.)
2. Which of the shapes would match criterion **c**? Why?
3. Which of the shapes would match criterion **d**? Why?

So a compromise has to be reached between criteria **c** and **d**.

4. What other criteria might the manufacturing firm have?
5. What shape would you use for the can?

The manufacturing firm decides to reject the first two of the following shapes for the cans, for the following reasons.

Cube: Although it will pack without any gaps, it would not roll out of a dispensing machine easily. The sharp corners might damage people. There would be too many edges to weld.

Sphere: Although it would roll easily and use very little material compared to a cube, it could not easily be put down without rolling around and spilling the drink.

Hexagonal prism: A possibility as if it is a regular hexagon it would roll and pack reasonably well. It would also stand upright without spilling the drink. It would need to have the edges creased and some welding done. To be considered.

Cylinder: Another possibility as it would roll and pack reasonably well. It would also stand upright without spilling the drink. No creasing required but some welding needed. To be considered.

Step 3: The mathematical problem

You now have to compute the surface area of a regular hexagonal prism and a cylinder that gives a volume of 330 cm^3 and find the best solution: the one that minimises the amount of metal used.

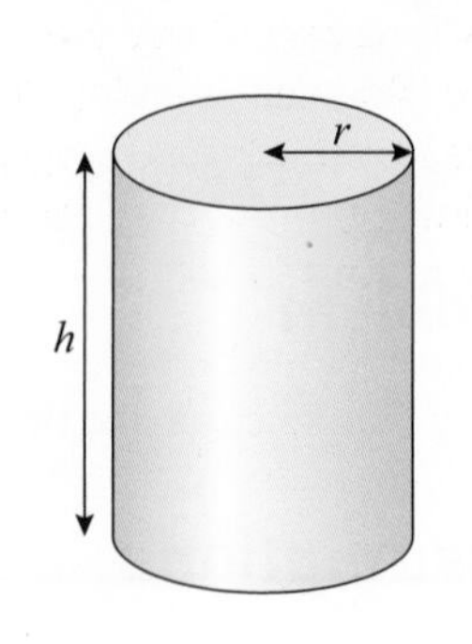

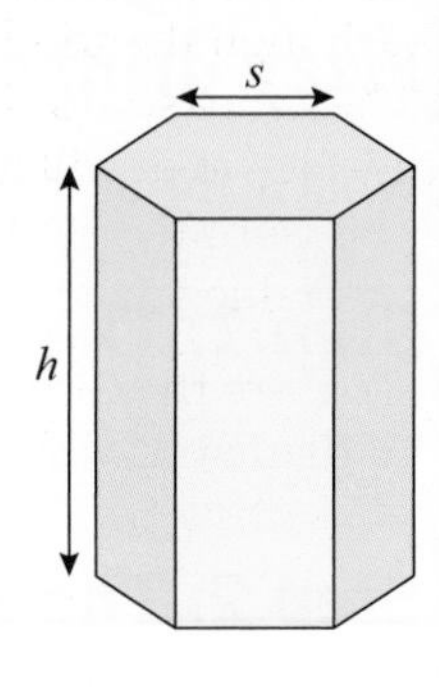

Step 4: Solving the mathematical problem

The case of a hexagonal prism will be worked through here and you are expected to do similar work for the cylinder.

The surface area of a prism is found by adding the area of the six rectangular sides to the area of the two hexagons on the top and bottom of the prism.

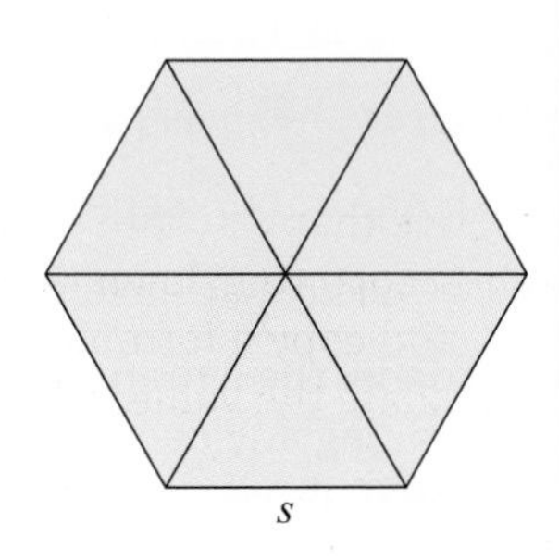

So you need to find the area of the hexagon.

A regular hexagon consists of six equilateral triangles as shown here.

The area of a triangle is $\frac{1}{2} \times \text{base} \times \text{height}$.

The height of the equilateral triangle can be found by Pythagoras' Theorem and this turns out to be $0.866s$, where the length of one side is s.

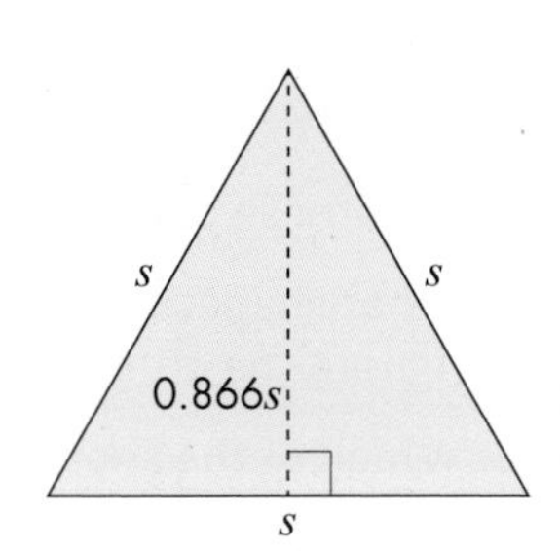

So the total surface area is $6 \times sh + 2 \times 6 \times \frac{1}{2} \times s \times 0.866s$.

This can be simplified to $6sh + 5.196s^2$.

So the total surface area, A, is given by $A = 6sh + 5.196s^2$.

The volume of any prism is found by multiplying the area of the cross-section by the length of the prism, so the volume, V, of the hexagonal prism is given by $V = 6 \times \frac{1}{2} \times s \times 0.866 \times h = 2.598s^2h$.

But you want the volume to hold 330 ml, so the volume must be 330 cm^3.

This means you need to find values of s and h such that $2.598s^2h = 330$.

This can be rearranged to find a formula for h: $h = \frac{330}{2.598s^2}$.

Exercise 3A

1 Find an expression that gives the total surface area of a cylinder that has radius r and height h.

2 Find the formula for the volume of the cylinder.

3 Write down the equation that the values of r and h have to satisfy for the cylinder.

4 How would the formulae change if the can had to hold 500 ml?

Now use a spreadsheet to calculate the height, h, and total surface area, A.

	A	B	C
1	Side of hexagon, s, in cm	Height, h, in cm	Total surface area, A, in cm^2
2		$=330 \div (2.598s^2)$	$=6sh+5.196s^2$
3	1	127.02	767.32
4	2	31.76	401.85
5	3	14.11	300.81
6	4	7.94	273.67
7	5	5.08	282.32
8	6	3.53	314.08
9	7	2.59	363.48
10	8	1.98	427.81

Notice that as the side of the hexagon increases, the height decreases, but, although the total surface area decreases at first, it then starts to increase. Therefore somewhere between the side being 3 cm and 5 cm, the total surface area will have its smallest value, called its minimum value, and if you want to use the least amount of metal in making the can, that will be the value you will try to find.

Here are the formulae used in the spreadsheet.

	A	B	C
1	Side of hexagon, *s*, in cm	Height, *h*, in cm	Total surface area, *A*, in cm^2
2		=330÷(2.598*s*^{2})	=6*sh*+5.196*s*^{2}
3	1	=330/(2.598*A3^2)	=6*A3*B3+5.196*A3^2
4	=A3+1	=330/(2.598*A4^2)	=6*A4*B4+5.196*A4^2
5	=A4+1	=330/(2.598*A5^2)	=6*A5*B5+5.196*A5^2
6	=A5+1	=330/(2.598*A6^2)	=6*A6*B6+5.196*A6^2
7	=A6+1	=330/(2.598*A7^2)	=6*A7*B7+5.196*A7^2
8	=A7+1	=330/(2.598*A8^2)	=6*A8*B8+5.196*A8^2
9	=A8+1	=330/(2.598*A9^2)	=6*A9*B9+5.196*A9^2
10	=A9+1	=330/(2.598*A10^2)	=6*A10*B10+5.196*A10^2

Notice that you start with 1 in cell A3 and in cell A4 you first put = A3 + 1 then copy this down the column. Similarly in cell B3 you replaced s with A3 and copy a formula down. Cell C3 uses the values in cells A3 and B3 to calculate the values of A.

To zoom in on the minimum value, change the increments from 1 to 0.1, starting with 3.

	A	B	C
1	Side of hexagon, *s*, in cm	Height, *h*, in cm	Total surface area, *A*, in cm^2
2		=330÷(2.598*s*^{2})	=6*sh*+5.196*s*^{2}
3	3	14.11	300.81
4	3.1	13.22	295.78
5	3.2	12.40	291.37
6	3.3	11.66	287.53
7	3.4	10.99	284.22
8	3.5	10.37	281.40
9	3.6	9.80	279.04
10	3.7	9.28	277.11
11	3.8	8.80	275.59
12	3.9	8.35	274.45
13	4	7.94	273.67
14	4.1	7.56	273.23
15	4.2	7.20	273.12
16	4.3	6.87	273.31
17	4.4	6.56	273.80

Now you see that when $s = 4.2$, A is at its lowest value.

You could now redo the spreadsheet taking smaller steps for s, of 0.01, but measuring to the nearest mm should be accurate enough.

Exercise 3A (continued)

5 Set up a spreadsheet like this for the formulae you found earlier in the exercise for the cylinder and find the best dimensions for a cylindrical fizzy drink can.

Step 5: Interpreting the solution

From the spreadsheet you find the minimum value of A occurred when $s = 4.2$ and $h = 7.20$.

This means the hexagonal prism has a side of 4.2 cm and height of 7.2 cm.

You could also use graph-plotting software to find the minimum value.

You know that $A = 6sh + 5.196s^2$ and $h = \dfrac{330}{2.598s^2}$ so you can replace h in the formula for A to get $A = 6s \times \dfrac{330}{2.598s^2} + 5.196s^2$ or $A = \dfrac{1980}{2.598s} + 5.196s^2$.

Here is the graph of this with the minimum point.

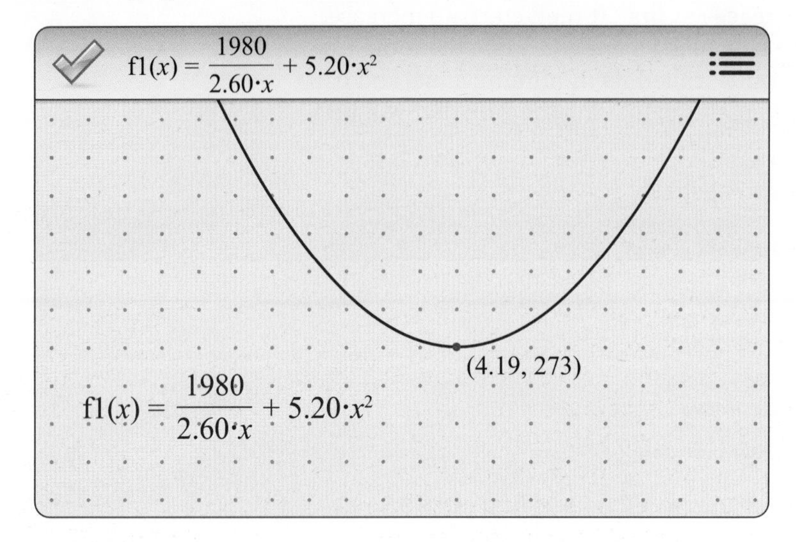

Notice that the software has done some rounding in places.

Step 6: Validating the model

You can now draw a net of the hexagonal prism, which has a side of 4.2 cm and height of 7.2 cm, and make it up.

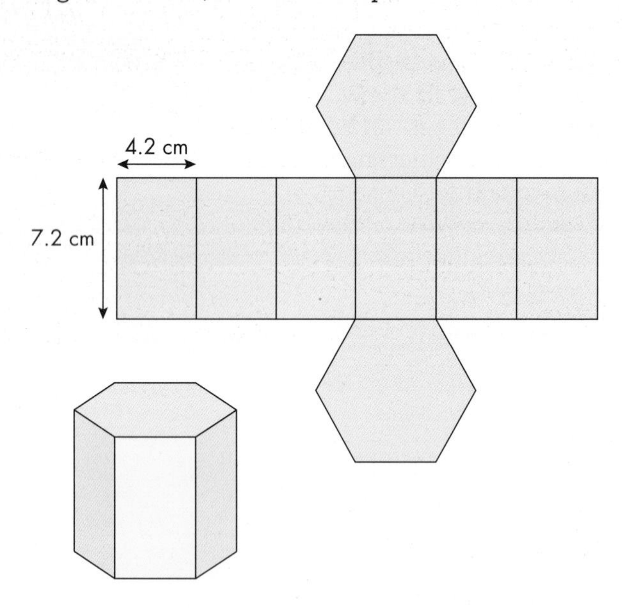

Making the solid from the net means you can check that it rolls. You know that it uses the least amount of metal from the work with the spreadsheets and graph, so you now check that this would work in the real world.

Of course this is a crude model. It has not taken into account whether the can would be watertight or whether it could be easily constructed. However, you would be able to see whether the cylinder gives a better result.

Step 7: Use the model to explain, predict, decide or design

To finish you can compare the model to the actual design of fizzy drink cans. Compare your result for the cylinder with an actual drinks can that contains 330 ml. How close were the dimensions of your model to reality?

Exercise 3A (continued)

6 When you are ready, summarise your findings on one sheet of A3 paper and split your group into two parts. One half should stay with your work to answer any questions and the other half should go round and see what the others have done, asking questions and making suggestions. When your group is back together consider how you could have improved your work.

Using modelling in sport

Nowadays mathematics and science play a big part in sport. To break records that seem to be on the limit of human achievement, sports scientists analyse the movement of various joints and the best way to maximise the potential of the human body. Sensors are put on people to measure their blood pressure, temperature, heart rate, movement, muscle contraction, air resistance, etc., then the data is analysed to see if any improvements are possible. Using high-speed cameras, sport scientists can follow the movement of the human body and then models are used to see if there are more efficient ways of running, throwing, rowing and cycling, etc. Mathematical modelling plays a major part in this.

British cycling has benefitted from sports science and all the mathematical modelling that is included in its programme. In 2012, British cyclists won eight gold medals at the London Olympics, matching the tally of the Beijing Olympics four years earlier. The Paralympics squad won medals in all 15 events they entered, winning five gold, seven silver and three bronze. In the 2012 Tour de France, team members Bradley Wiggins and Chris Froome finished first and second overall, while the World Champion Mark Cavendish won three stages. Of course in 2013 and 2015, Chris Froome won the Tour de France – an incredible achievement. In the past 20 years Great Britain has moved from mediocrity to become one of the most feared but well-respected cycling nations in the world.

You will now explore some running records to see whether there are any patterns that allow you to make predictions for the future.

Running a mile by men

The data below shows the records for running a mile since 1930. Each shows the date the record was set, and the time the runner achieved.

(Some records that are very close in dates or times have been omitted so that there is a manageable amount of data here.)

Date	4/10/31	15/7/33	16/6/34	28/8/37	1/7/42	4/9/42	1/7/43	18/7/44	6/5/54
Time	4:09.2	4:07.6	4:06.8	4:06.4	4:06.2	4:04.6	4:02.6	4:01.6	3:59.4

Date	21/6/54	19/7/57	6/8/58	17/11/64	9/6/65	17/7/66	17/5/75	12/8/75	17/7/79
Time	3:58.0	3:57.2	3:54.5	3:54.1	3:53.6	3:51.3	3:51.0	3:49.4	3:49.0

Exercise 3B

1 Use this data to plot a graph and use it to predict when the record might reach 3 minutes 43 seconds.

2 What assumptions have you made in question **1**?

3 When do you think the record will drop to 3 minutes 30 seconds?

Discussion

Discuss the following with your peers.

What do you think is the lower bound for the time to run a mile? Why?

Running long distances by women

Here are the distances and the times of the world records (July 2015) for increasing distances for women.

Distance on track (m)	100	200	400	800	1000	1500	2000
Time (h:mm:ss.ss)	10.9	21.34	47.60	1:53.28	2:28.98	3:50.07	5:25.36

Distance on track (m)	3000	5000	10 000	20 000	25 000	30 000	100 000
Time (h:mm:ss.ss)	8:06.11	14:11.15	29:31.78	1:05:26.6	1:27:05.8	1:45:50	
Distance on road (m)			10 000	20 000	25 000	30 000	100 000
Time (h:mm:ss)			30:21	1:01:54	1:19:53	1:38:23	6:33:11

Key

sprint   middle distance   long distance

Exercise 3C

1 Use this data to plot a graph and use it to predict what the long distance track record might be for 7500 m.

2 What assumptions have you made in question **1**?

3 Find a model that fits the sprint data and one that fits the long distance track data.

4 Find a model that fits the long distance road data.

Why do you think it is different from the long distance track data?

Braking distances

The Highway Code publishes a chart of the braking distances needed when travelling at various speeds. You will need to know these distances in order to pass your driving theory test and as mathematicians you should be interested in the modelling cycle and reasons behind these distances.

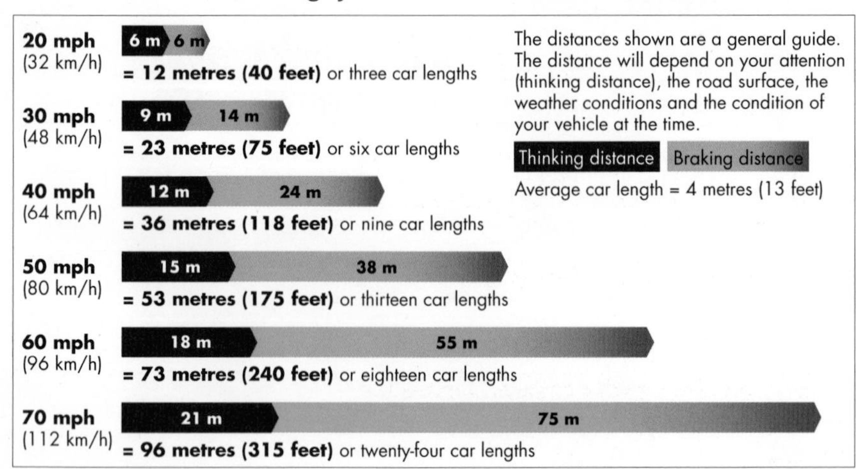

Speed, u (km/h)	Thinking distance, d_t (metres)	Braking distance, d_b (metres)	Stopping distance, d_s (metres)
32	6	6	12
48	9	14	23
64	12	24	36
80	15	38	53
96	18	55	73
112	21	75	96

Thinking distance is the distance travelled by a vehicle in the time it takes the brain to see the hazard to the application of the brakes.

Braking distance is the distance travelled by a vehicle once the brakes are applied before it stops.

The calculations behind the distances use a number of assumptions.

- The thinking distance is directly proportional to the speed.
- The thinking time is independent of the speed, the person's reaction times and road conditions.
- The braking distance is directly proportional to the square of the speed, independent of the condition of the car's braking system and road conditions.

(Steps 2 and 3 of the modelling cycle) You can see that they have listed some of those assumptions on the chart.

Exercise 3D

1 Use the first two columns of the table above to check that the assumed thinking time is $\frac{27}{40} = 0.675$ seconds.

2 Plot the graph of thinking distance, d_t, against speed, u.

Use the information to express d_t in terms of u.

(Step 4 of the modelling cycle)

3 Plot the graph of the braking distance, d_b, against the square of the speed, u^2.

What sort of a graph do you get?

Use the information to express d_b in terms of u^2.

(Step 4 of the modelling cycle)

4 Use the information from questions **1** and **2** to express d_s in terms of u.

(Step 5 of the modelling cycle)

5 What do you think are the limitations of this model?

(Step 6 of the modelling cycle)

Pizza delivery service

Modelling is important when setting up or running a business.

Large food franchises use modelling to offer standard sized portions at prices that maximise profits.

Imagine you are opening a pizza delivery business. Some of the questions you would ask might be:

- What size pizzas are best?
- What is the best price to charge, considering your costs and expected profit?
- Should there be a delivery charge?
- What is the competition charging and can that be bettered?

Modelling the situation before setting up a franchise store is an important part of any business plan.

For Exercise 3E, try to collect real data from local pizza shops. You can show your data in a table like the one shown here.

Diameter (in inches)	7	9.5	11.5
Number of slices	4	6	8
Price of basic pizza	£4	£8	£10
Price of moderate pizza	£5	£9	£12
Price of luxury pizza	£7	£11	£14

Exercise 3E

1 Imagine you are opening a pizza delivery business.

Rank the four questions given above in order of importance to you.

Based on achieving your prime objective, write down any assumptions you are going to make about the pizzas.

2 Using any relevant assumptions, find a formula that models the number of slices in a pizza based on the pizza's diameter.

3 Using any relevant assumptions, find a formula that models the cost of a basic pizza depending on its diameter.

Then find formulae that model the cost of a moderate pizza depending on its diameter and the cost of a luxury pizza depending on its diameter.

Comment on the similarities and differences of the three formulae.

4 Produce a mathematical model for the cost of how much a family will spend on this type of pizza in a year. Think about the number and ages of people in the family, how much they can eat, how often they might eat pizza, etc. These are your estimations based on your own experiences.

Remember that you need to write down your assumptions before you start. You might want to change them later, but that is all part of the modelling procedure.

Daniel in the lions' den

How did Daniel manage to survive overnight in the lions' den? (Step 1 of the modelling cycle)

(If you don't know the Daniel and the lions' den story, look it up online.)

One possibility could have been that he knew about mathematics and about lions.

Lions are very territorial and so will not intrude into another lion's territory. This presents a problem when they are kept in captivity as their territory is constantly changing. So if there were two lions, all Daniel had to do was to remain equidistant between them. Then neither lion would eat him as they would not see him as being in their territory. Of course, as the lions move around, that boundary is constantly changing, but Daniel is still able to move so that he is equidistant between them.

To begin with, model the lions' den as a rectangle. Also, to keep things simple, start by having just two lions and estimate that they move at the same speed and that Daniel can estimate distances so can keep the same distance between them. (Step 2 of the modelling cycle)

The **locus** (plural **loci**) is the path of an object under certain conditions.

So, you can build up the model and show the locus of Daniel at various positions of the lions. (Step 3 of the modelling cycle)

If you have access to a computer with dynamic geometry software, use it: that way you can move the points representing the lions around and see how the loci change.

Exercise 3F

You will need mathematical drawing instruments for this exercise.

1 Draw a rectangle and mark two points, L_1 and L_2 to represent the positions of two lions.

Construct the perpendicular bisector of L_1L_2. This is the locus of Daniel's safe positions.

(Step 4 of the modelling cycle)

2 Draw a new rectangle and mark points, L_1, L_2 and L_3 to represent the positions of three lions.

Construct the perpendicular bisectors of L_1L_2, L_1L_3 and L_2L_3.

Where is the locus of Daniel's safe positions now?

Remember that if he is closer to one lion than any other, that lion considers him to be in its territory.

(Steps 4 and 5 of the modelling cycle)

3 Is there any way the three lions can combine to manage to trap Daniel?

(Steps 6 and 7 of the modelling cycle)

4 Repeat the process with four lions if you want to practise your drawing skills.

5 Daniel was fortunate to be put in the lions' den rather than a lion's den. Why?

Siting a supermarket warehouse

Many of the big supermarket chains have set up smaller local supermarkets in towns and cities. One problem faced by supermarket owners is where to site the warehouses to minimise the travel required to supply the supermarkets with what they need. In Exercise 3G, you will work through a series of questions that will give you a feel for the modelling cycle in this situation.

Exercise 3G

1 Write a list of assumptions that you will consider for siting a warehouse to supply smaller local supermarkets.

  There are no right or wrong assumptions at the moment and not all assumptions will be used.

2 In the simplest case there is just one supermarket to be supplied by a warehouse.

  Where would you put the warehouse? (Yes, the answer is obvious!)

3 Now consider supplying two supermarkets a distance D km apart by one warehouse.

  Also assume that the transport costs are £C per km.

  a If the demand at each one is the same, where would you put the warehouse?

  b If the demand at one is twice the demand at the other, where would you put the warehouse?

4 Now consider three supermarkets (not in a straight line) with equal demands.

  Where would you put the warehouse?

5 Now consider four supermarkets (not in a straight line) with equal demands.

  Where would you put the warehouse?

One crude and simple way of solving the general case is to make a map of the supermarkets and then put weights on strings, the mass of each weight depending on the demand of the supermarket. Join the ends together and where this lands (A, the position of the minimum potential energy) is the optimum position.

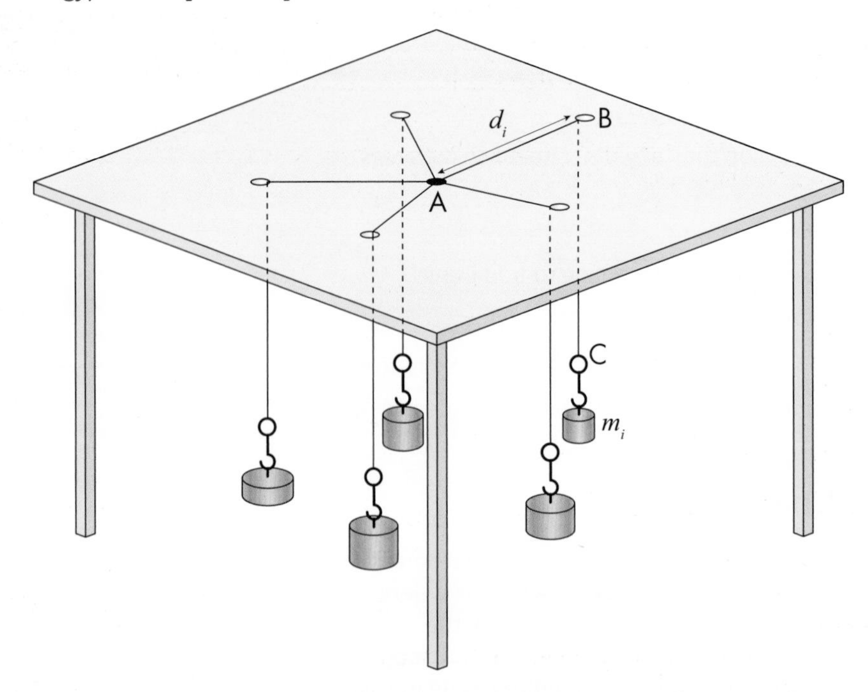

Of course this doesn't take into account things like the road system and the suitability of the land, but it does give a reasonable approximation to the site of the warehouse. Other considerations can come into play later.

Mathematics in the real world

You have probably seen reduced speed limits on motorways and roads where road works are taking place. At times a limit of 20 mph on a road in a built up area seems rather slow when all the traffic is progressing well and you think they could all go faster. The reason why the traffic flows well is because people keep the correct distance between each other and travel at that constant speed of 20 mph! Any faster and cars start to get too close (because they won't leave the correct distance between each other) and then brake. That makes the car behind brake and so on, and this causes traffic to slow even more and hold ups start to occur. If the speed limit is increased then the distance between cars needs to be greater and so fewer cars go past a given mark in a given time. Not only that, the damage done to pedestrians in accidents involving cars is very serious at speeds over about 25 mph.

Here is the modelling behind maximum flow rate.

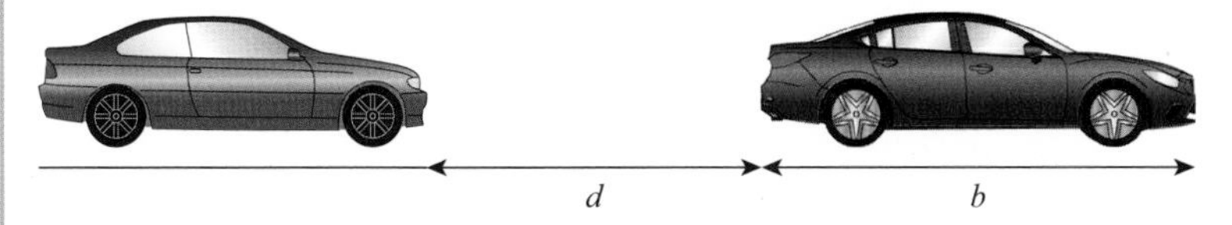

The aim is to find the optimum speed to get the biggest number of cars past a point in a given time.

If one car has passed a fixed point, the time taken for the next car to pass the same point is the total distance needed for that car (which is the length of the car, b, plus the stopping distance, d) divided by the speed, u, of the car.

Remember: $\text{time} = \dfrac{\text{distance}}{\text{speed}}$.

It is essential to be consistent with units, so if speeds are given in kilometres per hour and lengths in metres, you need to convert km/h to m/s.

If u is the speed in km/h, this is converted to m/s with the expression $\dfrac{u \times 1000}{60 \times 60} = \dfrac{5u}{18}$ m/s.

Replacing d, b and u in the equation for time gives:

$$\text{time} = \frac{d+b}{\frac{5u}{18}} = \frac{18(d+b)}{5u}$$

The flow rate f is the reciprocal of the time, $\dfrac{1}{t}$.

$$\text{flow rate, } f = \frac{5u}{18(d+b)}$$

Assume that the cars are travelling at the same speed and each car has the same stopping distance.

Now from Exercise 3D question 4 you know that $d = 0.1875u + 0.00598u^2$ and assume that the length of a car is 4 metres.

$$\text{So } f = \frac{5u}{18(0.1875u + 0.00598u^2 + 4)}.$$

Using a graph plotter you can find the maximum value of u, which is about 26 km/h or 16 mph.

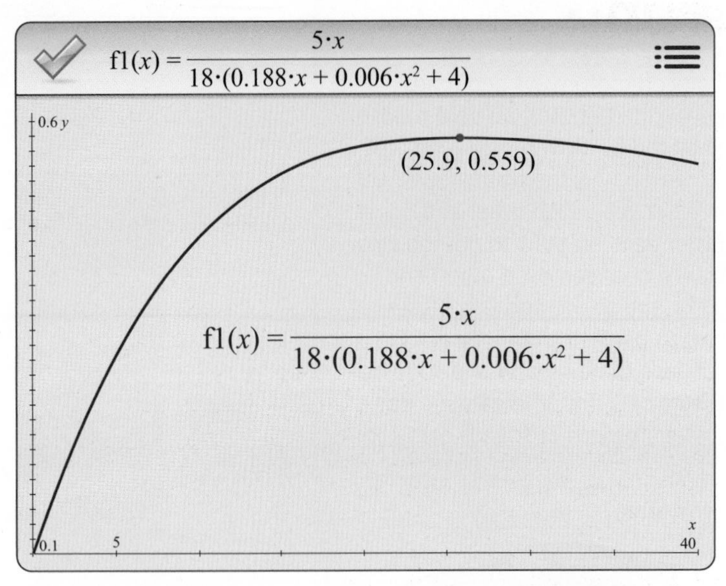

In real life people drive closer than the recommended distance and a more realistic model using experimental data is to take $d = 0.1875u + 0.003u^2$.

This gives a value of u of about 37 km/h or 23 mph, so setting a speed limit of 20 mph is a suitable compromise.

Discussion

Discuss with your peers the steps in the modelling cycle that have been used in the example above.

E2: Fermi estimation

Learning objectives

You will learn how to:

- Make fast, rough estimates of quantities which are either difficult or impossible to measure directly.

Introduction

Fermi estimation, or Fermi problems, are the sort of rough and ready, or 'back of the envelope', calculations that are made to get a reasonable estimate to a given problem when a lot of information is lacking.

Fermi estimations are rough and ready estimations used to get an idea of the size of the answer.

They are named after the physicist Enrico Fermi who was able to make good estimates with very little data. A classic example is Fermi's estimate of the strength of the atomic bomb detonated at the Trinity test on 16 July 1945. He based his estimate of 10 kilotons on the distance travelled by pieces of paper he dropped from his hand. The accepted value of the test is 20 kilotons, so Fermi's estimate was of the correct order of magnitude.

The essential part of any Fermi estimation is to state your assumptions clearly. Without them your work cannot be checked and will not gain any credit. You are not going to be judged on whether your assumptions are correct as there will often not be the data to check those assumptions. If you use numbers in your assumptions, you will usually be multiplying and dividing the numbers, rather than adding and subtracting.

Example 1

How many piano tuners are there in Liverpool?

To estimate an answer, assumptions are needed about the number of pianos in Liverpool. The number of pianos will depend on the number of people, so start by thinking about how many people there are in Liverpool. Here is a set of assumptions to be used in the estimate.

1 There are about 2 000 000 people in the Liverpool area.

2 There are on average five people in a household.

3 About one household in 50 has a piano that is tuned once a year.

4 A piano tuner can tune two pianos a day.

5 A piano tuner works 5 days a week for 50 weeks.

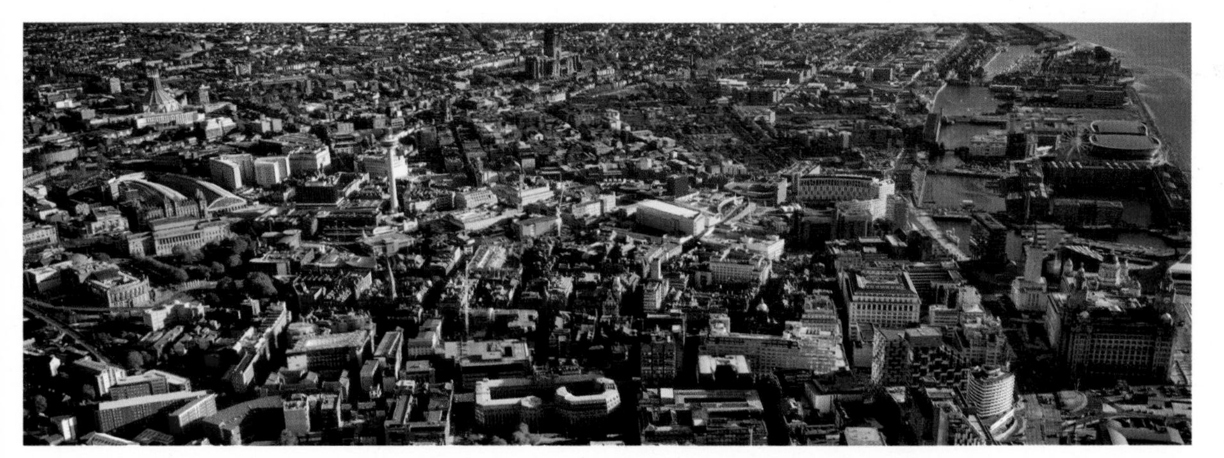

And here is the reasoning behind those estimates.

1 Liverpool is a big city but London, Birmingham and Manchester are probably larger. As a ball park figure, the population of London is about 8 000 000 and the populations are about 3 000 000 for Birmingham and Manchester. From this, estimate that the population of Liverpool is 2 000 000. Note that this is a nice round number which is easy to work with.

2 The number in a family varies. Estimate five people in each household. It might be more or less, but ten would be a reasonable upper bound and one is a lower bound so the figure in the middle seems reasonable.

3 Assume an upper bound for the number of households with a piano is two households out of every 50 and a lower bound is no households in 50. The middle of these assumptions is one household per 50 households, hence 1 in 50.

4 Take into account the need to travel and the time taken to tune a piano. Assume an upper bound of four in a day and a lower bound of zero, so the mean number is 2.

5 5 days a week is a standard working week and 50 is a convenient number with which to work.

Notice that the reasoning uses upper and lower bounds to help give reasonable estimates.

- From assumptions 1 and 2, there are $\frac{2\,000\,000}{5} = 400\,000$ households.
- From assumption 3, the number of pianos is $\frac{400\,000}{50} = 8000$
- From assumptions 4 and 5, one piano tuner can tune $2 \times 5 \times 50 = 500$ pianos in a year.

Therefore there should be about $\frac{8000}{500} = 16$ piano tuners in Liverpool.

An online search lists 18 piano tuners in Liverpool, which is very close to the estimate. However, this doesn't necessarily mean that all the estimates in the assumptions were correct.

Discussion

Discuss the following with your peers.

Are the assumptions made in Example 1 reasonable?

What would you estimate for the number of piano tuners in the nearest city to you?

Compare your Fermi estimate with other discussion groups. How close are your estimates?

Sometimes it is handy to have a feel for numbers. It can be helpful to learn and use the following approximations in some of the questions that follow.

- Population of the UK: 65 million
- Population of London: 8 million
- Height of a person: 2 metres
- Radius of the Earth: 6000 km

Using upper and lower bounds

To get some estimates where you do not really have a clue, use a sensible upper and lower bound and work from that. Imagine you need an estimate for the length of time a 17-year-old spends on social media each day. You can sensibly set upper and lower bounds for this. It is reasonable to think that a teenager will spend at least half an hour each day on Facebook, Twitter, Instagram, WhatsApp and so on, so this is the lower bound. Given other things that a teenager has to do in the rest of the day (8 hours asleep and 8 hours at work or college and travelling), it is reasonable to set an upper bound of 8 hours, so the range is between 0.5 and 8 hours.

The arithmetic mean of this is $\frac{0.5 + 8}{2} = 4.25$.

However, this mean is about 8 times higher than the lower bound and only about half of the upper bound, so it may not be a very good representation.

To get round this, you can use the geometric mean. This is calculated by multiplying the two numbers together and taking the square root:

$$\sqrt{0.5 \times 8} = \sqrt{4} = 2.$$

In this case, the geometric mean is a factor of 4 above the lower bound and the upper bound is a factor of 4 above the geometric mean. To get the final Fermi estimate within a scale factor of 10 then it is better to use the geometric mean.

Geometric mean is the nth root of a product of n numbers.

Exercise 3H gives you the chance to practise making estimates using lower and upper bounds with the geometric mean. Do not use a calculator. Estimate the result of multiplying the bounds if they seem to involve too much difficult multiplication and estimate the square root. You should know the following square roots.

Number	1	4	9	16	25	36	49	64	81	100	1000
Square root	1	2	3	4	5	6	7	8	9	10	Approximately 30

Exercise 3H

For each of these estimates write down your lower and upper bounds and use them to work out the geometric mean. Where possible, check the accuracy of your estimates, after you have completed the whole exercise.

1 The time spent texting by one of your parents in a week.

2 The time your mathematics teacher spends marking books each week.

3 The number of grams of sugar in a can of non-diet fizzy drink.

4 The population of Paris.

5 The number of hairs on your head.

Estimating areas and volumes

Sometimes you can estimate numbers using formulae that you have met before.

For example, use the method of the geometric mean to estimate the number of gooseberries in this bowl.

You can quickly estimate that there will be more than 40 and fewer than 1000 gooseberries. Using these numbers as lower and upper bounds gives a geometric mean of

$$\sqrt{40 \times 1000} = \sqrt{40\,000} = 200.$$

What difference does it make to the geometric mean if you change the upper bound to 500?

Another way is to use the formula for the area of a circle (πr^2).

There are about 8 gooseberries to a diameter, so the radius is 4 gooseberries.

Therefore the top layer contains approximately $\pi \times 4^2 \approx 3 \times 16 \approx 50$ gooseberries.

There are about 4 layers, so that makes about $4 \times 50 = 200$ gooseberries in total.

Notice that both estimates give the same figure.

In fact there are 192 gooseberries in the bowl.

Estimates won't always be as close as this and so the answers may be given as a range of values.

Exercise 3I

For each of these estimates write down any formulae you use as well as the assumptions you make.

Remember to calculate your answer approximately: do not use a calculator at any point.

1 Estimate the circumference, in centimetres, of a bicycle wheel.

2 Estimate the number of bricks in the exterior of the building you are in.

3 Estimate the volume, in cubic millimetres, of the pen you are holding.

4 Estimate the volume, in cubic millimetres, of your phone.

Fermi questions

Remember, the essential part of any Fermi estimation is to state your assumptions clearly and that they involve multiplying and dividing rather than adding and subtracting.

Example 2

A high street coffee shop uses wooden stirrers.

Make Fermi estimates to answer these questions.

1 How many stirrers are used by the coffee shop in a day?

2 How many stirrers can be made from a tree?

Assumptions

1 Every customer at the coffee shop uses one stick.

2 There is always a queue at the coffee shop.

3 There are 4 people taking orders at the same time.

4 It takes 3 minutes to serve each customer.

5 The coffee shop is open from 6 a.m. to 11 p.m. daily.

6 A tree is 30 metres tall and its trunk is a cylinder with an average diameter of 1 metre.

7 The branches of a tree are not used.

8 A typical stirrer is 150 mm by 5 mm by 2 mm.

Working

The coffee shop is open for 17 hours. For easy calculation, this can be rounded to 20 hours.

The number of people processed in an hour is $20 \times 4 = 80$.

So the number of sticks used in a day is approximately $80 \times 20 = 1600$.

The volume of a tree trunk is $\pi r^2 h \approx 3 \times 0.5^2 \times 30 = 90 \div 4 \approx 20\ \text{m}^3$.

$20\ \text{m}^3 = 20\,000\,000\,000\ \text{mm}^3$.

Volume of a stick is $150 \times 5 \times 2 = 1500\ \text{mm}^3$.

Number of sticks from a tree is $\dfrac{20\,000\,000\,000}{1500} \approx 10\,000\,000$.

Exercise 3J

Use Fermi estimation to estimate the following.

Before working as an individual, discuss the assumptions to be made on the first three questions and the reasons behind those assumptions.

The questions in this exercise have no exact answers. You will find that the answers are given as a range. Provided you are within that range you can consider that you have tackled the questions well (unless you have just guessed without any reasoning!).

1 The number of times your heart has beaten since birth.

2 The number of secondary schools in the UK.

3 The number of tweets tweeted in a week in the UK.

4 The area covered when a box of paper hankies is spread out without overlapping.

5 The number of golf balls needed to fill a wheelie bin.

6 The number of pies eaten at football matches on one Saturday.

7 The number of Smarties needed to fill a double decker bus.

8 The amount spent on foreign holidays in a year by UK residents.

9 The value of a line of £2 coins stretching around the equator.

10 The value of £20 notes filling an Olympic sized swimming pool.

Case study

How many taxis do you think operate in your area?

On my way into town recently I noticed the following licence numbers on five taxis: 45, 394, 61, 64 and 546.

I assume that the licence numbers start at 1 and are numbered sequentially. I imagine that there are probably fewer than 10 000 taxis because I would have expected to see at least one with a higher number than 1000 if there are near to 10 000 taxis.

There is a statistical formula that provides you with a likely answer in such cases.

$$T = s - 1 + \frac{s}{n}$$

where

T is the estimated total number of objects

s is the highest serial number seen

n is the number of objects seen.

Here $s = 546$ and $n = 5$, so

$$T = 546 - 1 + \frac{546}{5} = 654.2$$

so an estimate for the number of taxis is 654.

This formula played a crucial part in the D-Day preparations in 1944. The Allies were concerned about the numbers of German Panzer Mark V tanks. The planners needed to have an estimate of the numbers of Panzers they would be likely to meet when landing on the D-Day beaches.

Earlier in the war, two of these tanks had been captured. Different parts of the tanks had different serial numbers and the planners needed to work out which numbers to use. They considered using the chassis numbers, but the problem here was that they were not numbered sequentially as they were made by different manufacturers with big gaps in the numbers. The gearboxes were numbered from 1 upwards, but two is a very small sample upon which to base such a crucial decision!

Fortunately the rubber tyres on the bogie wheels had mould numbers and with each tank having eight axles with six bogie wheels on each axle, this meant there were a total of 96 different numbers with which to work. The result was that the estimate of the number of tanks produced each month was 270, which was many more than the planners expected. The plans were therefore revised and the rest is history.

How good was this estimate of 270? The exact figures were found after the war; in the month of February 1944, 276 Panzer tanks had been produced.

Project work

Project work – 1

The sketch map shows town centres with their populations and distances between towns and crossroads. A new sixth form college is to be built to accommodate all sixth formers in the area. No new roads are planned. Recommend a suitable site assuming that there are no objections to it being placed by a road anywhere in the area.

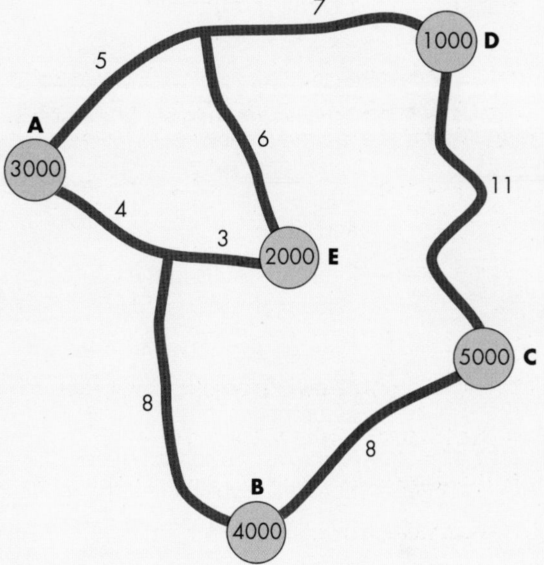

Project Work – 2

In the 18th century, Johann Bode and Johan Titus found a simple sequence of numbers that approximates to the average planetary distance from the Sun.

Their data is shown in the table. (Neptune and Pluto are not shown because they had not been discovered at that time).

Find a mathematical model (i.e. a formula) that gives the distance of a planet from the Sun using its number from the Sun and you will have discovered the Titus–Bode rule independently.

Planet	Mercury	Venus	Earth	Mars	Asteroids	Jupiter	Saturn	Uranus
Number from Sun	1	2	3	4	5	6	7	8
Distance from Sun in AU*	0.4	0.7	1	1.5	2.8	5.2	9.5	19.2

*The distance from the Sun is measured in Astronomical Units (AU) which is defined as the distance between the Sun and the Earth.

Kepler's law is an extension of this. It concerns the distance from the Sun and the orbital period (the number of years it takes a planet to revolve around the Sun). Due to the fact that some of the numbers involved get very large, Neptune and Pluto have again been omitted as well as the asteroids: once you find Kepler's law you can check that Neptune and Pluto obey it.

Planet	Mercury	Venus	Earth	Mars	Jupiter	Saturn	Uranus
Distance from Sun in AU (R)	0.4	0.7	1	1.5	5.2	9.5	19.2
Orbital period in years (T)	0.24	0.62	1	1.9	11.9	29.5	84.0

Start by plotting a graph of T^2 against R^3.

Check your progress

How confident are you feeling in your level of knowledge? What do you need to practise more?

Spec reference	Learning objective	▶▷▷	▶▶▷	▶▶▶
E1.1	Represent a situation mathematically, make assumptions and simplifications			
E1.2	Select and use appropriate mathematical techniques for problems and situations			
E1.3	Interpret results in the context of a given problem			
E1.4	Evaluate methods and solutions including how they may be affected by assumptions made			
E2.1	Make fast, rough estimates of quantities which are either difficult or impossible to measure directly			

4 Critical analysis of given data and models (including spreadsheets and tabular data)

Critical analysis: what does it mean?

In this chapter you will learn what is expected of you when you are asked to do a critical analysis of a piece of work or to write a report. You will be expected to draw on the mathematics you have learnt for GCSE as well as the mathematics in Chapters 1 and 2. You will start off by looking at how to critically analyse a report then deal with how to write a report.

Discussion

Discuss the following with your peers.

What do the words 'critical analysis' mean to you?

One dictionary describes the word 'critical' as "rigorously discriminating", another as "characterised by careful evaluation and judgement". Analysis in the sense of this chapter can be defined as "Look and identify errors, suggest improvements and evaluate whether a claim is justified."

Critical analysis is important because in everyday life you will be faced with many different scenarios. Sometimes you will need to make decisions based on what you read, for example, political literature in the run-up to a general election or referendum trying to persuade you which way to vote. Other times you will need it when you have to present a logical argument to others, for example, to obtain funding for a project for your business or in seeking support for a worthy cause.

Articles can be written descriptively or critically and the following table will help you distinguish between the two types.

Critical	Descriptive
identifies the significance of events	states what happened
argues a case according to the evidence	says how to do something
identifies whether something is appropriate or suitable	notes the methods used
indicates why something will work	explains how something works
identifies why the timing is of importance	says when something occurred
structures information in order of importance	lists in any order
shows the relevance of links between pieces of information	states links between items
draws conclusions	gives information

So critical analysis attempts to:

- get to the root of things
- discover reasons behind things
- break things down into logical steps and order things in appropriate formats
- examine strengths and weaknesses based on concrete evidence.

Do not confuse critical analysis with critical path analysis – that is a separate chapter!

Critical analysis

Exercise 4A

This exercise can be used as a discussion exercise.

For each of the following extracts, decide whether it is an example of critical analysis or descriptive writing, giving your reasons.

1 In 2016 Maria Sharapova failed a drug test at the Australian Open. She was provisionally suspended from 12 March.

She tested positive for meldonium, a substance she had been taking legally since 2006 under the name of mildronite.

Nike said they "decided to suspend our relationship with Maria" while the investigation proceeded.

In a press conference she admitted taking meldonium, which promotes blood flow, for ten years on medical grounds but did not notice that the World Anti Doping Authority had banned it on 1 January.

2 The two box and whisker plots shown compare the enjoyment of men and women when doing experiments that involve using a pendulum or investigating the equation formed by hanging a chain between two points.

The following abbreviations were used: **p** (for the pendulum activity), **c** (for the chains activity), **e** (for the enjoyment of the task), **l** (for the learning that took place), **f** (for a female response), **m** (for a male response).

Therefore **pef** refers to the number given on the pendulum activity for the enjoyment by the women.

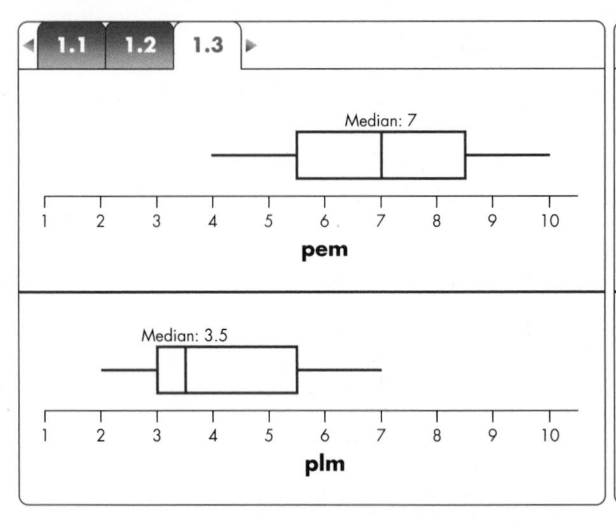

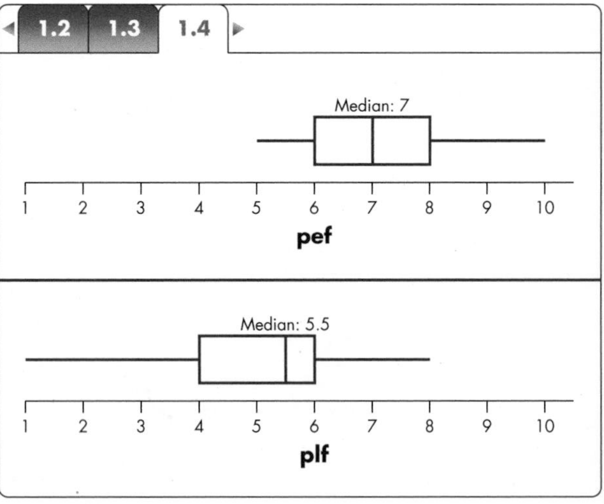

You see that the median for the enjoyment of the pendulum activity in each case is 7 and the dispersion for men is greater than that for women.

3 In a recent survey, the Money Advice Service found that 76% of shoppers spend more than they intended, with an extra £11 on average per shop, after being enticed by special offers in supermarkets.

A source said "The problem is that quite often we see a special offer at the supermarket and we don't want to miss out – so we put it in our trolley without really thinking about whether it is a good deal or we actually need it".

Supermarkets often make cheaper lines harder to reach and place them out of the natural eye line.

4 People with Type 1 diabetes can have low blood sugar levels at night. There are many causes for this including:

- too much insulin at bed time
- too much insulin and/or medication at dinner time
- not enough to eat at dinner or bed time
- being more active without making changes to insulin or food intake.

These can be prevented by:

- checking blood sugar levels during the night (the doctor will tell you the best time as this depends on the type of insulin)
- checking your medication doses with the doctor (you may be taking too much medication)
- eating more carbohydrate (you may not be eating enough for your medication dose)
- adjusting your insulin or food intake when exercising (less insulin is needed when exercising).

5 British mums who have just given birth stay in hospital the least amount of time in the developed world. A survey says that new mums (who have had just one child through a natural birth) spend an average of one-and-a-half days in hospital. In the US the average stay is two days, three days in Germany, 4.2 days in France and 6.2 days in the Ukraine. Experts say short stays mean there is insufficient time for checks on the mother and child, which increases health risks and means education or offering support is lacking.

C1: Presenting logical and reasoned arguments in context

Learning objectives

You will learn how to:

- Criticise the arguments of others.

The brown cow argument

A writer, a geographer and a critical analyst were travelling on a train in Derbyshire. They looked out of the window and saw a brown cow in a field. The writer said "Look! Derbyshire cows are brown!"

The geographer said "No, some Derbyshire cows are brown."

The critical analyst looked irritated and said "In Derbyshire there is at least one field, containing at least one cow, of which at least one side is brown."

From this you should realise that sometimes statements are made that could be true, but that you should not jump to conclusions. Each statement above as you read it seems fine, yet the conclusions you deduce from the evidence given can vary a lot.

In or out? Should the UK remain in the EU?

In 2016 there were many arguments for and against staying in the UK before a referendum took place in June. In March 2016 the Governor of the Bank of England, Mark Carney, told MPs that he believed that some of Britain's largest firms were planning to leave the country if the UK quit the EU. He said that Britain's economy would also be hit by uncertainty after a vote to leave in June. "It is the biggest domestic risk to financial stability", he said, qualifying it with "I would say that in my judgement the global risks, including from China, are bigger than the domestic risks."

He warned that "without question" the City of London would lose business. "Mutual recognition arrangements are possible, but they take a very long time to achieve. I can't give you a precise number in terms of institutions or jobs, because we don't know where we would be between full mutual recognition or third-country access."

The problem with Mark Carney's argument was that it is all based on supposition. Notice that he does not name any firms, quantify any risks or give any numbers. Arguments based on opinion, unsupported by facts, should be avoided when writing critically. There is also a certain amount of implicit emotional pressure: nobody wants to risk financial stability as people fear this would affect their standard of living. However, in 2016, Mark Carney held a position of great responsibility and influence in the UK. Only time will tell whether his opinion was correct.

Puncturing John's argument

John always carried a punctured tyre when he went out on a bike ride. Janet could not understand this, so she asked him why he carried the punctured tyre. John said "I read that the probability of a bike having a puncture when being ridden is 0.03. Therefore, the probability of a bike having two punctured tyres is 0.03×0.03, or 0.0009. So I have managed to lower the probability that I get a puncture in one of my good tyres from 0.03 to 0.0009."

Janet soon critically analysed John's argument. She said "Your reasoning is false, John. Since you know you have a punctured tyre with you, the probability of that happening is 1, not 0.03. Therefore, the probability of another punctured tyre in this case is not 0.03×0.03, but 1×0.03, which is still 0.03."

John ditched the punctured tyre and had less cumbersome rides after that.

Similar WWI examples

In World War I soldiers were encouraged to seek refuge in shell holes where bombs or shells had just landed because it was thought that the probability a second bomb would land there was far less than before because that spot had already been hit. In fact, this was a fallacious argument, just like John's argument above. It may even have been more likely to be hit again. This is because there was no guarantee of independence: if the guns that fired the shells were static, then perhaps it would be more likely that a shell would land in the same place.

Is it true that 2 = 1?

Here is a proof that $2 = 1$.

Follow the argument through and see if you can identify where the error in the argument lies.

Step 1: Let $a = b$

Step 2: Multiply both sides by a: $a^2 = ab$

Step 3: Add a^2 to each side of the equation: $a^2 + a^2 = ab + a^2$

Step 4: Simplify: $2a^2 = ab + a^2$

Step 5: Subtract $2ab$ from each side: $2a^2 - 2ab = ab + a^2 - 2ab$

Step 6: Simplify: $2a^2 - 2ab = a^2 - ab$

Step 7: Factorise: $2(a^2 - ab) = 1(a^2 - ab)$

Step 8: Divide each side by $(a^2 - ab)$: $2 = 1$

Discussion

Now clearly there is something wrong here. Discuss with your peers where the argument has gone wrong.

Martingale: a winning way?

The martingale was a betting strategy popular in the 18th and 19th centuries. It was a doubling-up system in which the gambler is fairly sure of a small gain. However, to compensate for this, there is a small risk of a large loss.

Here is how it works.

You bet some amount, say £1, to start with and bet that amount every time you win. However, each time you lose you double your bet the next time. Provided you quit and go home after a win, you will be in profit.

For example, consider a simple game that consists of 10 tosses of a coin and winning a toss if it shows heads (H).

The tables show the results of three such games.

Game 1

Bet (£)	1	1	1	2	1	2	1	2	4	1
Result	H	H	T	H	T	H	T	T	H	T
Won (£)	1	1		2		2			4	
Lost (£)			1		1		1	2		1
Total (£)	1	2	1	3	2	4	3	1	5	4

So after Game 1 the gambler is £4 ahead.

Game 2

Bet (£)	1	1	2	4	8	1	2	4	1	1
Result	H	T	T	T	H	T	T	H	H	H
Won (£)	1				8			4	1	1
Lost (£)		1	2	4		1	2			
Total (£)	1	0	−2	−6	2	1	−1	3	4	5

So after Game 2 the gambler is £5 ahead.

Game 3

Bet (£)	1	2	4	8	16	1	2	4	8	1
Result	T	T	T	T	H	T	T	T	H	T
Won (£)					16				8	
Lost (£)	1	2	4	8		1	2	4		1
Total (£)	−1	−3	−7	−15	1	0	−2	−6	2	1

So after Game 3 the gambler is £1 ahead.

So is this infallible?

Martingale and other doubling up systems are called d'Alembert systems after an 18th century French mathematician. He pointed out the fallacy of what was then known as the Law of Equilibrium. He wrote that this law supposes a balancing of events over an infinite time interval, not in a brief series of events limited by man. For example, if the probabilities of two events are equal and there are no other possible events, after a very, very long time each event will happen nearly the same number of times. Now when tossing a coin you can only toss it a small number of times and although you might argue the probability of heads and tails are equal, it is likely that there is some slight difference due to various factors.

Exercise 4B

This exercise can be used as a discussion exercise.

1 Play the game described above a couple of times.
Compare your winnings or losses with a friend.
Critically analyse the game based on your results.

2 A clock strikes four in 3 seconds.
Ben says it will take 6 seconds to strike eight.
Why is he wrong?

3 Jeremy drove his car for 1 mile at a speed of 30 miles per hour then drove the next mile at a speed of 60 miles per hour.

He claimed he had averaged 45 miles per hour.

James said he was right and Rachel said he was wrong.

Who is correct and why?

4 Perhaps fears of overpopulation are ungrounded.

It is possible to show that there were many more people around years ago than now.

To be born you must have had two parents, four grandparents, eight great grandparents and so on.

So, one generation back you had 2 ancestors, two generations back you had 4, or 2^2 ancestors, three generations back you had 8, or 2^3 ancestors.

Carrying on like this, n generations ago you had 2^n ancestors.

Now suppose that there are 25 years to a generation.

Therefore 500 years ago, or 20 generations back, you had 2^{20} or 1 048 576 ancestors.

In February 2016 the world population reached 7.4 billion (7 400 000 000) and since each of these people had 1 048 576 ancestors 500 years ago, the population of the world 500 years ago was about 7 thousand trillion or 7×10^{15}.

Why is this a fallacious argument?

Mathematics in the real world

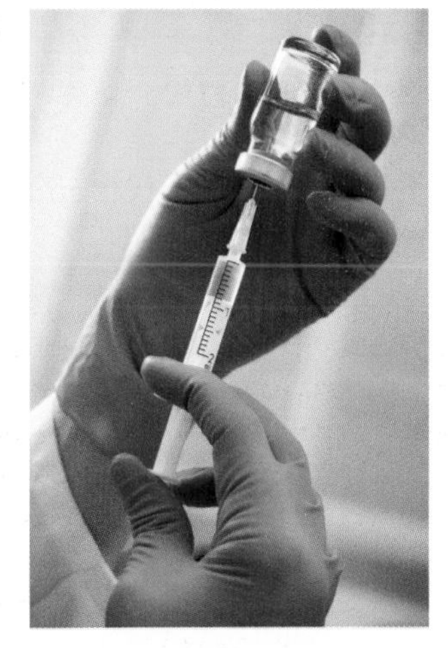

In 1998 Andrew Wakefield published a medical research paper claiming that there was a link between the MMR (measles, mumps and rubella) vaccine routinely given to young children and medical conditions including autism and bowel disease. The research received widespread media interest and, as result, the inoculation rate in the UK dropped from 92% to less than 80%. This led to an increase in the numbers of cases of measles, mumps and rubella. In 1998 there were 56 measles cases in the UK; by 2008 there were 1348 cases, with two confirmed deaths.

Wakefield's research was investigated and much of the methodology was criticised. The reasons for this criticism included facts or data being misused, small sample size and self-selection of the data. The media also distorted the scientific evidence, reporting selectively on the evidence suggesting that MMR was risky and repeatedly ignoring evidence to the contrary. (See Ben Goldacre's book *Bad Science* for more information about the media coverage.) In January 2010 a tribunal of the General Medical Council found that Wakefield had acted "dishonestly and irresponsibly" in his published research and *The Lancet* retracted the publication, stating the paper was "utterly false" and noting that elements of the manuscript had been falsified. The scientific consensus now is that there is no evidence to link the MMR vaccine with autism and the vaccine's benefits greatly outweigh the risks due to the death and morbidity rates of measles, mumps and rubella if inoculation does not take place.

The Wakefield case shows how easy it can be for a piece of research to go badly wrong and have serious implications in the wider world.

C2: Communicating mathematical approaches and solutions

Learning objectives

You will learn how to:

- Summarise and write reports.

Example 1

Sam has collected data from the college garden about the number of plants that have grown from the number of seeds planted. Sam works out the percentage of the seeds that have grown and enters all the data into a table. Sam's results and report are shown below.

	A	B	C	D	E	F
1	**Type of seed**	**carrot**	**Lettuce**	**Beetroot**	**Tomataoe**	**Total**
2	**Number planted**	100	80	50	20	250
3	**Number of seeds that grow**	74	72	40	17	213
4	**Percentage (%)**	74	90	20	85	269

> The best results came from the lettuce seeds: this is fantastic because I love eating lettuce! I was pleased to see that beetroot did not do well as nobody likes beetroot as it makes your tongue turn red. I think there may be something wrong with the percentage of 269 as it seems a bit big.

a Identify any errors in the table and suggest ways that the table could be improved.

b Criticise Sam's report and write an improved short report based on the data.

---

a There are a number of errors in the table:

1 The number in cell F3 should be 203, not 213.

Sam should use a formula like =SUM(B3:E3) in cell F3 to calculate the total automatically.

If a formula was used in cell F2 then it could be copied down the F column.

2 The percentage in cell D4 is incorrect.

It should be 80% (because 40 out of 50 is the same as 0.8), not 20%.

Again, Sam could use a formula in cell B4 like =100*B3/B2 and copy that along the row.

3 In cell F4 Sam has added the percentages along the row, which is incorrect.

Sam should have used the figures in cells F3 and F2 to calculate the result (which would be 81.2% using 203 and 250).

4 Although the column headings are clear, Sam should be consistent and always use lower case letters for the plant names.

The spelling should be 'tomato', not 'tomataoe'.

b You lack too many facts about the data: whether the quality of the college garden might have varied from one part to another, whether any fertiliser that was used, what was grown previously, whether the crops were properly watered, etc. However based on the data you are given you can criticise Sam's report.

To produce a good report Sam needs to concentrate on the facts provided by the data. Since different numbers of seeds were planted for each crop, if any comparison is to be made it makes sense to use the percentages. If you look at what Sam has written, you can comment on it as follows:

Sam has used emotive language ("this is fantastic because I love eating lettuce!", "nobody likes beetroot") and has made statements that do not relate to the data in the table ("it makes your tongue turn red"). Sam has noticed a possible error, but has made no attempt to correct it. An improved report follows.

The seeds that grew best were the lettuce seeds as 90% of those seeds grew. The worst seeds were carrots as only 74% of those seeds grew.

Overall the success rate was 80% to the nearest 10%. I have rounded the results to the nearest 10% because I might have made some counting errors when I counted the seeds or when I counted the plants that grew.

Although I am happy with my results I need to consider whether the differences in the percentages are due to factors such as soil conditions, chemicals already in the ground, the quality of the seeds, etc.

Exercise 4C

1 As a travel agent you want to know what kind of holidays people want.

You downloaded the data shown in the spreadsheet from the Office for National Statistics.

	A	B	C	D	E
1	**The Number Of Holiday Visits to GB and Overseas Destinations, 2007 - 2013**				
2					
3	**Outbound visits (000s)**	**Domestic overnight visits (000s)**			
4	45000	52000			
5	46000	50000			
6	38000	59000			
7	36000	55000			
8	36000	58000			
9	36000	58000			
10	38000	57000			
11					

Data from: How does the UK economy affect where the British go on holiday?/ Office of National Statistics

You have to write a short report for your staff about what the data means.

Here are some points to guide you.

- Use the skills you developed in Chapter 1 for representing data numerically and diagrammatically.
- Which data is for which year? (Assume row 4 is for 2007 down to row 10 for 2013).
- Are people staying at home or going abroad?

2 Read a report by someone else who has tackled question **1**.

Do you agree with what they have written?

How could they improve their report?

On a separate piece of paper, give them some constructive feedback.

3 Look at the feedback from the person who has commented on your report.

Amend your report addressing the points they have made.

4 As a manager of health services you have access to the data shown in the spreadsheet.

It shows the median age of the UK population based on the ages of the population in the middle of each year.

	A	B	C	D	E	F	G	H	I	J	K	L	M	N	O	P	Q	R	S	T	U	V
1	**Mid-year**	1974	1975	1976	1977	1978	1979	1980	1981	1982	1983	1984	1985	1986	1987	1988	1989	1990	1991	1992	1993	
2	**Median age**	33.9	33.8	33.8	33.9	34.1	34.2	34.3	34.5	34.8	35.1	35.3	35.4	35.5	35.5	35.6	35.7	35.8	35.8	36.0	36.1	
3																						
4	**Mid-year**	1994	1995	1996	1997	1998	1999	2000	2001	2002	200	2004	2005	2006	2007	2008	2009	2010	2011	2012	2013	2014
5	**Median age**	36.3	36.5	36.6	36.9	37.1	37.3	37.6	37.9	38.1	38.4	38.6	38.7	38.9	39.0	39.1	39.3	39.5	39.6	39.8	39.9	40.0

Data from: Ageing of the UK population / Office of National Statistics

Write a one-page report for the government about the trend.

Here are some points to guide you.

- Use the skills you developed in Chapter 1 for representing data numerically and diagrammatically.
- Why is the median age being used?
- Can you find a formula that fits the growth? If so, what assumptions are you making?

5 Read a report by someone else who has tackled question **4**.

Do you agree with what they have written?

How could they improve their report?

On a separate piece of paper, give them some constructive feedback.

6 Look at the feedback from the person who has commented on your report.

Amend your report addressing the points they have made.

7 The data below shows quarterly recycling rates in one Local Authority area.

Table 3, Dry recycling (in 1000 tonnes)	**Jan–Mar 2013**	**Apr–Jun 2013**	**Jul–Sep 2013**	**Oct–Dec 2013**	**Jan–Mar 2014**
Dry recycling rate (excludes organic recycling from calculation)*	**32.4%**	**31.9%**	**32.1%**	**31.7%**	**32.5%**
total collected dry (includes organic waste in the residual stream)	4313	4561	4515	4326	4531
Sent for dry (recycling, reuse)	1398	1455	1450	1372	1473
of which glass	283	274	283	263	286
of which paper and card	614	580	588	611	623
of which metals	55	54	55	54	58
of which plastic	91	95	103	99	104
of which textiles	31	28	29	26	32
of which WEEE & other scrap metals	103	134	125	101	116
of which other materials	220	288	266	218	254

Statistical data set ENV18 – Local authority collected waste: annual results tables/ Department for Environment, Food & Rural Affairs 8 November 2012

As a prospective MP who wants to make a mark in the area, write a two-page report for your local paper.

Here are some points to guide you.

- Use the skills you developed in Chapter 1 for representing data numerically and diagrammatically.
- Is the recycling increasing, decreasing or staying constant in all types of materials?
- Is there any period where recycling rates fell? When? Why do you think that is? How could this be improved in the future?
- How much is not recycled? How does this concern you?

8 Read a report by someone else who has tackled question **7**.

Do you agree with what they have written?

How could they improve their report?

On a separate piece of paper, give them some constructive feedback.

9 Look at the feedback from the person who has commented on your report.

Amend your report addressing the points they have made.

C3: Analysing critically

Learning objectives

You will learn how to:

- Compare results from a model with real data.
- Critically analyse data quoted in media, political campaigns, marketing, etc.

In Chapter 3 you will have compared some results from a model with real data, such as the braking distances at various speeds and the best speed to travel at when traffic is restricted. In this section you will look at some other results from mathematics and the media. The first example is from a verbal report.

HM Submarine Ocelot

HM Submarine Ocelot is on show at the Historic Dockyard, Chatham. It was the last warship built for the Royal Navy at Chatham Dockyard. Launched in 1962, it saw service with the Royal Navy throughout the height of the Cold War until 1991.

At the front of the submarine there are six torpedo tubes, arranged in two columns of three.

It is said that three torpedoes were fired at the same time because the chance of one hitting the intended target was one in three, so by firing three at once it was nearly certain that at least one would hit the target.

The following examines this claim.

Let H be the event that a torpedo hits the target.

Then you can write $P(H) = \frac{1}{3}$.

So the probability that a torpedo misses the target is

$$1 - P(H) = 1 - \tfrac{1}{3} = \tfrac{2}{3}.$$

Now assume that the event H is independent for each torpedo, i.e. the outcome of one torpedo does not affect the outcome of another one.

Also assume that the torpedoes do not emerge simultaneously, but that there is a slight delay between each one.

The probability that the first two torpedoes both miss is

$$\tfrac{2}{3} \times \tfrac{2}{3} = \tfrac{4}{9}.$$

And the probability that the first three torpedoes all miss is

$$\tfrac{4}{9} \times \tfrac{2}{3} = \tfrac{8}{27}.$$

The probability that at least one torpedo hits the target is therefore 1 minus the probability they all miss and is

$$1 - \tfrac{8}{27} = \tfrac{19}{27} = 0.7037...$$

Therefore the claim that by firing three at once it was nearly certain that at least one would hit the target is incorrect – it is about 70%.

However you have no evidence to support the assumption that the event H is independent for each torpedo, so you might be wrong.

If the events are not independent, then it could mean that if the first one misses, the second one could be more likely to miss.

Modelling: comparing results

In 2002 *New Scientist* reported that the Belgian one euro coin was thought to be unfair: it was thought that the probability of landing heads or tails was not $\frac{1}{2}$. This, of course, has serious consequences for those sports that toss a coin to decide important decisions.

Two Polish teachers, Gliszczynski and Zawadowski, had their students spinning Belgian euros. (Gliszczynski believed that spinning is a more sensitive way of revealing whether a coin is biased than tossing in the air.)

They found that out of 250 spins, 140 landed heads up. This gave an experimental probability of 0.56. They felt that because the head of King Alfred of Belgium was so large it weighted that side more than the tails side, so the coin was more likely to land heads up.

Howard Grubb, a statistician, noted that this sample was relatively small and that anything between 0.438 and 0.562 could not be said to be biased. This is because when modelled, a result of 0.56 is expected in about 7 out of 100 experiments of this size with a fair coin. Therefore, in this case, the assumption that the coin is biased cannot be said to be true or false: the real data suggests that it might be biased, but the model tells us that it is not necessarily the case.

If you go on to study Chapters 5 and 6 you will do more on this sort of work.

Misleading statistical diagrams

At the start of this chapter you looked at some data extracted from media reports, so you will build on that now.

You may have dealt with some of these for GCSE Mathematics: here is a summary of the common ways the media tries to mislead you with inappropriate diagrams. They are illustrated with just two pieces of data that compare the average house price according to the Office for National Statistics in the Southeast (£365 000) and the Northeast (£155 000) in March 2016.

Omitting the start of a scale

This can exaggerate the difference between the two prices.

Here the visual impression makes it seem as if the average house price in the Southeast is more than four times that in the Northeast.

Irregular scale intervals

Again, this can exaggerate the difference between the two, but includes a zero so you think it's not deceiving like the first case.

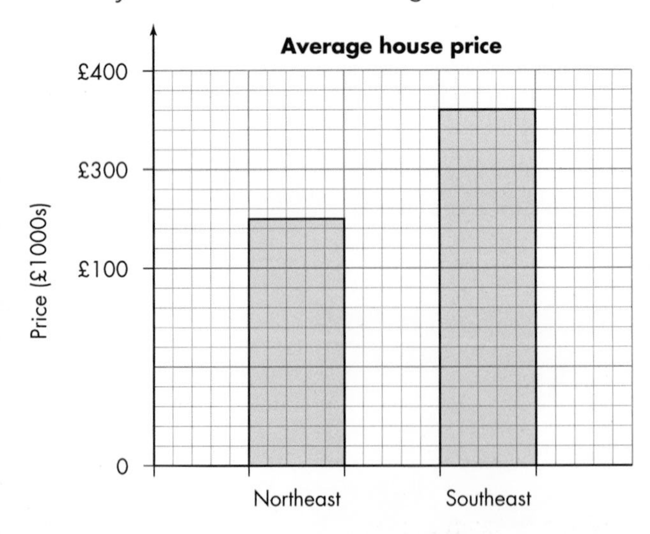

Now the visual effect is that there is not a lot of difference when in fact the average house price in the Southeast is more than twice that in the Northeast.

Using two (or even three)-dimensional pictures

Here your eyes are drawn to the area, where there is a big difference, yet you should be considering just one dimension (usually the height).

Here the visual impression makes it seem as if the house prices in the Southeast are more than four times that of houses in the Northeast, yet the heights and the widths of the houses are in proportion to the prices.

Speculative effects

Beware of spurious data based on what might happen in the future!

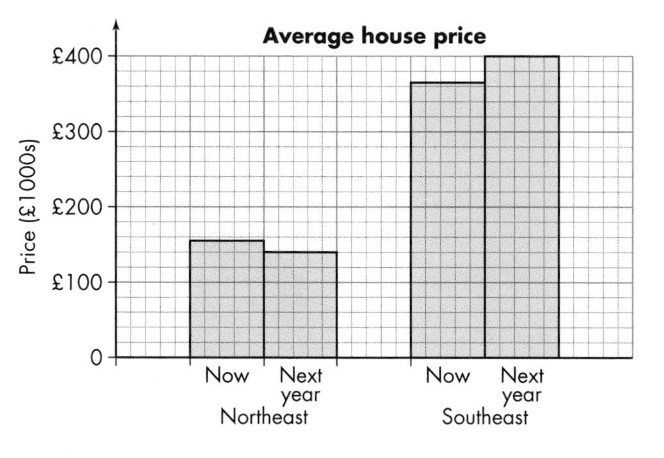

Here the scale is fine, but it looks as though prices will decrease in the Northeast and increase in the Southeast, but you do not know what will happen!

Discussion

The following chart comes from a report written in 2015 concerning the lack of students choosing STEM (Science, Technology, Engineering, Mathematics) subjects at A-level, despite the fact that when they enter secondary school, 74% were interested in and enjoying science. Britain's place as a leading technology economy means it needs more students to carry on with STEM subjects.

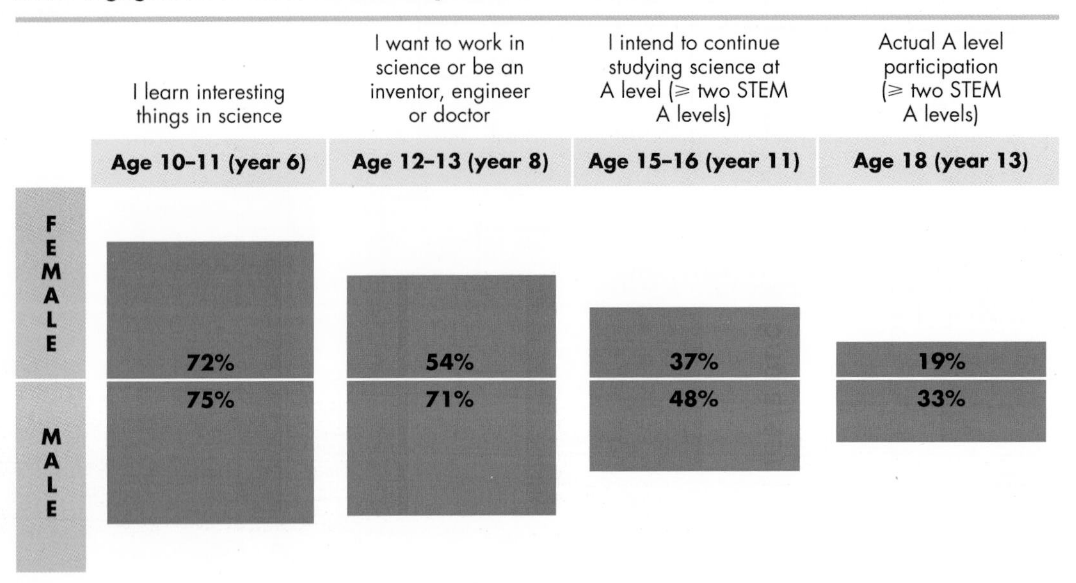

Study the diagram and then discuss the following questions with your peers.

1. Look at the diagram then cover it up.

   Which features do you remember most?
2. Look at the proportion of boys and girls aged 18.

   Discuss what you notice about this part of the diagram.
3. Look at the proportion of boys and girls in Year 6.

   Which parts of the diagram represent the 72% and 75%? The start height, the end height, the width, the area of the shape, the 'volume' of the shape or what?
4. How would you display the data more effectively?
5. How does your school or college compare with the Year 11 and Year 13 data?

   You may need your teacher's help in obtaining relevant data if it is available.
6. Discuss any other features of the diagram that concern you.

Exercise 4D

1 You know that if you do the same thing to each side of an equation, the resulting equation is true. So if

$$x^2 = 36$$
$$\sqrt{x^2} = \sqrt{36}$$
$$x = 6$$

Now

$$\tfrac{1}{4}\text{ pound} = 25\text{ pence}$$
$$\sqrt{\tfrac{1}{4}\text{ pound}} = \sqrt{25\text{ pence}}$$
$$\tfrac{1}{2}\text{ pound} = 5\text{ pence}$$

What is wrong here?

2 Cancelling is a way to simplify fractions.

Fred has misunderstood how to cancel fractions. His work is shown below.

$$\frac{19}{95} = \frac{1\not{9}}{\not{9}5} = \frac{1}{5} \qquad \frac{16}{64} = \frac{1\not{6}}{\not{6}4} = \frac{1}{4}$$

$$\frac{12}{24} = \frac{1\not{2}}{\not{2}4} = \frac{1}{4} \qquad \frac{26}{65} = \frac{2\not{6}}{\not{6}5} = \frac{2}{5}$$

He has cancelled incorrectly in all these examples, but three of them coincidentally give the correct result!

a State which fraction does not give the correct result.

b Explain what Fred did wrong and how he should cancel fractions.

3 In 2016 the cost of the cheapest adult season ticket for Manchester City was £299 and for Arsenal was £1035.

Here is a diagram comparing the costs.

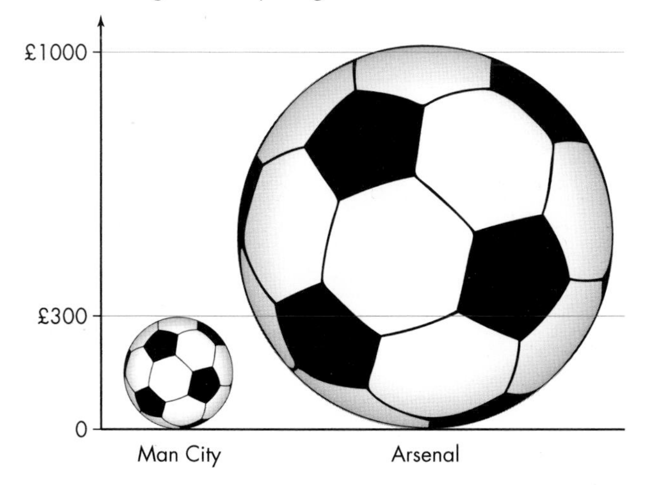

a Critically analyse the diagram.

b Draw a more appropriate diagram using the data.

4 Pierre was very interested in pie charts and what they look like, so he asked students in his college "What do you think a pie chart looks like?" and presented the results as shown.

Critically analyse the diagram.

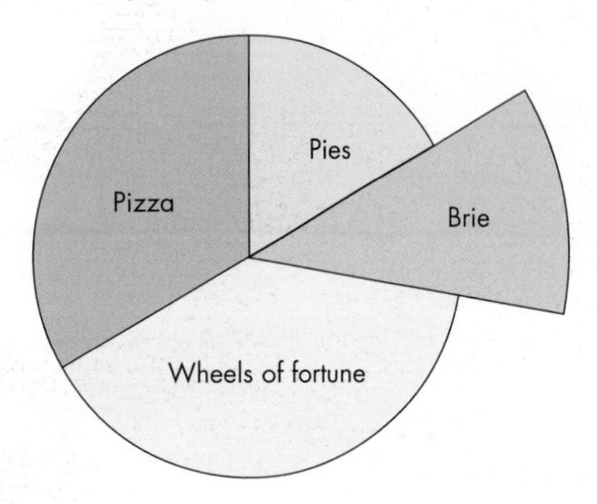

5 The diagram shows the growth in the number of GCSE students aged 17 and over.

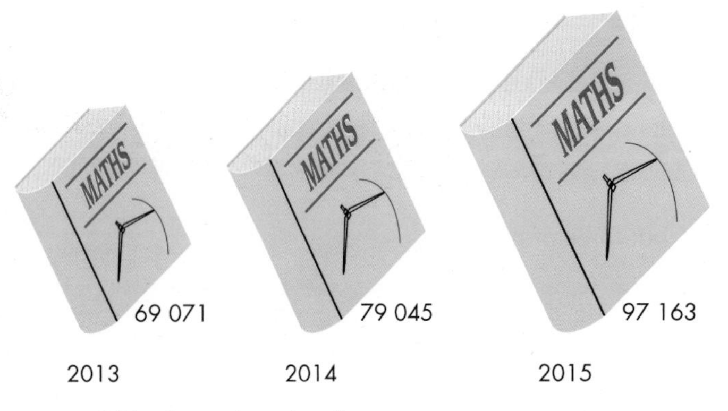

a Critically analyse the diagram.

b Draw a more appropriate diagram using the data.

6 The chart shows the number of GCSE grades challenged in appeals.

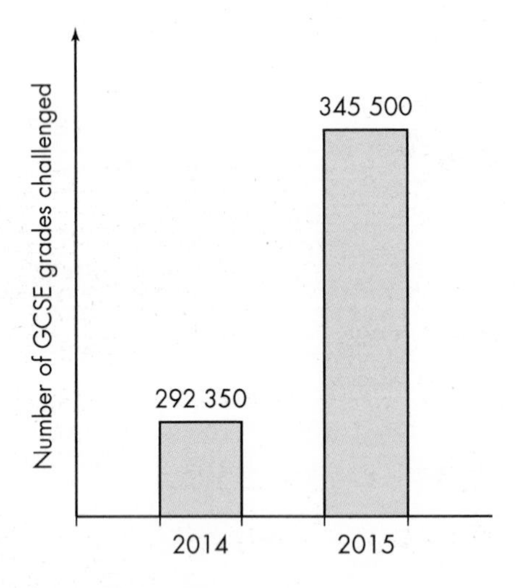

a Critically analyse the diagram.

b Draw a more appropriate diagram using the data.

7 The chart below shows the number of drivers in fatal crashes one year.

A newspaper reported that "16-year-olds are safer drivers than people in their twenties, and that octogenarians are very safe".

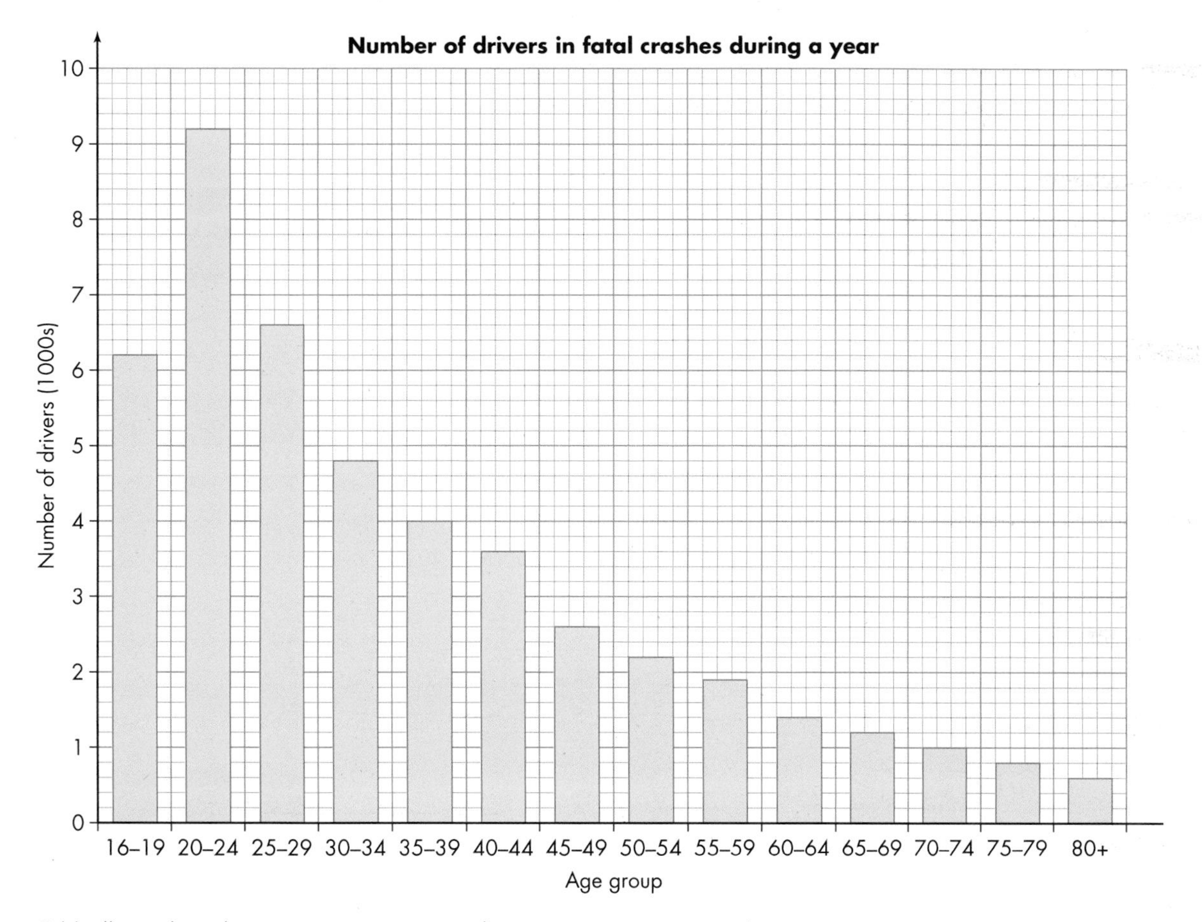

Critically analyse the newspaper quote and suggest a more appropriate diagram they should use.

Critically analysing calculations

Example 2

Steph wants to buy a house.

Her annual salary is £50 000 and her partner's is £25 000.

The bank will lend her four times her annual salary plus her partner's salary, or three times their combined salary.

The house she wants to buy is £300 000 and the bank requires Steph to provide a deposit of 15%.

Steph makes the following notes:

300,000 ÷ 15 = 20,000
So I need to find a deposit of £20k.
4 × 50,000 + 25,000 = 4 × 75000
= 300,000
So I can buy the house!

What has Steph done wrong?

---

1 She has worked out the deposit incorrectly.

   It should be $300\,000 \times 0.15 = 45\,000$.

2 To calculate the amount she can borrow, she should multiply her own salary by 4 and add her partner's salary:

   $50\,000 \times 4 + 25\,000 = 200\,000 + 25\,000 = 225\,000$

3 The bank will not loan the money as the £225 000 is less than the mortgage required:

   £300 000 − £45 000 = £255 000

Discussion

Discuss the following with your peers.

Burt is a landscape gardener.

He wants to know if gardens are getting smaller and compares the areas, in square metres, of gardens he designed for clients in 2015 and 2016.

He displayed his results in this stem and leaf diagram.

Year

2015		2016
0	25	2
6 4 1	26	0 5 8
6 5 2 2	27	3 4 9
3 0	28	
3 1	29	2
7	30	8

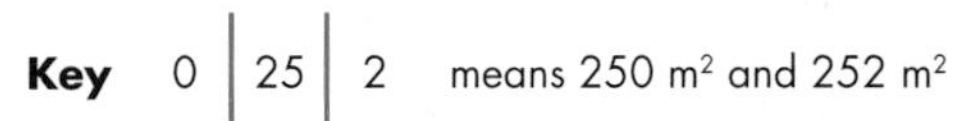

He makes two statements:

"The median area for 2016 is 279 square metres and the median for 2015 is 272 square metres so gardens are getting bigger."

"The mode is 2 square metres."

Critically analyse Burt's statements making corrections where necessary.

Note that:

1 There are 13 values for 2015, so the median will be the 7th value when they are arranged in order.

2 There are 9 values for 2016, so the median will be the 5th value when they are arranged in order.

3 Although 2 is the single digit that appears most often, the value that occurs most often is 272 metres.

Exercise 4E

1 Carys decides to save £10 a month in a building society that pays 2% compound interest per annum.

She decides to see how much that will be worth at the end of the year.

She uses a spreadsheet to show her results.

	A	B	C	D	E	F	G	H	I	J	K	L	M
1	Month number	1	2	3	4	5	6	7	8	9	10	11	12
2	Amount in account at start of the month	£10.00	£22.00	£36.40	£53.68	£74.42	£99.30	£129.16	£164.99	£207.99	£259.59	£321.50	£395.81
3	Interest	£2.00	£4.40	£7.28	£10.74	£14.88	£19.86	£25.83	£33.00	£51.92	£51.92	£64.30	£79.16
4	Amount in account at end of month	£12.00	£26.40	£43.68	£64.42	£89.30	£119.16	£154.99	£197.99	£311.50	£311.50	£385.81	£474.97
5													

Critically analyse what Carys has done.

2 Jude negotiates a 10% discount with a company because he wants to buy three toner cartridges for his laser printer.

The total amount he has to pay after 15% carriage charge is added to the price is £120.

Jude wants to know the original price of each toner cartridge before carriage is added and how much he has saved.

He makes these notes:

£120 ÷ 3 = £40
40 × 0.15 = 6
40 − 6 = 34
34 ÷ 0.10 = 3.40 original price = £34 + £3.40 = £37.40

Critically analyse Jude's calculations.

3 Dale spends time on Twitter each day and he wants to find out about how much other people use it and make a presentation to his peers.

Here is his PowerPoint slide.

a Analyse Dale's report, identifying any errors.

b Suggest any improvements he could make.

Twitter transactions

To study how much time is spent on Twitter,
I asked my friends to count how many tweets they sent and received.
I entered the data into a spreadsheet like this:

	A	B	C	D
1	Person	Tweets sent	Tweets read	Total number of tweets
2	Ben	4	15	19
3	Charles	9	23	32
4	Dick	18	17	45
5	John	11	19	30
6	Total	42	84	126

4 Margaret does a survey of how much students at her college earn in a week during term time.

Here are her results.

Amount earned, £e	$0 < e \leqslant 10$	$10 < e \leqslant 15$	$15 < e \leqslant 20$	$20 < e \leqslant 40$	$40 < e$
Number of students	8	12	10	7	2

She draws the following chart.

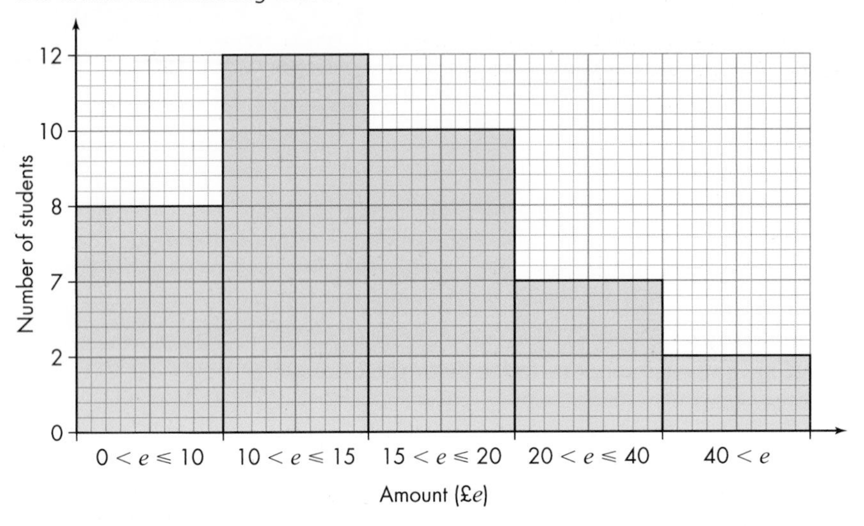

a Critically analyse what Margaret has done and suggest improvements.

b Margaret decides to estimate the mean amount earned and jots down her calculation.

$$\frac{8 \times 5 + 12 \times 12.5 + 10 \times 17.5 + 7 \times 30 + 2 \times 40}{5} = \frac{655}{5} = £131$$

What has she done wrong?

c She realises the answer she got for part **b** is too large, so decides to estimate the median amount instead.

She says this is £17.50.

What could she have done wrong?

5 Sam sets out a spreadsheet showing his progress in short revision tests on the first four chapters of *Mathematical Studies*.

	A	B	C	D	E	F
1	**Question number**	**Test 1**	**Test 2**	**Test 4**	**Test 3**	**Median**
2	**1**	8	4	6	9	7
3	**2**	3	6	5	-	5
4	**3**	9	8	8	7	8
5	**4**	4	4	4	5	4
6	**2.5**	6	5.5	5.75	7	**Mean**

a Identify any errors in Sam's table and suggest some improvements that he could make.

b Sam said "I did best in Test 3" and "Question 4 was the worst of all the questions."

Critically analyse these statements, showing working to justify your comments where appropriate.

Case study

"CT scans in childhood can triple the chance of developing brain cancer or leukaemia."

What does this mean?

It is a statement of the **relative risk** of developing cancer. It tells us how much more, or less, likely brain cancer is in one group, compared to another.

It does not tell us anything about the overall likelihood of brain cancer happening at all – that is what is known as the **absolute risk**.

In 2012 a study in *The Lancet* indicated that brain tumours in later life were three times as common among people who had two to three CT head scans in childhood.

But the researchers who carried out the study emphasise that the overall absolute risk of people developing cancer after receiving CT scans remains small.

If you dig deeper, you find that the chances of developing these cancers are very small (0.4 per 10 000 children aged 0–9 develop brain tumours and 0.6 per 10 000 children aged 0–9 develop leukaemia). So the increased risk would mean about one additional case of brain cancer and one of leukaemia for every 10 000 children **given the scans**. Of course being given the scans is probably beneficial because any early symptoms can be identified and problems treated at an early stage.

Therefore, a frightening looking headline needs to be taken in context and you need to look carefully at whether it is the relative risk or absolute risk that is being considered.

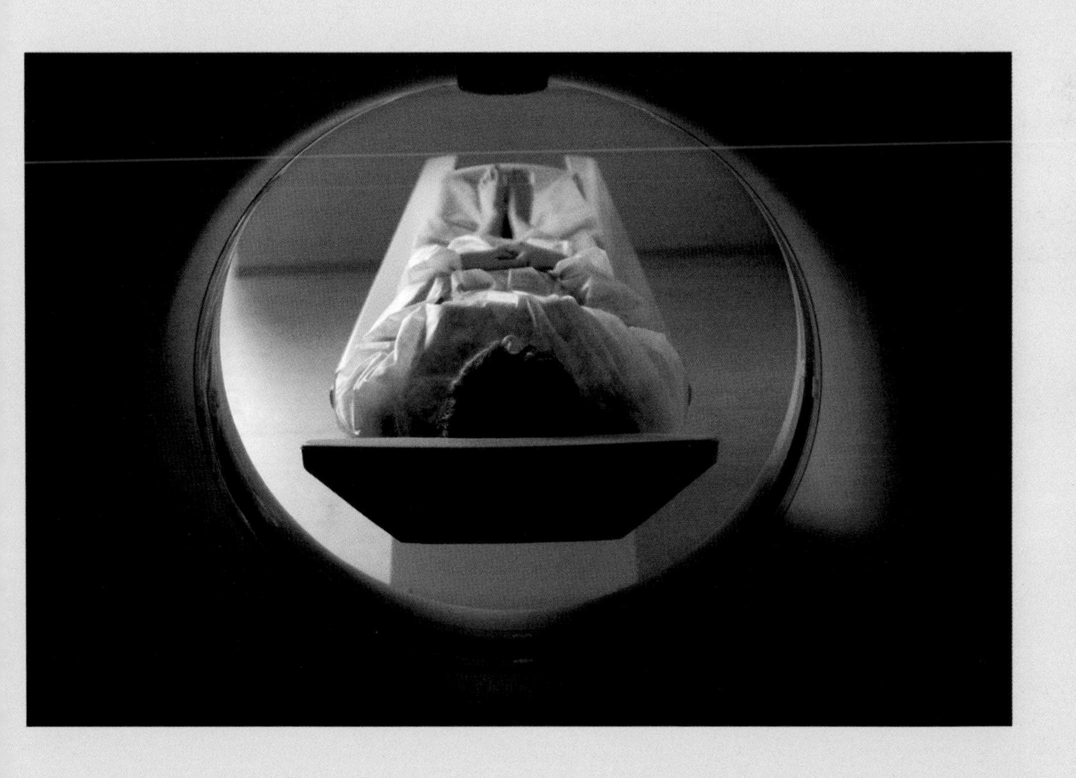

Project work

Collect some examples of statistical diagrams or claims from magazines and journals and critically analyse each one. Some diagrams or claims may contain many dubious facts, but others might not contain any, so if that is the case, comment on why you think it is a good representation.

Your project should be a minimum of four A4 pages, four PowerPoint slides or a four-minute video.

When you have completed your project, give it to a friend for them to check what you have done and ask them to critically analyse it. In return, you can do the same for their project.

You can give your feedback in written or oral form.

Check your progress

Critical analysis will be examined in all the Paper 2s of Mathematical Studies. The questions on critical analysis will be the same in Paper 2A, Paper 2B and Paper 2C because they examine part of the compulsory content of the qualification.

How confident are you feeling in your level of knowledge? What do you need to practise more?

Spec reference	Learning objective			
C1.1	Criticise the arguments of others			
C2.1	Summarise and write reports			
C3.1	Compare results from a model with real data			
C3.2	Critically analyse data quoted in media, political campaigns, marketing, etc.			

5 The normal distribution

Are you friends with any of the adults in your family on Facebook? Do they follow you on Twitter? Do they follow you on any other social media sites?

How many friends do you have on Facebook? Do your friends at school or college have more Facebook friends than you? Is the number of Facebook friends you have typical? Do the adults in your family have more Facebook friends than you?

On Facebook, Parents Are Friends with Their Teens

YES NO

Are you friends with your teenager on **Facebook?**
Among adult Facebook users with teen on Facebook

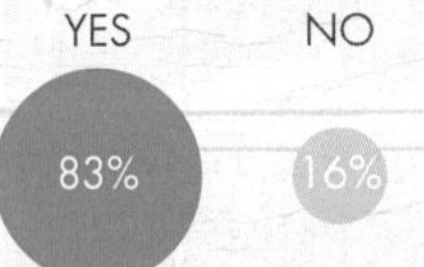

Do you follow your teenager on **Twitter?**
Among all adults with a teen on Twitter

Are you connected with your teenager on any (other) social media sites?
Among all adults with teens

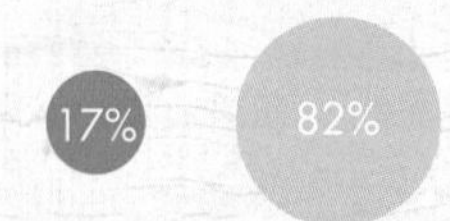

You can do comparisons between your peers and your family members by asking them questions and exchanging information, but how do you know if your number of Facebook friends is typical?

Facebook friend counts
Median number of friends by age

Age	Median number of friends
18–29	300
30–49	200
50–64	75
65+	30

According to a survey carried out in 2014, Facebook users differ greatly when it comes to the number of friends in their networks.

- 39% of adult Facebook users have between 1 and 100 Facebook friends.
- Half of all adult Facebook users have more than 200 friends in their network.
- 23% have 101–250 friends.
- 20% have 251–500 friends.
- 15% have more than 500 friends.

If the mean number of Facebook friends worldwide is 176 people and the standard deviation is 33, (standard deviation is a measure of spread expressing how much some items differ from the mean) how do you work out what proportion of people have the same number of Facebook friends as you?

S1: Properties of a normal distribution

Learning objectives

You will learn how to:

- Recognise normally distributed data.
- Describe the properties of a normal distribution.
- Use your knowledge about specific standard deviations and their probabilities.

Introduction

If you had been sitting in the Olympic stadium in London on 27 July 2012 watching the spectators take their seats for the opening ceremony of the Olympic Games, you would have noticed that there was a variation in the heights of the people. Some people would have been quite short while others would have been quite tall. (Hopefully one of the tall people wasn't in front of you!) Generally, most people would have tended to have been around a similar height.

If you measured the heights of all the people in the stadium that night and plotted the results, the graph would have looked something like the histogram here.

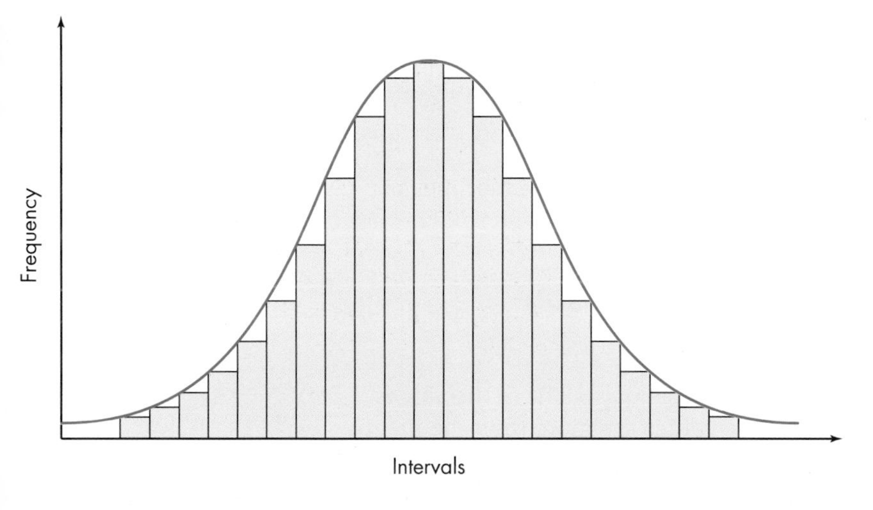

The middle column of the graph is the highest and so contains both the median and the modal height. As the curve is symmetrical, it also shows the mean height. The extreme left-hand column represents the shortest people and the extreme right-hand column represents the tallest people. If you had plotted the results for men and women separately, the histograms would have been a similar shape but the men's heights would have centred around the men's mean height of 175 cm and the women's heights around the women's mean height of 162 cm (assuming most of the crowd was from the United Kingdom).

The graph shows a bell curve distribution of data. The distribution is symmetrical about the centre with fewer values as the distribution moves away from this middle value. The statistical distribution that is used as a model for continuous variables distributed with these properties is a **normal distribution**.

A **normal distribution** is symmetrical about the mean.

Mathematics in the real world

An EU Directive in 2012 required that sales of traditional incandescent light bulbs be phased out. The bulbs had to be replaced by more energy-efficient technologies such as compact fluorescent lamps (CFLs), halogen lamps and light-emitting diodes (LEDs). It was predicted that this would save 39 terawatt-hours of electricity across the EU annually by 2020 and the UK government said the ban would bring an average annual net benefit of £108m to the UK between 2010 and 2020 in energy savings. So how have these huge energy and financial savings been calculated? How sure can you be that they are correct?

One way is to examine the claims on the packaging for the new types of bulb. A manufacturer produces a new fluorescent light bulb and claims that it has an average life of 4000 hours with a standard deviation of 375 hours. What is the probability that any one light bulb will last for more than 3500 hours? Assuming the lifetimes of the light bulbs are normally distributed, you can use a normal distribution to answer this question.

Discussion

Choose one of the following and discuss it with your peers.

1. Imagine you had been sitting in the Olympic stadium in London on the opening night in July 2012. Talk about the other things you could have measured that are normally distributed.
2. Imagine you watched a day of athletics in the Olympic stadium during London 2012. Talk about whether the results in different disciplines, such as the high jump, shot put or 100 metres, are normally distributed.
3. Imagine you are in your kitchen at home. Talk about the lifetimes of other household items that are normally distributed.

The properties of a normal distribution

The properties of a normal distribution are:

- the distribution is symmetrical about the mean (so 50% of the values are less than the mean and 50% of the values are more than the mean)
- the mode, median and mean are all equal
- the area under the curve represents probability and the total area under the curve is 1
- the distribution is defined by two parameters (defining characteristics), the mean and the standard deviation.

The area under a normal distribution curve is defined as 1 and represents probability.

Standard deviation: A measure of the spread of a distribution.

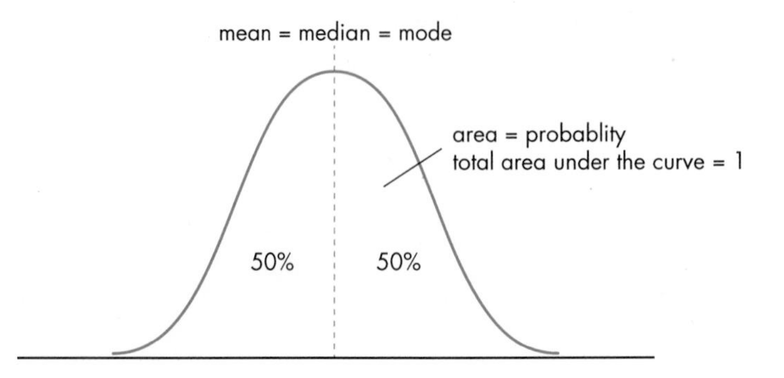

The mean is the line of symmetry in the middle of the distribution. If you move outwards from the mean, to both the left and the right, by 1 standard deviation, the area contains approximately $\frac{2}{3}$ of the data. So a large proportion of the whole data is within 1 standard deviation of the mean. Remember a normal distribution is a bell curve with fewer values as the distribution moves away from the middle value.

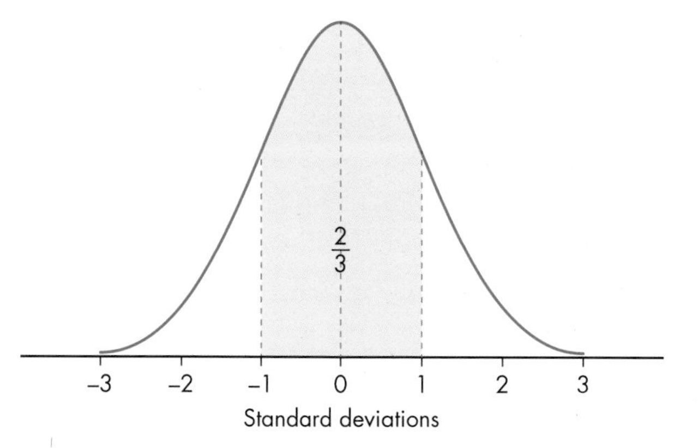

If you move outwards again from the mean, approximately 95% of the data is within 2 standard deviations of the mean. So there are few values more than 2 standard deviations from the mean.

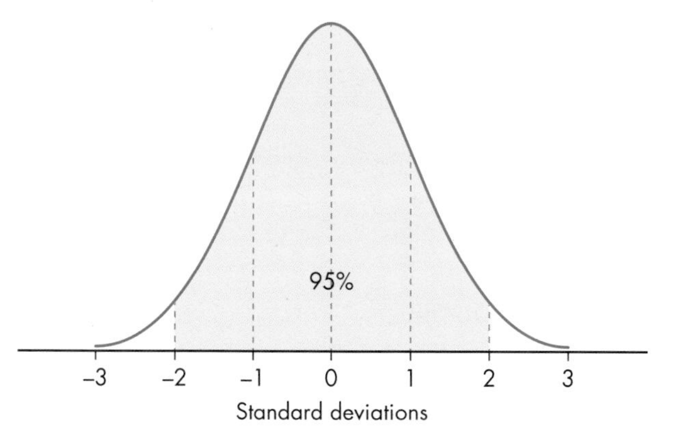

Exercise 5A

1 Correct the following list of properties of a normal distribution. (There is something wrong in each statement.)

a The distribution is symmetrical about the standard deviation.

b The mode, median and mean are all different.

c The total area above the curve is 1.

d The distribution is defined by three parameters: the mean, the median and the standard deviation.

2 Design a crossword with at least five clues to test a classmate's knowledge of a normal distribution.

3 What can you say about the two distributions in the diagram?

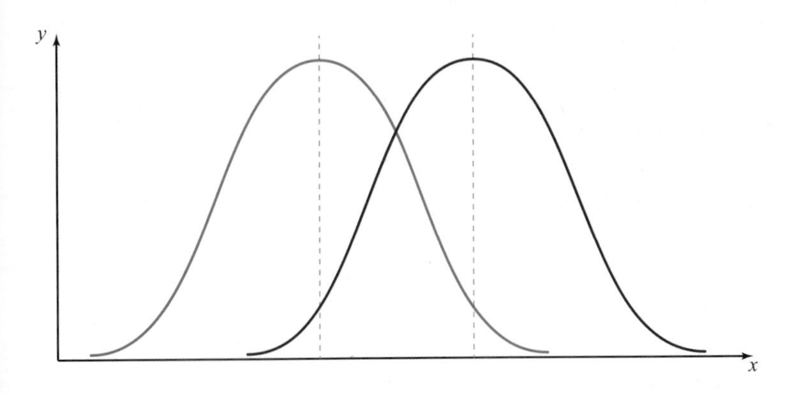

4 What can you say about the two distributions in the diagram?

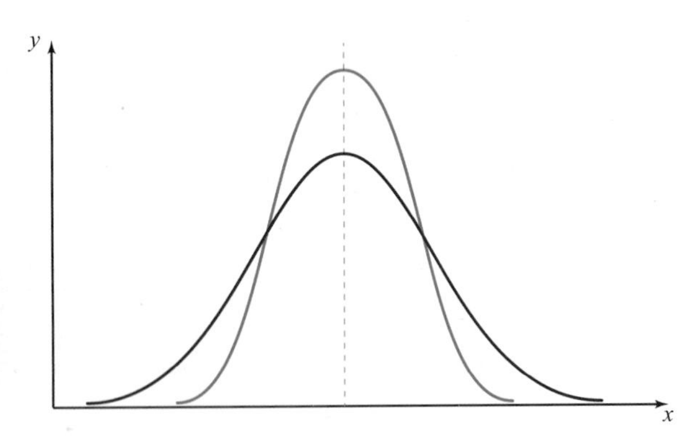

5 The heights of Year 12 students are normally distributed. Imagine you line up all the Year 12 students in your school or college in order of height from shortest to tallest.

a Where in the line would you expect to find the Year 12 student(s) with the mean height?

b What percentage of the Year 12 students would be above the mean height?

c What fraction of the Year 12 students would have heights between the mean and +1 standard deviation?

d What percentage of the Year 12 students would have heights between the mean and +2 standard deviations?

e What fraction of the Year 12 students would have heights between −1 standard deviation and +2 standard deviations?

S2: Notation

Learning objectives

You will learn how to:

- Understand and write the correct notation to describe a normal distribution.
- Understand and write the correct notation to describe the standardised normal distribution.

Introduction

What is notation? The definition from the Collins English Dictionary is "... representation of numbers or quantities in a system by a series of symbols ...". Notation can be formal, such as the mathematical notation you are familiar with (such as the operator symbols +, −, ×, ÷, and so on), or informal such as the emoticons and textspeak you might use in messaging.

Emoticons		Phrases		Abbreviations for words	
:)	happy face for humour, laughter, friendliness, sarcasm	LOL	laughing out loud	M8	mate, boy or girl friend
		L8R	later		
{{{{hug}}}}	really big hug	WYSIWYG	what you see is what you get		
ALL CAPS	yelling				

A normal distribution has its own notation which is used to describe a specific normal distribution. A normal distribution is defined by two parameters, the mean and the standard deviation, and it is important to be able to understand and write the notation to describe a normal distribution with the correct symbols.

Mathematics in the real world

The diagram shows some facts about Twitter usage and trends. What mathematical notation can you see? The obvious mathematical symbol is the percentage sign, %. But what about the numbers? A symbol that represents a number is a numeral. So all the numerals that make up the numbers are also mathematical notation.

40%
OF TWEETS COME FROM MOBILE DEVICES

70%
OF TWITTER ACCOUNTS ARE OUTSIDE THE U.S.

50%
OF USERS ACCESS TWITTER ON MORE THAN ONE PLATFORM

LESS THAN **25%** OF USERS GENERATE **90%** OF TWEETS WORLDWIDE

Discussion

Choose one of the following and discuss it with your peers.

1. Imagine you have taken a classmate out to a pizzeria. Talk about the mathematical notation you would see during your visit to the restaurant.
2. Imagine you have got to explain the quadratic formula $\left(x = \frac{-b \pm \sqrt{b^2 - 4ac}}{2a}\right)$ to someone who has never seen it before. Talk about the mathematical notation in the quadratic formula.
3. Imagine you are in a GCSE science class and the topic is our solar system. Talk about all the mathematics and associated notation that might be included in the lesson.

The notation used to describe a normal distribution

If a variable X has a normal distribution then you can write this as follows using mathematical notation:

$$X \sim \mathrm{N}(\mu, \sigma^2)$$

where

- $\sim$ N tells you that X has a normal distribution
- μ is the mean
- σ is the standard deviation
- σ^2 is the square of the standard deviation and is called the variance.

If $X \sim \mathrm{N}(100, 15^2)$, this tells you that

- the variable X has a normal distribution
- with a mean $\mu = 100$
- and a standard deviation $\sigma = 15$.

(This means that 68% of the values lie between 85 (100 – 15) and 115 (100 + 15).)

It's important to be clear about the difference between standard deviation and variance. If you have

$$B \sim \mathrm{N}(28, 9).$$

This tell you that

- variable B has a normal distribution
- a mean $\mu = 28$
- a standard deviation $\sigma = \sqrt{9} = 3$.

(This means that 68% of the values lie between 25 (28 – 3, the mean minus one standard deviation) and 31 (28 + 3, the mean plus one standard deviation).)

Don't forget the second parameter is the variance and the standard deviation is the square root of this number.

$\sim$: $\sim$ means 'is distributed as'

μ: μ is the Greek letter mu, pronounced 'mew'. It is used to represent the mean of a normal distribution.

σ: σ is the Greek letter sigma. It is used to represent the standard deviation of a normal distribution.

σ^2: σ^2 is the square of the standard deviation and is called the variance.

The standardised normal distribution

To work out probabilities for variables with a normal distribution, the standardised normal distribution is used. The standardised normal distribution is defined as follows:

$$Z \sim N(0, 1).$$

This tells you that

- the standardised normal random variable is denoted by Z
- Z has a normal distribution
- with a mean $\mu = 0$
- and a standard deviation $\sigma = 1$.

The diagram shows the standardised normal distribution.

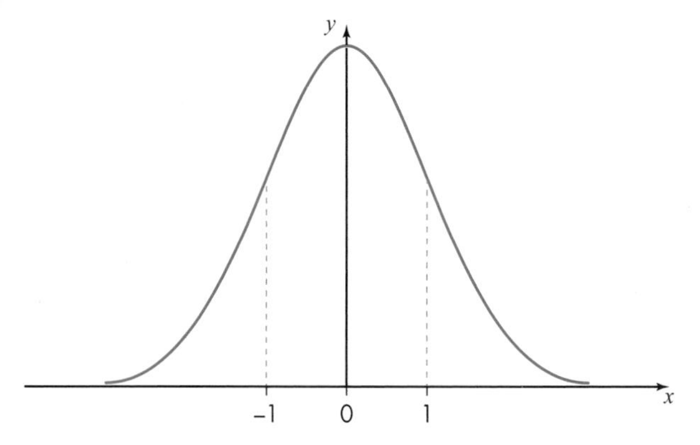

The most important element of the standardised normal distribution is that any normal distribution can be made to fit the standardised normal distribution. The process is called standardising, and is explained later in the chapter.

Exercise 5B

1 During a series of Strictly Come Dancing, two different contestants had scores with normal distributions as follows.

Contestant A: mean = 3, variance = 4

Contestant B: mean = 2, standard deviation = 3

Use the correct mathematical notation to describe the distribution of each of the contestant's scores.

2 During two different series of Stephen Fry's QI, Alan Davies had scores with normal distributions as follows.

Series 1: $\mu = 0$, $\sigma^2 = 5$

Series 2: $\mu = -1$, $\sigma = 2$

Use the correct mathematical notation to describe the distribution of his scores for each series.

3 Two mathematics students sat a set of tests and had results with normal distributions as follows.

Student A: $X \sim \text{N}(5, 7^2)$

Student B: $Y \sim \text{N}(6, 81)$

Write down mean, variance and standard deviation of the students' scores.

4 A mathematics student was asked to write down the mathematical notation for the scores on a new PS4 game which have a normal distribution with mean $\mu = 7$ and standard deviation $\sigma = 11$.

The student wrote $\text{N} \sim X(11, 49)$.

Write down everything that is wrong with what the student wrote and then write down the correct notation.

5 A mathematics student was asked to write down the mathematical notation for the standardised normal distribution.

The student wrote $\text{N} \sim Z(1, 0^2)$.

Write down everything that is wrong with what the student wrote and then write down the correct notation.

S3: Calculating probabilities

Learning objectives

You will learn how to:

- Use a calculator or tables to find probabilities for normally distributed data with known mean and standard deviation.
- Use a calculator or tables to find values for normally distributed data with known mean, standard deviation and probabilities.

Introduction

The average height of men has increased from 167.05 cm to 177.37 cm, an increase of over 10 cm, since the mid-19th century. A company that made beds in the mid-19th century made beds 173.00 cm long so that the average Victorian man could easily sleep in it without his feet sticking out. If the same company was still making beds the same size today, what proportion of men in 2015 could easily sleep in the beds without their feet sticking out?

The heights of men are normally distributed so you can answer this question. In this section you will discover how to calculate probabilities for normally distributed data.

Mathematics in the real world

Researchers at King's College in London have been studying the rise in IQ (intelligence quotient) test scores worldwide.

Many psychological tests (of which IQ tests are an example) are constructed so that the test scores will be approximately normally distributed when used with an appropriate population.

So what is average IQ and how do you calculate what proportion of the population have IQs above a given level? What proportion of the population have a high IQ (around 140 or better)?

On the majority of modern IQ tests the mean μ is set at 100 and the standard deviation σ at 15 so that the distribution models a normal distribution. Using a normal distribution you can answer the questions above.

Discussion

Choose one of the following and discuss it with your peers.

1. Imagine that you have been asked to estimate the proportion of the population with a high IQ (> 140). Talk about what you think the proportion might be (bearing in mind that the mean IQ, μ, is 100 and the standard deviation, σ, is 15) and justify your reasoning.
2. Imagine that you have been asked to estimate the proportion of the population with a genius IQ (> 160). Talk about what you think the proportion might be (bearing in mind that the mean IQ, μ, is 100 and the standard deviation, σ, is 15) and justify your reasoning.
3. Imagine that you have been asked to estimate the proportion of the population with an immeasurable genius IQ (> 200). Talk about what you think the proportion might be (bearing in mind that the mean IQ, μ, is 100 and the standard deviation, σ, is 15) and justify your reasoning.

Calculating probabilities

The area under a normal distribution curve represents probability, and the total area under the curve is defined as 1. This is used to calculate the probabilities of given values.

The standardised normal distribution is denoted as $Z \sim N(0, 1)$, which means it has mean $\mu = 0$ and standard deviation $\sigma = 1$. When calculating probabilities for normally distributed data, it is important to always draw a clear diagram to help your understanding of what you are being asked to find.

There are three different cases to consider.

- Z is less than a value a: $(Z < a)$.
- Z is greater than a value a: $(Z > a)$.
- Z is between values a and b: $(a < Z < b)$.

If you are given $Z \sim N(0, 1)$, you can say that the probability that Z is less than a value a (where $a > 0$) is shown by

$$P(Z < a)$$

where P is the probability.

You know that the mean of the standardised normal distribution is 0 so the diagram will look like this:

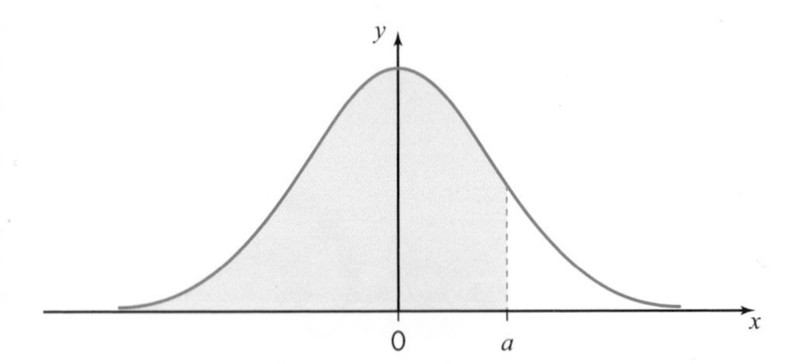

If you are given $Z \sim N(0, 1)$, you can say that the probability that Z is greater than a value a (where $a > 0$) is shown by

$$P(Z > a)$$

where P is the probability.

You know that the mean of the standardised normal distribution is 0 so the diagram will look like this:

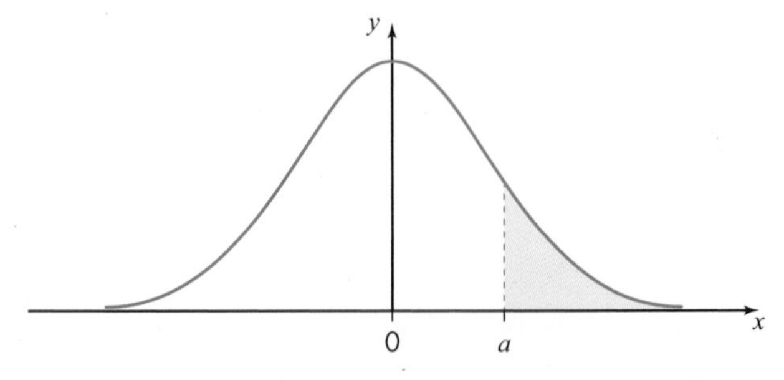

If you are given $Z \sim N(0, 1)$, you can say that the probability that Z is between the value a and b (where $a, b > 0$ and $b > a$) is shown by

$$P(a < Z < b)$$

where P is the probability.

You know that the mean of the standardised normal distribution is 0 so the diagram will look like this:

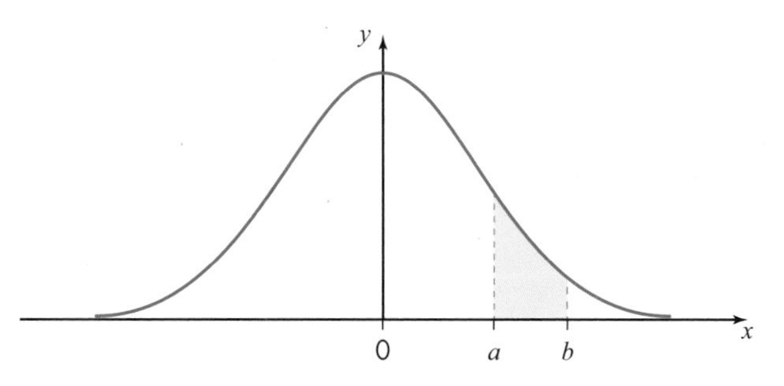

Exercise 5C

Given that $Z \sim N(0, 1)$, draw a clear diagram for:

1 $P(Z > a)$ where $a < 0$.

2 $P(Z < a)$ where $a < 0$.

3 $P(a < Z < b)$ where $a < 0$, $b > 0$.

4 $P(a < Z < b)$ where $a, b < 0$, and $b > a$.

5 $P(0 < Z < a)$ where $a > 0$.

Understanding and using statistical tables for the normal distribution

The probability (area) under the curve has been calculated using the equation of the curve (beyond the scope of this book) and tabulated for values of Z. This table can be found at the back of this book. At first glance the table looks complicated, but it is actually quite straightforward to use once you know how.

Before you look at some specific values there are two key points to note.

- For any particular value the table gives the probability (area) to the left of the value.
- The table only gives probabilities (areas) for positive values.

Given that $Z \sim N(0, 1)$, find $P(Z < 0.31)$.

First draw a clear diagram.

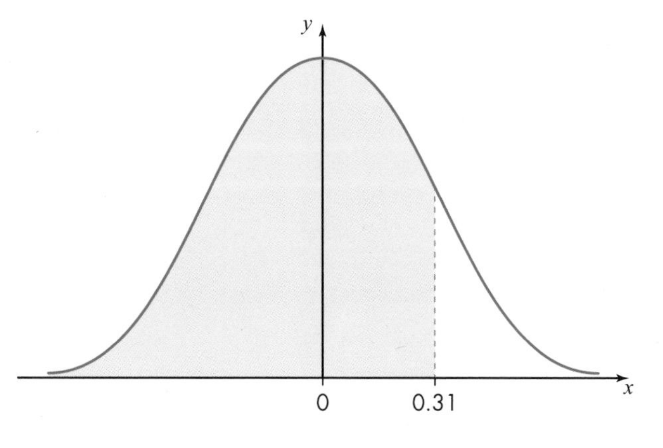

To use the tables correctly, you need to work out the value to look for in the tables. The expression for the probability is:

$$P(Z < 0.31) = \Phi(0.31)$$

where Φ represents the area to the left of the value in brackets. i.e. any given value of z. (Φ is the capital Greek letter phi, pronounced 'fi').

The table gives values for Φ, so you look for the value 0.31 in Table 1: The normal distribution function at the back of this book, part of which is shown below.

You need to find 0.31 then find the corresponding probability.

- First look down the far left column to find 0.3.
- Then look along the top row to find 0.01. (0.31 = 0.3 + 0.01)
- Where the column and row meet you find the required probability as shown below.

z	**0.00**	**0.01**	**0.02**
0.0	0.500 00	0.503 99	0.507 98
0.1	0.539 83	0.543 80	0.547 76
0.2	0.579 26	0.583 17	0.587 06
0.3	0.617 91	0.621 72	0.625 52
0.4	0.655 42	0.659 10	0.662 76

So $P(Z < 0.31) = 0.621\ 72$.

This means the probability of Z having a value < 0.31 is 0.621 72

How do you find $P(Z > 0.31)$? Remember the table only gives the probabilities to the left of a value i.e. less than the value. However, the total area under the curve is 1.

$$\begin{aligned} \text{So } P(Z > 0.31) &= 1 - \Phi(0.31) \\ &= 1 - 0.621\ 72 \\ &= 0.378\ 28 \end{aligned}$$

This means the probability of Z having a value > 0.31 is 0.378 28.

Note that $P(Z < 0.31) + P(Z > 0.31) = 1$.

How do you find $P(Z < -0.31)$?

First draw a clear diagram.

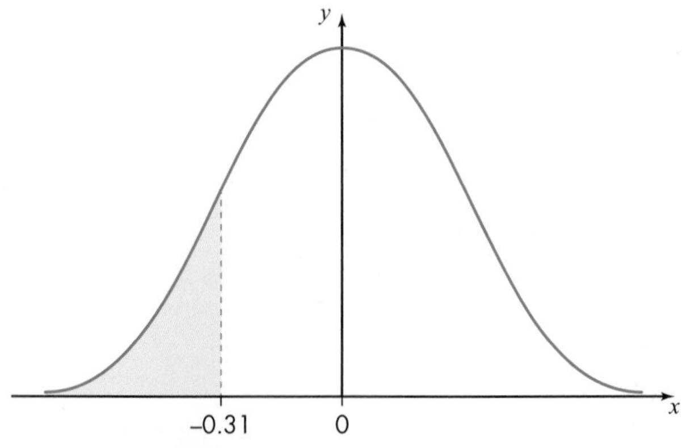

The table does not have values for negative numbers, so it is necessary to use known properties of the normal distribution.

You know that the normal distribution is symmetrical about the mean, so:

$$\begin{aligned}P(Z < -0.31) &= P(Z > 0.31)\\ &= 1 - P(Z < 0.31)\\ &= 1 - \Phi(0.31)\\ &= 1 - 0.621\,72\\ &= 0.378\,28\end{aligned}$$

This means the probability of Z having a value < -0.31 is 0.378 28.

Note this is the same as $P(Z > 0.31)$.

Finally how do you find $P(0.1 < Z < 0.31)$?

First draw a clear diagram.

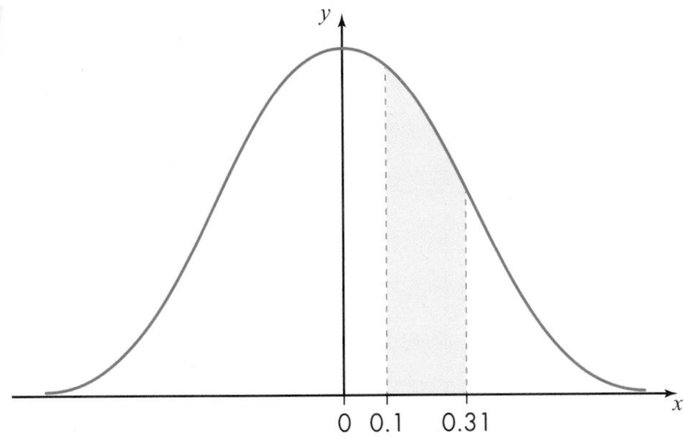

To find the probability (area) between 0.31 and 0.1 you need to find the probability less than 0.31 and subtract the probability less than 0.1.

$$\begin{aligned}P(0.1 < Z < 0.31) &= P(Z < 0.31) - P(Z < 0.1)\\ &= \Phi(0.31) - \Phi(0.1)\\ &= 0.621\,72 - 0.539\,83\\ &= 0.081\,89\end{aligned}$$

What is the probability that Z will have a value greater than the mean but less than 1.5?

As always, first draw a clear diagram.

To find the probability (area) between the mean, 0, and 1.5 you need to find the probability that Z will have a value less than 1.5 and subtract the probability that Z will have a value less than 0.

$$P(0 < Z < 1.5) = P(Z < 1.5) - P(Z < 0)$$

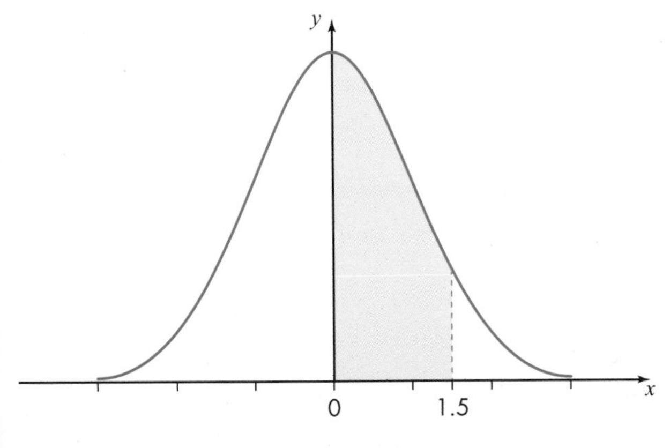

The probability that Z is less than 0 is 0.5 since zero is the mean and half of a normal distribution lies below the mean and half above it.

So

$$P(0 < Z < 1.5) = P(Z < 1.5) - P(Z < 0)$$
$$= \Phi(1.5) - 0.5$$
$$= 0.933\,19 - 0.5$$
$$= 0.433\,19$$

Exercise 5D

For this exercise, present your working as shown in the examples above and use the correct notation.

1. Find the probability that Z will be less than 1.5.
2. Find the probability that Z will be less than 0.57.
3. Find the probability that Z will be greater than 2.06.
4. Find the probability that Z will be less than -1.
5. Find the probability that Z will be greater than -1.
6. Find the probability that Z will be between 1 and 2.
7. Find the probability that Z will be between -1 and 1.
8. Find the probability that Z will be between -2 and -1.

Standardising

The standardised normal distribution has a mean of 0 and a standard deviation of 1. Any normal distribution can be transformed into a standardised normal distribution. This means that you don't need statistical tables for every possible combination of mean and standard deviation. Instead you can transform any normal distribution to the standardised normal distribution and just use one set of statistical tables. You do this by subtracting the mean μ and dividing by the standard deviation σ as follows:

Standardising: Standardising is a mathematical process to transform any normal distribution to the standardised normal distribution with mean = 0 and standard deviation 1.

$$Z = \frac{X - \mu}{\sigma}$$

Example 1

You are given the random variable $X \sim N(60, 10^2)$, find $P(X < 75)$.

Draw a clear diagram.

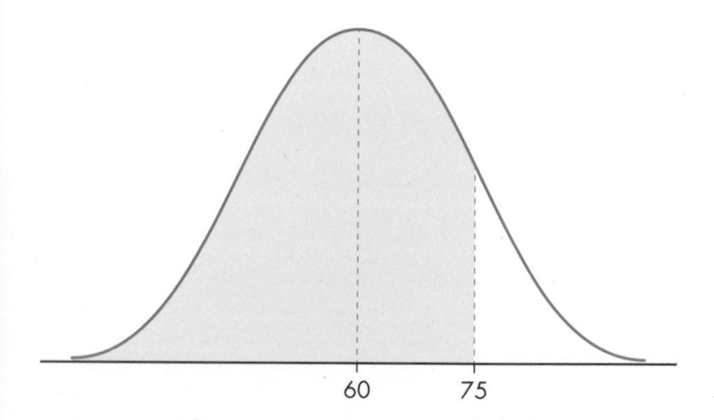

$$P(X < 75) = P\left(Z < \frac{75-60}{10}\right)$$

$= P(Z < 1.5)$

$= \Phi(1.5)$

$= 0.933\ 19$

So, the probability of X having a value < 75 is 0.933 19.

Example 2

For the same random variable X, find $P(X > 75)$.

Draw a clear diagram.

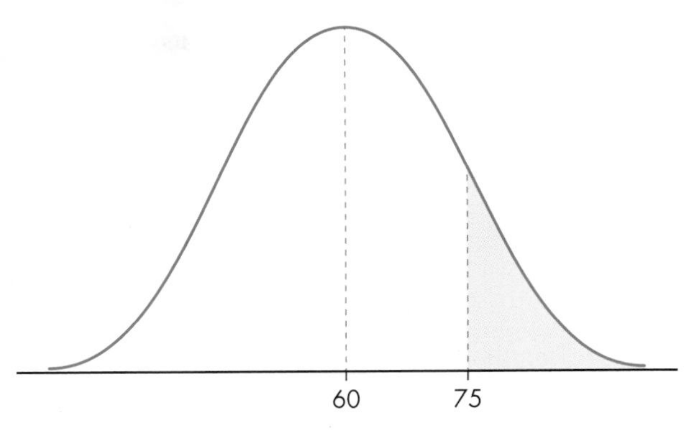

$P(X > 75) = 1 - P(X < 75)$

$$P(X > 75) = 1 - P\left(Z < \frac{75-60}{10}\right)$$

$= 1 - P(Z < 1.5)$

$= 1 - \Phi(1.5)$

$= 1 - 0.933\ 19$

$= 0.066\ 81$

So, the probability of X having a value > 75 is 0.066 81.

Example 3

For the same random variable X, find $P(65 < X < 75)$.

Draw a clear diagram.

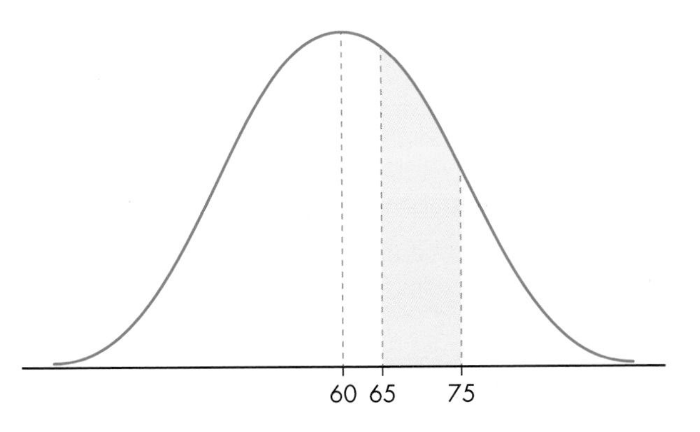

$$P(65 < X < 75) = P\left(\frac{65-60}{10} < Z < \frac{75-60}{10}\right)$$
$$= P(0.5 < Z < 1.5)$$
$$= P(Z < 1.5) - P(Z < 0.5)$$
$$= \Phi(1.5) - \Phi(0.5)$$
$$= 0.933\ 19 - 0.691\ 46$$
$$= 0.241\ 73$$

So, the probability of X having a value between 65 and 75 is 0.241 73.

Exercise 5E

For this exercise, present your working as shown in the examples above and use the correct notation.

1 Given that $X \sim N(30, 100)$, find the following probabilities.

- **a** $P(X < 35)$
- **b** $P(X > 38.6)$
- **c** $P(X > 20)$
- **d** $P(35 < X < 40)$
- **e** $P(15 < X < 32)$
- **f** $P(17 < X < 19)$

2 Find the proportion of the population with a high IQ (> 140) given that the mean IQ is 100 and the standard deviation is 15.

3 A Victorian bed manufacturer made beds with a standard length of 1.73 m.

The heights of men are normally distributed and at the start of the 21st century the mean height is 1.77 m and standard deviation is 0.07 m.

Find the proportion of 21st century men that are shorter than the Victorian bed length.

4 A particular breed of hens produces eggs which have a mean mass of 60 grams with a standard deviation of 4 grams and mass is found to be normally distributed.

The eggs are classified as small, medium, large and extra large depending on their weight, as follows.

Classification	Weight
Small	Less than 55 grams
Medium	Between 55 grams and 65 grams
Large	Between 65 grams and 70 grams
Extra large	Greater than 70 grams

- **a** Find the proportion of eggs that are classified as small.
- **b** Find the proportion of eggs that are classified as medium.
- **c** Find the proportion of eggs that are classified as large or extra large.
- **d** If the classification for medium eggs is changed to between 55 grams and 68 grams, by how much does the proportion of eggs classified as medium increase?

5 The police regularly monitor the speeds of cars on a section of motorway and it is found that the speeds are normally distributed. The mean speed is 68.5 mph with a standard deviation of 5 mph.

a Find the proportion of motorists that break the speed limit (70 mph).

b A motorist believes that people aren't fined unless they are 10% over the speed limit.

What proportion of motorists will be fined on this basis?

c A motorist is given a fixed penalty fine for doing 85 mph.

What percentage of motorists exceed this speed?

Finding values of variables from known probabilities

Sometimes you may know the probability of an event and then need to work backwards to find the corresponding value X.

Example 4

A random variable X is normally distributed with a mean 9 and standard deviation 0.27.

Find the value of x so that $P(X < x) = 0.751\ 75$.

First you need to transform the distribution in the question to the standardised normal distribution.

$$P\left(Z < \frac{x-9}{0.27}\right) = 0.75175$$

Using Table 1: The normal distribution function, you then need to find the Z value that corresponds to a probability of 0.751 75. It is a case of looking in the body of the table until you find the required probability.

$$\Phi(0.68) = 0.751\ 75$$

Therefore:

$$\frac{x-9}{0.27} = 0.68$$

Rearranging gives

$x = (0.27 \times 0.68) + 9$

$x = 9.1836$

Percentage points of the normal distribution

You can also find the value of x when a probability is given by using Table 2: Percentage points of the normal distribution at the back of the book.

This table has column and row headings corresponding to probabilities and the values in the cells are x values.

To find the Z value that corresponds to a probability of 0.61, you can use Table 2, part of which is shown below.

- First look down the far left column to find 0.6.
- Then look along the top row to find 0.01.
- Where the column and row meet, you find the required probability as shown below.

p	0.00	0.01	0.02
0.5	0.0000	0.0251	0.0502
0.6	0.2533	0.2793	0.3055
0.7	0.5244	0.5534	0.5828
0.8	0.8416	0.8779	0.9154
0.9	1.2816	1.3408	1.4051

So $P(Z < 0.2793) = 0.61$.

Example 5

A random variable X is normally distributed with mean 25 and standard deviation 1.25.

Find the value of x so that $P(X < x) = 0.75$.

First you need to transform the distribution in the question to the standardised normal distribution.

$$P\left(Z < \frac{x-25}{1.25}\right) = 0.75$$

Using Table 2: Percentage points of the normal distribution, you then need to find the Z value that corresponds to a probability of 0.75.

$$\Phi(0.6745) = 0.75$$

Therefore:

$$\frac{x-25}{1.25} = 0.6745$$

Rearranging gives:

$x = 25.843\,125$

It is important to know when to use the two different tables of values.

- Table 1: The normal distribution function is used when you are given (or can calculate) z values and are asked to find the area under the curve (the probability).
- Table 2 is used when you are given specific probabilities and you need to find the corresponding z values.

You need to be confident using both tables.

Exercise 5F

For this exercise, present your working as shown in the examples above and use the correct notation.

1 Given that $X \sim N(30, 100)$, find the value of x so that:

a $P(X < x) = 0.99$

b $P(X < x) = 0.979\,82$

c $P(X > x) = 0.194\,89$

d $P(X < x) = 0.75$

e $P(X > x) = 0.99$

f $P(X > x) = 0.05$

2 A manufacturer produces a new fluorescent light bulb and claims that it has an average lifetime of 4000 hours with a standard deviation of 375 hours.

Find the lifetime such that 90% of light bulbs will last for less than this duration.

3 A Victorian bed manufacturer made beds with a standard length of 1.73 m.

The heights of men are normally distributed and at the start of the 21st century the mean height is 1.77 m and standard deviation is 0.07 m.

Find the bed length such that 10% of 21st century men will be longer than the bed length.

4 A particular breed of hens produces eggs which have a mean mass of 60 grams with a standard deviation of 4 grams and mass is found to be normally distributed.

Find the mass such that 75% of eggs are greater than this mass.

5 The police regularly monitor the speeds of cars on a section of motorway and it is found that the speeds are normally distributed. The mean speed is 68.5 mph with a standard deviation of 5 mph.

Find the range (symmetrical about the mean) within which 80% of the speeds lie.

6 A dairy produces cartons of milk with a mean capacity of 1 litre.

Cartons of milk containing less than 960 ml are sold at a reduced price.

The capacity of milk in the carton is normally distributed with a standard deviation of 25 ml.

a What proportion of cartons are sold at a reduced price?

b Cartons of milk are produced with a normal retail price per carton of £1.05.

The price of cartons containing less than 960 ml of milk is 80p.

How much revenue will the dairy expect to get when they sell 1000 cartons of milk?

Case study

At the start of this chapter, some questions were asked about numbers of Facebook friends. In this chapter, you have learnt about normally distributed continuous random variables. Is 'number of Facebook friends' a discrete or continuous variable? What do you think?

If you're not sure, discuss the differences between discrete and continuous data with a classmate. You should have reached the conclusion that 'number of Facebook friends' is a discrete variable. (You can't have half a friend!) So can you still use the normal distribution? The simple answer is yes, if the pattern of frequencies for each data value (number of people who have each specific number of Facebook friends) satisfy all the conditions for a normal distribution. You are going to assume they do.

Using the data given at the start of the chapter, that the mean number of Facebook friends worldwide is 176 people and the standard deviation is 33, you can now work out what proportion of users have less than or more than the number of Facebook friends that you do. You could also work out what proportion of users have, for example, more than 5000 Facebook friends, as some people claim.

Project work

On the internet find the IQ of:

- a pop star
- a famous scientist
- a well known actor
- a famous painter
- a TV personality.

Given that the mean IQ is 100 and the standard deviation is 15, find the proportions of the population that have higher and lower IQs than the individuals you have selected.

Present your calculations in the way that has been shown in this chapter.

You might find Question **2** in Exercise 5E helpful.

Check your progress

How confident are you feeling in your level of knowledge? What do you need to practise more?

Spec reference	Learning objective			
S1.1	Recognise normally distributed data			
S1.1	Describe the properties of a normal distribution			
S1.1	Use your knowledge about specific standard deviations and their probabilities			
S2.1	Understand and write the correct notation to describe a normal distribution			
S2.1	Understand and write the correct notation to describe the standardised normal distribution			
S3.1	Use a calculator or tables to find probabilities for normally distributed data with known mean and standard deviation			
S3.1	Use a calculator or tables to find values for normally distributed data with known mean, standard deviation and probabilities			

6 Probabilities and estimation

If you follow Formula One, you will be familiar with the theme tune for the BBC's television coverage. The opening music features an instrumental from a Fleetwood Mac song called *The Chain*, which was sampled and built into the opening music for the TV show. In music, sampling is the act of taking a portion, or sample, of one sound recording and reusing it as an instrument or a sound recording in a different song or piece.

Sampling is now very common in music, and samples are often taken from unexpected sources. Dizzee Rascal and Jay Z have both sampled a 1980 song by Billy Squier, *The Big Beat*, which is now one of the most-sampled songs. Kanye West is a prolific sampler, well known for building on an existing song or picking the best bits from several songs.

But what has this got to do with probability and estimation? All these artists are doing something in their profession that statisticians do in theirs – sampling. The music artists have back catalogues of songs from which to take samples. Musicians take samples for many different reasons, while statisticians generally use samples when it isn't practical (due to limitations such as cost and time) to gather data on an entire population.

S4: Population and sample

Learning objectives

You will learn how to:

- Understand what is meant by the term 'population' in statistical terms.
- Develop ideas of sampling to including the concept of a simple random sample from a population.

Introduction

Imagine that a market research company had offered you free tickets to the opening ceremony of the London 2012 Olympics. This company had been employed by a T-shirt manufacturer to determine how many of the crowd are wearing their brand of T-shirt. Before you can sit down and enjoy the atmosphere and the entertainment you have to survey the opening ceremony crowd to provide data for the market research company. You are their only employee that night so you have to do the whole job yourself.

You don't want to miss anything and quickly realise that you can't possibly look at everybody in the crowd who are wearing T-shirts so you decide to just look at a sample of the crowd. You decide that you will stand at the entrance to the VIP seats and do the survey there, and you might get to see a few celebrities. Do people who sit in the VIP seats wear T-shirts? If they do, are they likely to buy the brand you have been asked to look out for or will they be wearing designer T-shirts? How confident are you that your sample will be representative of the whole crowd? What might have been a better way to select a sample of the crowd?

Mathematics in the real world

Every 10 years, censuses are held across the UK, run by the Office for National Statistics (ONS) in England and Wales, the General Register Office for Scotland and the General Register Office for Northern Ireland. The last census in 2011 was the 21st such census.

Census statistics help paint a picture of the nation and how we live. They provide a detailed snapshot of the population and its characteristics, and governments use them to decide how to allocate money for public services.

Discussion

Choose one of the following and discuss it with your peers.

1. Imagine that the chief executive of the Formula 1 group has two VIP tickets for the next British Grand Prix to give away to students at your school or college. Talk about fair ways that two students could be picked to receive the tickets.
2. Imagine you had been the person offered the free ticket to the London 2012 Olympics opening ceremony by the market research company. Talk about other, better ways you could have chosen a sample of the crowd.
3. Imagine you are a statistician at the Office for National Statistics (ONS) and you have been asked to select data for a 1% sample of the people in the census. Talk about how you might go about selecting the 1% sample.

Populations and samples

In statistics, a **population** can be defined as all the objects or individuals under consideration. The population is analysed using samples. It is important to note that 'populations' are not just people. Populations may consist of, but are not limited to, people, animals, objects or events. When looking at data, it is important to clearly identify the population being studied or referred to, so that you can understand *who* or *what* is included in the data. For example, if you were looking at data for Formula One races, you would need to understand whether the population of data is for all Grand Prix races or just those for a particular year or continent, whether it is for all teams or just specific teams.

Population: All the objects or individuals under consideration.

Sample: A subset of a population which can be used to represent the whole population.

In statistics, a **sample** can be defined as a subset of a population which can be used to represent the whole population.

How do you go about choosing a sample? A sample should be representative of the whole population and have the same characteristics as that population.

If you had been the person offered free tickets to the opening ceremony of the London 2012 Olympics by the market research company, you would have had to survey the crowd to see if they were wearing a particular brand of T-shirt. The impracticalities of trying to survey the whole crowd and the drawbacks of choosing a sample of VIP ticket holders have already been mentioned. How else could you have chosen a sample? You could have chosen an area of the stadium where the general public were going to sit and done the survey as they took their seats. But what if tickets for this area had been allocated to a country where the brand of T-shirt you were looking for wasn't sold? This wouldn't be a representative sample.

Alternatively you could have decided to stand at one of the entrances to the Olympic Park for one hour and just survey the people under the age of 30. But how would you know who is under 30? Do people under 30 buy this brand of T-shirt or is it marketed to people over 30 or to all ages?

What you could have done instead is chosen a **random sample**. A random sample is one in which each member of the population has an equal chance of being selected. You need to use a procedure for selecting a sample, such as drawing names from a hat, throwing a die or tossing a coin, ensuring that the procedure is unbiased. You could have asked the Olympic organising committee to suggest a section of the stadium that would be representative of the whole crowd where there was no bias of country, cost of ticket, accessibility, etc. You could have then put all the seat numbers for this section of the stadium into a hat and drawn a random sample of seat numbers out of the hat. Just before the opening ceremony you could have walked up and down each row in the section making a note of whether or not the people in the seat numbers that you had chosen were wearing the brand of T-shirt you are interested in.

Random sample: A sample in which each member of the population has an equal chance of being selected.

As another example, suppose you have 200 Facebook friends and you want to choose a random sample of 20 of these friends to contact. You can use random numbers. To do this you need to make a list, in any order, of all your 200 Facebook friends and number the names from 1 to 200. You could then use the random number generator on a calculator and the total of 200 to select the 20 names of the people to contact.

A calculator generated the following 20 random numbers.

50	168	86	183	16
190	186	96	27	29
75	128	144	115	59
161	116	74	121	4

So the first name you would choose would be name 50, then name 168, etc.

Exercise 6A

1 Draw a diagram to illustrate the words 'population' and 'sample'.

2 Would the following situations produce a random sample? If not, why not?

a Conducting a survey about methods of transport to a football match by asking part of a section of the crowd sitting in the stadium

b Conducting a survey of hospital waiting times in a particular hospital by asking the patients in the Accident and Emergency department

3 Write down two different ways to randomly select a sample of 10 people from a population of 100 (without using random numbers).

4 The head teacher of a school wants to talk to eight students out of the 50 students in the sixth form about the sixth form common room.

She asks the head of the sixth form to make a list of all the students and number them from 1 to 50.

Use the list of random numbers below to identify eight students.

1 2 0 0 5 8 4 0 0 0 5 1 0 5 1 9 2 6 7 4 7 6 5 3 5 7 8 9 0 4
1 8 0 7 5 0 2 9 3 2 4 9 0 3 9 9 4 9 9 8 8 5 9 0 9 8 8 4 4 5
2 7 5 0 9 4 6 7 9 3 0 2 6 3 2 9 1 2 1 3 9 4 1 2 3 2 0 6 6 2
4 1 9 6 7 7 7 6

5 Use the random number generator on your calculator to select a sample of 10 people from a population of 100.

How does your list compare to those of your classmates?

S5: The mean of sample size n

Learning objectives

You will learn how to:

- Know that the mean of a sample is called a 'point estimate' for the mean of the population; appreciate that accuracy is likely to be improved by increasing the sample size.

Introduction

If you take a sample from a population, you can calculate the **sample mean,** $\bar{x}$.

Sample mean: The mean of a sample of a population, denoted by $\bar{x}$.

$$\bar{x} = \frac{\Sigma x}{n}$$

where

- $\bar{x}$ is the sample mean
- Σx is the sum of all the relevant data values (Σ is the Greek letter sigma and is used in maths to mean 'the sum of'.)
- n is the number of items in the sample.

The **population mean,** μ, is the mean of the population. If you don't know the population mean, then the value of the sample mean, $\bar{x}$, can be used as an estimate for the population mean. You call the value of $\bar{x}$ a **point estimate** of μ.

Point estimate: A single estimated value of a parameter of a given population.

For example, if you wanted to find the mean number of Facebook friends people have, you could take a random sample of people on Facebook and find the mean number of Facebook friends for this sample of Facebook users. You would calculate the sample mean, for the sample of people you selected. This would give you an estimate of the population mean.

Mathematics in the real world

Best-selling author Bill Bryson has written:

"Your pillow alone may be home to 40 million bed mites. (To them your head is just one large oily bon-bon). And don't think a clean pillow-case will make a difference … Indeed, if your pillow is six years old – which is apparently about the average age for a pillow – it has been estimated that one-tenth of its weight will be made up of sloughed skin, living mites, dead mites and mite dung."

How was the number of bed mites in a pillow estimated? Nobody would have laboriously sat and counted them all, and it would have been very prone to error as the bed mites would move during the counting process. Instead, a very small sample section of a number of different pillows would have been taken and the number of bed mites in each section counted. The sample mean of the number of bed mites in each section would then have been multiplied by the number of sections in a whole pillow to calculate an estimate of the total number of bed mites in a pillow. So a number from a sample was used to estimate a number for a population.

Discussion

Choose one of the following and discuss it with your peers.

1. Imagine that you have been asked to count the number of bed mites in a number of different samples of a pillow. Talk about whether the mean number of bed mites in each sample will be the same.
2. Imagine that you have been asked to estimate the mean number of hours of sleep 17- and 18-year-olds have each night. Talk about how you might change the sample size, n, to make the sample mean a better estimator of the population mean.
3. Imagine that you have been asked to estimate the mean number of hours per day that secondary school students spend playing on their Xbox, Wii or PlayStation. Talk about how close the mean of the sample will be to the population mean.

Point estimation

Point estimation involves using a statistic from a random sample, in this case the sample mean, $\bar{x}$, to find an estimate for the population mean, μ. The value of $\bar{x}$ that you get from any one sample will be dependent on the data values in the sample and consequently you will get different values of $\bar{x}$ from different samples. The larger the sample size (the greater the value of n), the more likely it is that that $\bar{x}$ will be an accurate representation for the population mean. Another way of saying this is that $\bar{x}$ will be a better estimator of the population mean.

Exercise 6B

1 A friend decides to test Bill Bryson's suggestion that a pillow has around 40 million bed mites. They put a tiny section of their own pillow under a microscope and count the number of bed mites that they see. They calculate how many of the sections there are in the pillow and multiply their count by the number of sections. They declare that Bill Bryson's number is wrong and the number of bed mites is a lot less.

How could your friend's method be improved to give a better estimate of the number of bed mites in a pillow?

2 A study estimated the mean number of hours of sleep 17- and 18-year-olds have each night.

The sample means (in hours) from six different samples were:

10.55 11.02 12.0 10.76 11.54 10.85.

How should these sample means be used to estimate the population mean? Give reasons for your answer.

3 A study estimated the mean number of hours per day that secondary school students spend playing on their Xbox, Wii or PlayStation.

The results for a class of students were as follows:

2.5	3.0	3.2	4.0	3.5	3.4
0.0	0.5	0.0	3.7	4.2	1.5
2.2	3.3	3.8	3.9	0.0	0.3
2.6	0.7	0.7	3.6	4.3	2.9
2.0	0.2	3.2	3.3	2.8	0.1

a Take three different random samples, each with five data values and calculate the sample mean for each of the three samples.

b By summing the sample totals from part **a** and using the total number of data values in the three samples (i.e. 15), calculate a more accurate point estimate.

c For the data above, another random sample of eight data values was taken and the sum of these is 19.1. Calculate the mean of this sample.

d Describe the limitations of this method, and explain how the method could be improved to give a better estimate for the population of secondary students as a whole.

S6: Confidence intervals

Learning objectives

You will learn how to:

- Calculate confidence intervals for the mean of a normally distributed population of known variance using $\frac{\sigma^2}{n}$.

Introduction

The Office for National Statistics states:

Confidence intervals are indicators of the extent to which the estimate may differ from the true population value. The larger the confidence interval, the less precise the estimate.

The 2011 Census estimate for the England and Wales population had a 95 per cent confidence interval width of plus or minus 0.15 per cent (plus or minus 83,000 people).

Census estimates used the Census Coverage Survey to measure coverage of the census and to provide estimates of the population, including people missed by the census. A basic requirement of any estimate is a measure of its precision or uncertainty. A 95 per cent confidence interval, which provides a measure of accuracy, can be interpreted as the interval within which 95 times out of 100 the true value will lie if the sample were repeated 100 times.

You know that the sample mean can be used as a point estimate for the population mean. This can be developed further and for any sample mean you can find a range of values, or interval, within which the population mean is expected to be found with a given level of confidence. The interval is called a **confidence interval**.

Confidence interval: The range of values within which the population mean will be found, with a given level of confidence.

Mathematics in the real world

> The Queen has topped the Christmas Day television ratings, as more than 7.8 million viewers tuned in to watch her festive broadcast across the BBC and ITV.
>
> Bawdy BBC One sitcom *Mrs Brown's Boys* drew the highest viewing figures of any single programme for a second year running, but with an average audience of 7.61 million was well down on last year when it attracted 9.4 million viewers.
>
> The Telegraph, 26 December 2014

The Broadcasters Audience Researchers Board, BARB, recruit private homes so that they can estimate TV watching patterns across all TV households. The type of households they select is based on data from the Ipsos MORI Establishment Survey as BARB wants to ensure that the households on their panel reflect all households in the country. The BARB panel consists of over 5000 households which are representative of similar ones across the country.

When a household is selected to join the BARB panel all the TV viewing devices, including laptops and desktop computers, in their home are fitted with a meter. Each member of the household has a dedicated button on a special remote control. When they enter a room where TV is being watched they have to press their button to indicate that they are there and press it again when they leave the room.

So is the headline from The Telegraph correct? Was the Queen's broadcast the most watched broadcast over the Christmas period? Did *Mrs Brown's Boys* really have the single biggest audience?

Discussion

Choose one of the following and discuss it with your peers.

1. Imagine that you have been told that the mean number of hours of sleep for a sample of 17 and 18-year-olds is 6 hours. Talk about how confident you are that this is the best estimator for the population mean of all 17 and 18-year-olds.
2. Imagine that you have been asked to write an article for the school magazine about Christmas 2014 viewing figures. Talk about how the figures have been calculated and how confident you are that the information in The Telegraph is correct.
3. Imagine that you have been asked to survey a Year 7 class to determine the mean number of hours per day they spend playing on their Xbox, Wii or PlayStation. Talk about your level of confidence that this sample mean number of hours can be used as a mean for all secondary school students.

Calculating confidence intervals

If you know the mean and variance of a set of n values with a normal distribution, you can calculate confidence intervals.

If a variable X is normally distributed with a mean μ and a variance σ^2 then you can write:

$$X \sim N(\mu, \sigma^2).$$

If you take a sample of n data values from a population that is normally distributed with a mean μ and variance σ^2, then the sample mean will be μ and the variance will be $\frac{\sigma^2}{n}$ and you can write:

$$\bar{x} \sim N\left(\mu, \frac{\sigma^2}{n}\right)$$

As the standard deviation is the square root of the variance this gives a sample standard deviation of $\frac{\sigma}{\sqrt{n}}$.

If you know or can calculate the sample mean, $\bar{x}$, and are given the variance, σ^2, the sample size n and the confidence level required then you can work out what values the mean, μ could be for the required confidence level. Because the normal distribution is symmetrical about the mean, the confidence interval will be symmetrical.

To calculate the values that are within a specific confidence interval, you need to find the upper and lower limits. To start, draw a diagram.

The diagram shows a normal distribution with a 95% confidence interval.

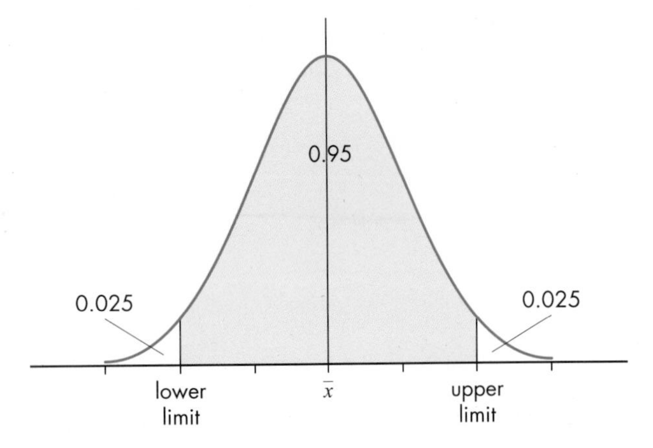

The whole area under the normal distribution curve is defined as 1 so the area outside the 95% confidence interval is:

$$1 - 0.95 = 0.05.$$

The normal distribution is symmetrical about the mean, so the areas above and below the upper and lower limit respectively are $0.05 \div 2 = 0.025$.

So how do you calculate the upper and lower limits? The distribution is symmetrical, so you start by calculating the upper limit.

The area under the curve to the **left** of the upper limit (remember the statistical tables always tell you the area to the left of a value) is

$$\text{area} = 0.025 + 0.95 = 0.975 \text{ (or } 1 - 0.025\text{)}.$$

You now use the standardised normal distribution to find the value that corresponds to an area of 0.975.

$0.975 = 0.97 + 0.005$, so use Table 2: Percentage points of the normal distribution to find a value of $z = 1.9600$. The relevant part of the table is shown below.

p	0.004	0.005	0.006
0.97	1.9431	1.9600	1.9774

Remember you use Table 2 when you are given a probability and have to find the corresponding z value.

This means that the upper limit is 1.96 and, by symmetry, the lower limit is –1.96. (Remember the mean of the standardised normal distribution is 0.)

Let's think about Alan Davies and his performance on the TV programme *QI*. If you had a sample of his scores from some of the *QI* programmes on which he had been a contestant (rather than all of his scores) and convert them to a standardised normal distribution, you can now say that the 95% confidence interval (the interval such that the probability that it includes the mean from the sample of his scores is 0.95) is (–1.96, 1.96).

You can repeat the calculations to find values for a 90% confidence interval.

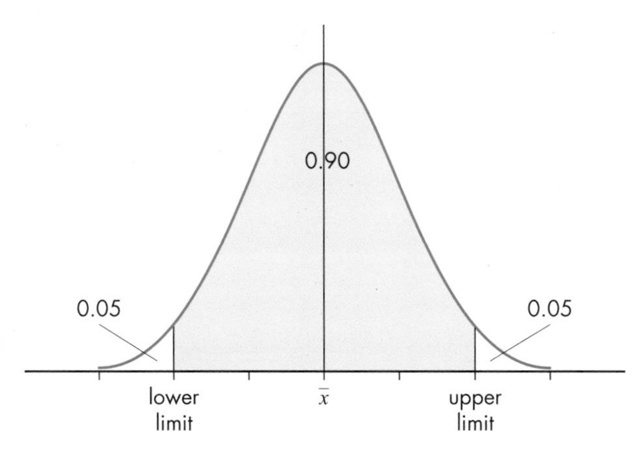

The whole area under the normal distribution curve is defined as 1 so the area outside the 90% confidence interval is:

$1 - 0.90 = 0.10.$

The normal distribution is symmetrical about the mean, so the areas above and below the upper and lower limit respectively are $0.10 \div 2 = 0.05$.

To calculate the upper and lower limits, you repeat the procedure above, using the fact that the distribution is symmetrical, so you start by calculating the upper limit.

The area under the curve to the left of the upper limit is

area = $0.05 + 0.90 = 0.95$ $(1 - 0.5)$.

You use the standardised normal distribution to find the value that corresponds to an area of 0.95.

$0.95 = 0.95 + 0.000$, and from Table 2 you find a value of $z = 1.6449$. The relevant part of the table is shown below.

p	0.000	0.001	0.002
0.95	1.6449	1.6546	1.6646

So the upper limit is 1.6449 and, by symmetry, the lower limit is –1.6449.

Let's return to Alan Davies. If you again had a sample of his scores from some of the QI programmes on which he had been a contestant and convert them to a standardised normal distribution, you can now say that the 90% confidence interval (the interval such that the probability that it includes the mean of all of his scores is 0.90) is (–1.645, 1.645).

Exercise 6C

1 Draw a diagram to illustrate a 80% confidence interval for a normal distribution. Clearly label all areas of the diagram.

2 Given that $Z \sim N(0, 1)$ determine the upper and lower limits for a 80% confidence interval.

3 Draw a diagram to illustrate a 98% confidence interval for a normal distribution. Clearly label all areas of the diagram.

4 Given that $Z \sim N(0, 1)$ determine the upper and lower limits for a 98% confidence interval.

5 Draw a diagram to illustrate a 99% confidence interval for a normal distribution. Clearly label all areas of the diagram.

6 Given that $Z \sim N(0, 1)$ determine the upper and lower limits for a 99% confidence interval.

Confidence intervals for μ

You have already seen that if you take a sample of n data values from a population that is normally distributed with a mean μ and variance σ^2, then the sample mean will follow the distribution:

$$\bar{x} \sim N\left(\mu, \frac{\sigma^2}{n}\right)$$

You have also already seen that the standard deviation for the sample mean will be:

$$SD = \sqrt{\frac{\sigma^2}{n}} = \frac{\sigma}{\sqrt{n}}.$$

You know that any normal distribution $X \sim N(\mu, \sigma^2)$ can be transformed into a standardised normal distribution, by subtracting the mean μ and dividing by the standard deviation σ as follows:

$$Z = \frac{X - \mu}{\sigma}$$

So for the sample mean you have:

$$Z = \frac{\bar{x} - \mu}{\sigma / \sqrt{n}}$$

If you now consider the 95% confidence interval:

$$P\left(-1.96 \leqslant \frac{\bar{x} - \mu}{\sigma / \sqrt{n}} \leqslant 1.96\right) = 0.95$$

You can calculate the interval such that the probability that it includes μ is 0.95 if you have:

- the sample mean, $\bar{x}$
- the variance σ^2 or standard deviation σ
- the number of data values n.

It is:

$$\left(\bar{x} - 1.96\frac{\sigma}{\sqrt{n}}, \bar{x} + 1.96\frac{\sigma}{\sqrt{n}}\right)$$

Using these formulae, you can determine the interval in which you can be 95% confident that that population mean, μ, will appear.

For example, if you calculated the mean GCSE mathematics result from a sample of 100 students (so $n = 100$) to be 54% with all the results having a standard deviation, σ, of 1.4, you could then calculate $\bar{x} - 1.96\frac{\sigma}{\sqrt{n}}$ to be 53.73% and $\bar{x} + 1.96\frac{\sigma}{\sqrt{n}}$ to be 54.27%. So you can now say that you are 95% confident that the mean of all the GCSE mathematics results will be between 53.73% and 54.27%.

Exercise 6D

Copy and complete the table. (It may help to use your answers to Exercise 6C.)

Confidence level	Confidence interval
80%	
90%	
95%	$\left(\bar{x}-1.96\frac{\sigma}{\sqrt{n}}, \bar{x}+1.96\frac{\sigma}{\sqrt{n}}\right)$
98%	
99%	

Example 1

IQ (Intelligence Quotient) is a measure of intelligence. The IQs of students at a sixth form college may be assumed to follow a normal distribution with mean μ and variance 22.

A random sample of 10 students at the sixth form college was chosen. Their IQs were measured and are recorded as follows:

112	97	119	103	108
122	96	115	107	99

Calculate a 90% confidence interval for μ.

$\bar{x} = (112 + 97 + 119 + 103 + 108 + 122 + 96 + 115 + 107 + 99) \div 10$

$= 1078 \div 10$

$= 107.8$

$\sigma = \sqrt{22}$

$n = 10$

90% confidence interval is given by $\left(\bar{x}-1.645\frac{\sigma}{\sqrt{n}}, \bar{x}+1.645\frac{\sigma}{\sqrt{n}}\right)$

$\bar{x}-1.645\frac{\sigma}{\sqrt{n}} = 107.8-1.645\left(\frac{\sqrt{22}}{\sqrt{10}}\right)$

$= 105.36$

$\bar{x}-1.645\frac{\sigma}{\sqrt{n}} = 107.8+1.645\left(\frac{\sqrt{22}}{\sqrt{10}}\right)$

$= 110.24$

So the 90% confidence interval (the interval such that the probability that it includes μ is 0.90) is (105.36, 110.24).

But what does this mean in reality? It means you can be 90% confident that between the IQs of 105.36 and 110.24 you can find μ, the mean IQ of all the students at the sixth form college.

Confidence intervals are used in quality control in manufacturing processes.

Example 2

The masses of 200 g bars of chocolate from a factory are normally distributed with a standard deviation of 2.5 g.

A sample of ten 200 g bars of chocolate is taken and their total mass is 2004 g.

Calculate a 98% confidence interval for μ and interpret your result.

$\bar{x} = 2004 \div 10 = 200.4$

$\sigma = 2.5$

$n = 10$

98% confidence interval is given by $\left(\bar{x} - 2.33\frac{\sigma}{\sqrt{n}}, \bar{x} + 2.33\frac{\sigma}{\sqrt{n}}\right)$

$$\bar{x} - 2.33\frac{\sigma}{\sqrt{n}} = 200.4 - 2.33\left(\frac{2.5}{\sqrt{10}}\right)$$

$$= 198.56$$

$$\bar{x} - 2.33\frac{\sigma}{\sqrt{n}} = 200.4 + 2.33\left(\frac{2.5}{\sqrt{10}}\right)$$

$$= 202.24$$

So the 98% confidence interval (the interval such that the probability that it includes μ is 0.98) is (198.56, 202.24).

This means you can be 98% confident that between the masses of 198.56 g and 202.24 g you can find μ, the mean mass of the 200 g chocolate bars produced by the factory.

Exercise 6E

1 A study was performed to estimate the mean number of hours per day that secondary school students spend playing on their Xbox, Wii or PlayStation.

The results for a class of students are shown.

2.5	3.0	3.2	4.0	3.5	3.4
0.0	0.5	0.0	3.7	4.2	1.5
2.2	3.3	3.8	3.9	0.0	0.3
2.6	0.7	0.7	3.6	4.3	2.9
2.0	0.2	3.2	3.3	2.8	0.1

Given that the number of hours per day that secondary school students spend playing on their Xbox, Wii or PlayStation is normally distributed with unknown mean μ and variance 1.69, calculate the sample mean and calculate the 98% confidence interval for μ.

2 A random sample of 15 data values, with mean 100, is chosen from a normal distribution with unknown mean μ and variance 36.

Calculate the 99% confidence interval for μ.

3 The masses of 250 g bars of chocolate are normally distributed with a standard deviation of 3.2 g.

A sample of ten 250 g bars of chocolate is taken and their total mass is 2510 g.

Calculate the 80% confidence interval for μ.

4 The amount of time that people spend watching TV each day is normally distributed with unknown mean μ and variance 8100 minutes.

A random sample of 3000 people is asked how many hours of TV they watch on average per week.

The sample mean is 17 hours (1020 minutes) of TV watched per week.

Calculate the 95% confidence interval for μ.

5 A random sample of 1000 1 litre cartons of milk were found to have a mean of 0.985 litres.

The estimated standard deviation was 25 ml.

Assume the amount of milk in a carton is normally distributed and calculate the 98% confidence interval for μ. Interpret your result.

6 Charlotte buys a pack of 24 batteries.

Her toddler has a toy which uses one battery at a time.

Out of curiosity Charlotte makes a note of how long each battery lasts before it needs to be replaced.

In total the 24 batteries last for 188 hours.

After some research on the internet Charlotte discovers that the standard deviation of the battery life is 20 minutes.

Calculate the 99% confidence interval for μ and interpret your result.

7 A random sample of eight test results from a group of maths students totalled 55.

The test results are normally distributed with a standard deviation of 4.

Find an 80% confidence interval for the mean of the population and interpret your result.

8 A random sample of 10 data values, with mean $\bar{x}$, is chosen from a normal distribution with unknown mean μ and variance 25.

Calculate the 90% confidence interval for μ.

Case study

The following is from a report by Ofcom, the UK regulator for the communications industries, produced in February 2015. Nearly one in three UK broadband connections are now superfast, up from around one in four in November 2013.

The growing take-up of superfast cable or fibre services – connections delivering 30 Mbit/s and above – has resulted in average UK broadband speeds increasing by a fifth in the six months to November 2014.

The average UK broadband speed is now 22.8Mbit/s, up from 18.7Mbit/s in May 2014, marking the largest absolute rise (4.1Mbit/s) in broadband speeds Ofcom has recorded.

Faster cable and fibre services have lower availability in rural areas, and rural broadband speeds are typically slower, delivering around one third of urban speeds on average.

Fastest download and upload speeds by provider

Virgin Media's 'up to' 152Mbit/s service achieved the fastest download speeds of the broadband services covered in this research, averaging 132.6Mbit/s over 24 hours.

Upload speeds, which are particularly important when sharing large files or using real-time video communications, were also examined. The research found that Plusnet's 'up to' 76Mbit/s package delivered the fastest upload speeds at 17.0Mbit/s on average.

This is a real-life example of the use of confidence intervals and many people, who have never studied statistics, will select their broadband provider based on these figures.

By choosing examples of lower and upper limits for particular service providers from the Ofcom table below, try writing a simple explanation of confidence intervals and what they mean in this instance.

Summary of average download speed by ISP package: November 2014

	Average download speed during period		
Package	**Maximum**	**24 hours**	**8-10pm weekdays**
BT ADSL2+	9.6Mbit/s to 12.6Mbit/s	8.9Mbit/s to 11.8Mbit/s	8.9Mbit/s to 11.7Mbit/s
KC ADSL2+	9Mbit/s to 11.3Mbit/s	8.6Mbit/s to 10.8Mbit/s	8.5Mbit/s to 10.8Mbit/s
Plusnet ADSL2+*	9.9Mbit/s to 12.8Mbit/s	9.3Mbit/s to 12.1Mbit/s	9.2Mbit/s to 12Mbit/s
Sky ADSL2+	9.3Mbit/s to 11.4Mbit/s	8.6Mbit/s to 10.6Mbit/s	8.5Mbit/s to 10.5Mbit/s
TalkTalk ADSL2+	8.1Mbit/s to 10.2Mbit/s	7.5Mbit/s to 9.5Mbit/s	7.5Mbit/s to 9.5Mbit/s
BT 'up to' 38Mbit/s	33.1Mbit/s to 35.4Mbit/s	32.1Mbit/s to 34.4Mbit/s	31.9Mbit/s to 34.2Mbit/s
EE 'up to' 38Mbit/s	34.3Mbit/s to 36.3Mbit/s	29.5Mbit/s to 31.5Mbit/s	29Mbit/s to 30.9Mbit/s
Plusnet 'up to' 38Mbit/s*	32.2Mbit/s to 36.5Mbit/s	30.1Mbit/s to 34.2Mbit/s	30Mbit/s to 34.1Mbit/s
Sky 'up to' 38Mbit/s	35.3Mbit/s to 37.1Mbit/s	34.3Mbit/s to 36.4Mbit/s	34.2Mbit/s to 36.3Mbit/s
Virgin Media 'up to' 50Mbit/s	53.9Mbit/s to 54.2Mbit/s	52.5Mbit/s to 53.3Mbit/s	51.5Mbit/s to 52.8Mbit/s
BT 'up to' 76Mbit/s	62Mbit/s to 65.1Mbit/s	59.9Mbit/s to 63.1Mbit/s	59.3Mbit/s to 62.6Mbit/s
Plusnet 'up to' 76Mbit/s	61.3Mbit/s to 64Mbit/s	57.7Mbit/s to 60.5Mbit/s	57.1Mbit/s to 59.7Mbit/s
Virgin Media 'up to' 100Mbit/s	104.4Mbit/s to 106.5Mbit/s	94.3Mbit/s to 99.5Mbit/s	86.5Mbit/s to 95Mbit/s
Virgin Media 'up to' 152Mbit/s	159.1Mbit/s to 159.5Mbit/s	129.5Mbit/s to 135.8Mbit/s	115Mbit/s to 124.5Mbit/s

Note: Ranges show the 95% confidence interval of the average speeds measured on the ISP packages.

Project work

Ofcom reported in 2013 that people aged 4+ watch 232 minutes of TV each day, which is 3 hours and 52 minutes.

Ask the head of your year group to provide you with list of the names of all the students in your year group. Select a random sample of 20 students and ask them to record how many minutes of TV they watch over a period of a week.

Make a frequency table of the data.

Calculate the mean and variance.

Find a 95% confidence interval for the mean number of minutes of TV watched per day of the population sampled.

Check your progress

How confident are you feeling in your level of knowledge? What do you need to practise more?

Spec reference	Learning objective			
S4.1	Understand what is meant by the term 'population' in statistical terms			
S4.2	Develop ideas of sampling to including the concept of a simple random sample from a population			
S5.1	Know that the mean of a sample is called a 'point estimate' for the mean of the population; appreciate that accuracy is likely to be improved by increasing the sample size			
S6.1	Calculate confidence intervals for the mean of a normally distributed population of known variance using $\frac{\sigma^2}{n}$			

7 Correlation and regression

Tyler Vigen is a student at Harvard Law School in the USA. As part of a university project he has created a website called Spurious Correlations (tylervigen.com) as

"… a fun way to look at correlations and to think about data … the charts on this site aren't meant to imply causation nor are they meant to create a distrust for research or even correlative data … I hope this project fosters interest in statistics and numerical research."

For example, on Tyler's website you can see the strong, positive correlation between online revenue on Black Friday (traditionally the Friday after Thanksgiving in the USA when the Christmas shopping season begins) and the number of visitors to Disney World's Animal Kingdom. You can see that the two curves are very similar, and a mathematical analysis of the data gives a correlation coefficient of 0.9932. The correlation coefficient is a measure of the strength of the correlation, and a coefficient of 1 would be a perfect correlation, so you can see the correlation between the two variables is very strong.

Another example on his website shows the negative correlation between total US crude oil imports and the number of honey producing bee colonies in the US. In this case, the correlation coefficient is –0.9152. This isn't quite as strong as the first example, but it still suggests a good correlation between the two variables.

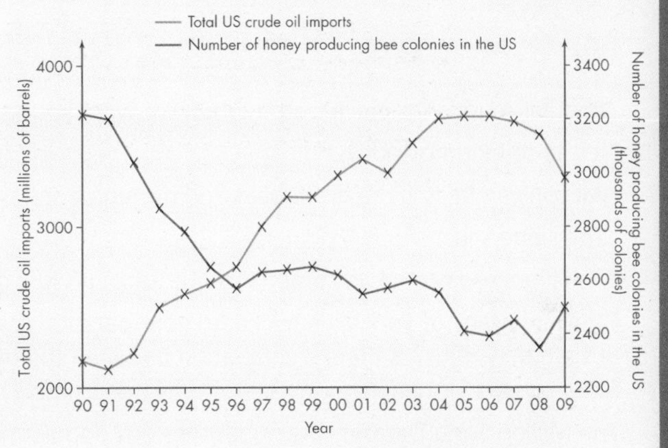

It is safe to say that there is no real correlation between online revenue on Black Friday and the number of visitors to Disney World's Animal Kingdom nor between total US crude oil imports and the number of honey producing bee colonies in the US. Correlation does not imply causation. In other words, a correlation between two variables does not necessarily mean changes in one cause changes in the other. But how did Tyler find these correlations?

S7: Correlation

Learning objectives

You will learn how to:

- Recognise when pairs of data are uncorrelated, correlated, strongly correlated, positively correlated and negatively correlated.
- Appreciate that correlation does not necessarily imply causation.
- Plot data pairs on scatter diagrams and draw, by eye, a line of best fit through the mean point.
- Understand the idea of an outlier; identify and understand outliers and make decisions whether or not to include them when drawing a line of best fit

Mathematics in the real world

"Popular music was much better when we were young."

"It's all so loud nowadays."

"All you can hear is the drum beat and nothing else."

"The singers just shout."

"The lyrics used to mean something but today it's just drivel."

Familiar? Many of you will have heard your parents, grandparents, or both say one or more of these things. But are they right? Is there a correlation between the musicality of popular music and the passage of time? Is it a positive or negative correlation or is there simply no connection at all?

The Million Song Dataset catalogues western popular music from 1955 to 2010. Using algorithms researchers were able to see what changes have taken place over this time. They focused on three variables: timbre, pitch and loudness.

- Timbre is a measure of texture and tone quality. It describes the variety and fullness of a sound.
- Pitch relates to how the chords progress, and the diversity of chord transitions.
- Loudness is a measure of the amplitude or volume of a recording.

The graph shows the relationship between timbre and pitch. Since 1965, chord transitions have become narrower while timbral variety has decreased a great deal.

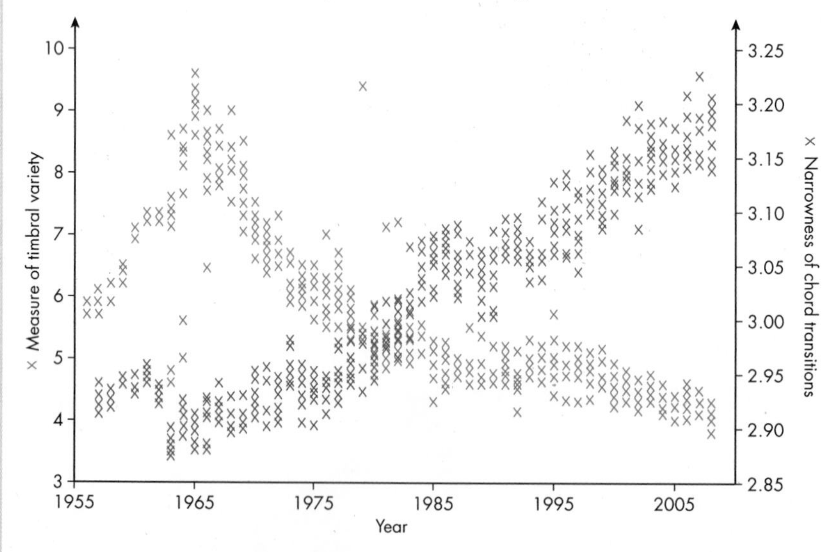

Does this mean that your parents and grandparents are actually right?

Discussion

Choose one of the following and discuss it with your peers.

1. Imagine that you have been asked by Sebastian Coe, chairman of the London 2012 Olympics Organising Committee, to draw a scatter diagram to illustrate the relationship between the number of countries participating and the number of medals won since 1896 (when the modern Olympic Games started).

   Talk about what variables will be on the axes and the type of correlation you think you might see.

2. Imagine you have been asked by the UK Met Office to draw a scatter diagram to illustrate the relationship between sea levels and world temperature since the Battle of Hastings in 1066.

   Talk about what variables will be on the axes and the type of correlation you think you might see.

3. Imagine you are a statistician at the Office for National Statistics (ONS) and you have been asked to draw a scatter diagram to illustrate the relationship between daily screen time (any device) and A-level results.

   Talk about what variables will be on the axes and the type of correlation you think you might see.

Correlation in scatter diagrams

Correlation describes a relationship or connection between two or more variables. When you a draw scatter diagram, you are plotting points to examine whether or not there is a relationship between two variables.

If the values of both variables are increasing, then you say that they are positively correlated.

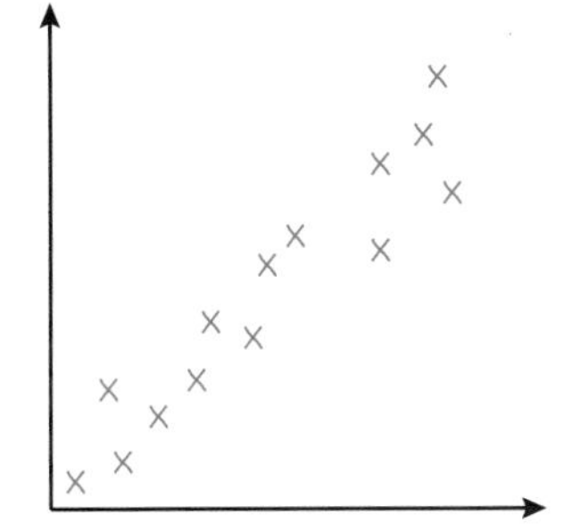

Correlation: A relationship or connection between two or more variables.

Positive correlation: As one variable increases, so does the second variable.

If you draw a line at the mean value of x, $\bar{x}$, and a line at the mean value of y, $\bar{y}$, and label the quadrants created as 1st through to 4th, starting at the top right-hand corner and going anticlockwise, you can analyse the correlation. You can see that for a positive correlation most points lie in the 1st and 3rd quadrants.

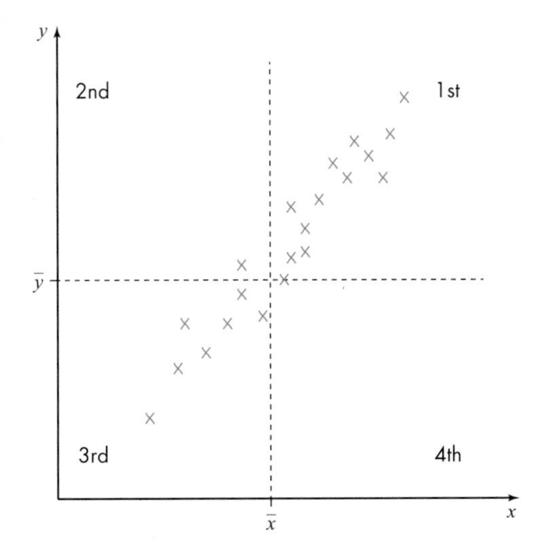

S7: Correlation

If the points are closely packed together, then you say there is a strong positive correlation. If they are spread apart, then you say there is a weak positive correlation.

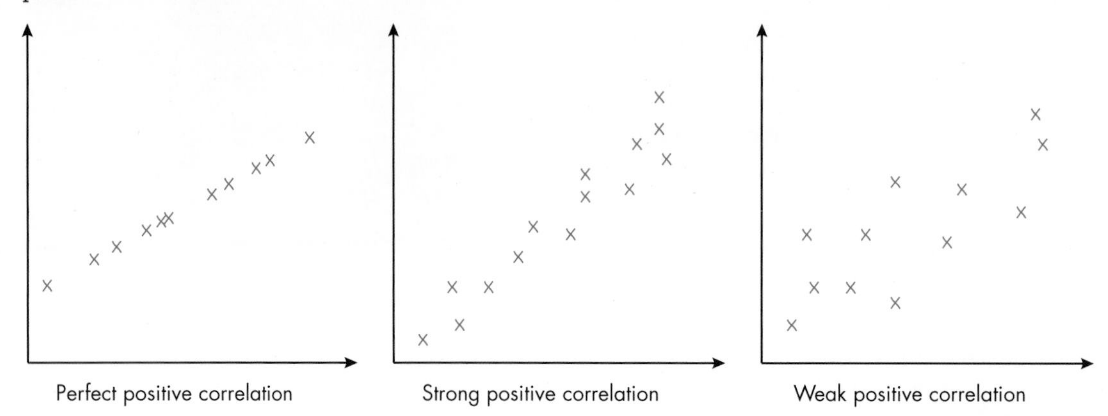

If the value of one variable is increasing as the other is decreasing, then you say that they are negatively correlated. For a negative correlation most points lie in the 2nd and 4th quadrants.

Negative correlation: As one variable increases, the second variable decreases.

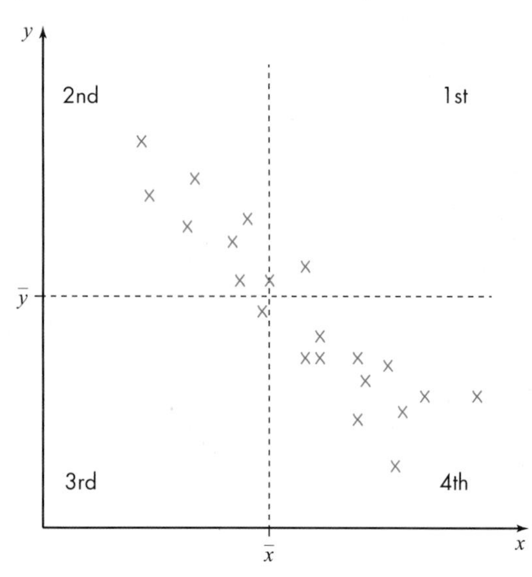

If the points are closely packed together, then you say there is a strong negative correlation. If they are spread apart, then you say there is a weak negative correlation.

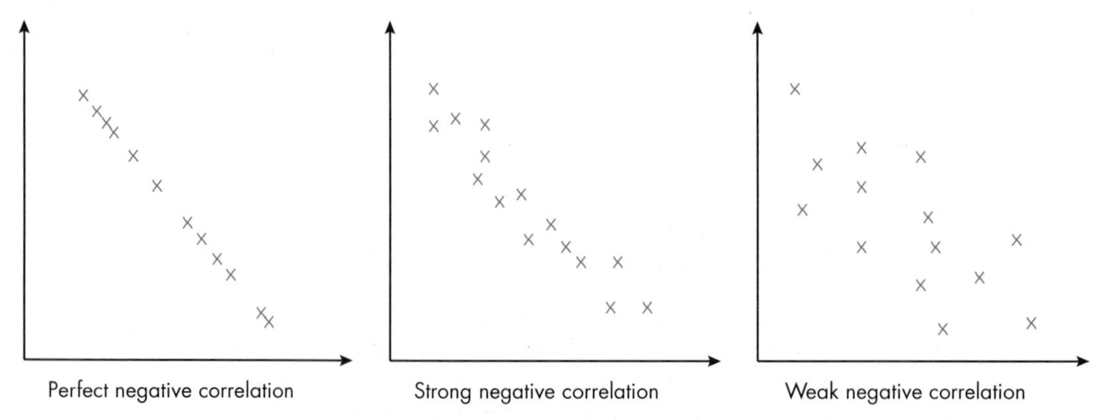

If there is no pattern, then you say that there is no correlation. You can see that the data points are approximately equally distributed in the four quadrants.

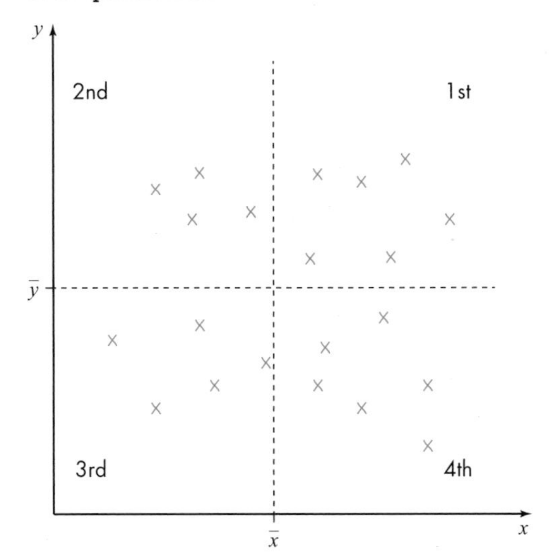

When you analyse the correlation, you can say what type of correlation it is and also, broadly, the strength of the correlation. Deciding on the strength of a correlation based on looking at the graph is subjective (your personal opinion) and consequently may not be very accurate. Later in the chapter you will look at how you can precisely measure the degree of correlation. It is important to note that you are only looking at linear (straight-line) correlations. There are other types of correlation, such as quadratic correlations, which produce different patterns of scatter, but these are beyond the scope of this book.

Correlation does not imply causation

Tyler Vigen has plotted the number of maths doctorates awarded and the amount of uranium stored at nuclear power plants in the US and found that there is a positive correlation between them.

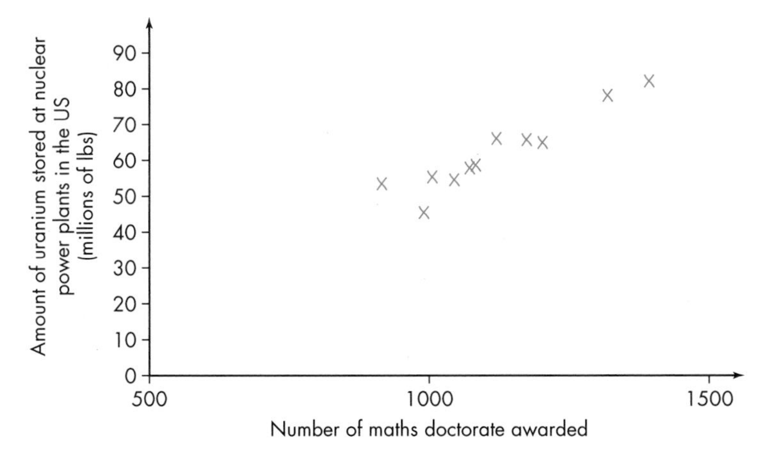

Does this mean that more uranium is stored at nuclear power plants because the number of maths doctorates awarded increases? Think about it. Just because there is a correlation between two variables, it does not mean that they necessarily affect each other. The existence of a correlation does not necessarily mean that one variable is causing a change in the other variable. Correlation does not imply causation. The correlation between the number of maths doctorates awarded and the amount of uranium stored at nuclear power plants in the US is a nonsense correlation.

Outliers and drawing a line of best fit

How do you draw a line of best fit on a scatter diagram when there is an outlier (extreme value)?

To start you need to ask the question why draw a line of best fit? You use a line of best fit to determine values, either x or y, that may not have been included in the scatter diagram. In the data you will encounter during this course, the line of best fit is always linear (straight) and later in this chapter, in the section on regression, you will look at a technique to determine the equation of the line of best fit.

As a line of best fit is straight, it isn't possible for it to touch all the points on the scatter diagram and it doesn't always touch as many points as possible either. The position of the line of best fit is determined by:

- the type of correlation between the two variables
- and, if present, the positions of any outliers.

The line of best fit is a representation of all the data and so it must include any outliers even though the rest of the data may be congregated in another area of the scatter diagram. There is, however, one point on the scatter diagram that the line of best fit *must* go through and that is the point $(\bar{x}, \bar{y})$ where $\bar{x}$ is the mean of x and $\bar{y}$ is the mean of y.

Look at this example. You have three pairs of coordinates, (1, 2), (2, 3) and (3, 4). First you calculate the mean of the x values and the mean of the y values in order to find the mean point, $(\bar{x}, \bar{y})$. The mean point in this case is (2, 3). In fact, the line of best fit will go through all of the points in this case because the points lie in a straight line, as illustrated in the graph on the right.

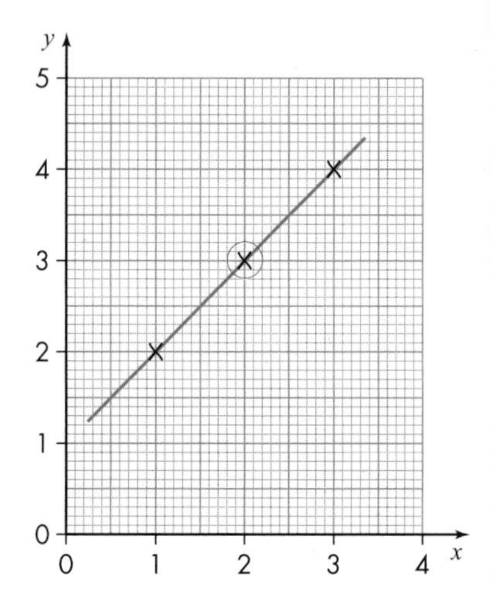

If you add in an outlier at (8, 2), the mean point becomes (3.5, 2.75) (calculated as above) and the line of best fit will now change position as illustrated in the following graph.

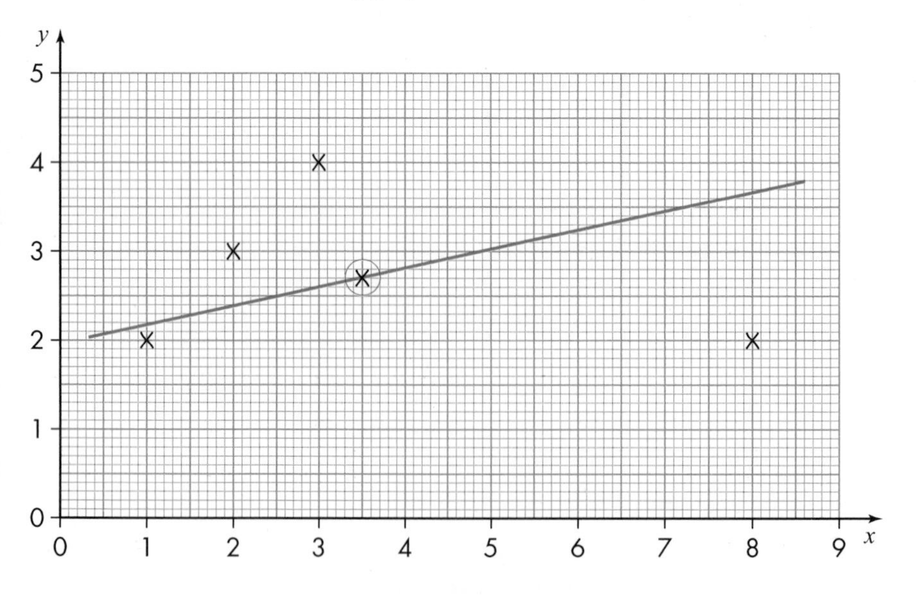

One final thing you need to remember: you need to be careful with outliers because they may be a result of incorrect experimental data. If they are, they should be discarded.

Exercise 7A

1 Mr and Mrs Brown have started keeping chickens, having never done so before.

They are very interested in egg production over time.

Study the scatter diagram and write down your observations.

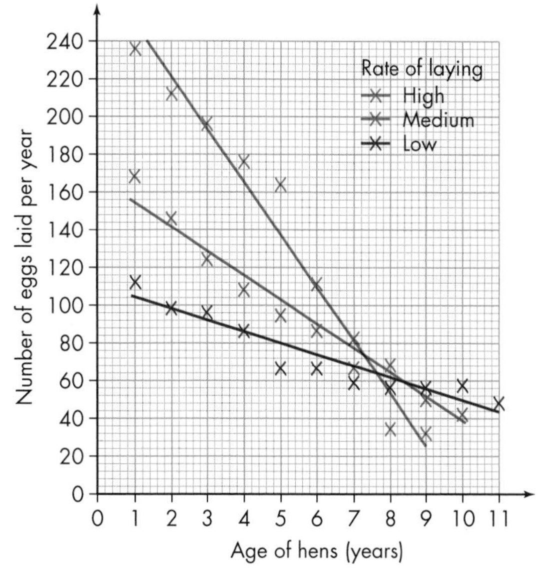

Data source: Optimal Reproductive Tactics, Eric R. Pianka

2 Draw a scatter diagram for the following set of data.

	2003	2004	2005	2006	2007	2008
Number of UK citizens who immigrated to the US	9527	14 915	19 800	17 207	14 545	14 348
US Uranium exports (millions of lbs)	13.2	13.2	20.5	18.7	14.8	17.2

Assess whether there appears to be a correlation between the two variables and, if there is, state the nature of the correlation.

3 Write down a real-life example and sketch a scatter diagram for each of the following scenarios.

a A strong positive correlation

b A weak negative correlation

c No correlation

d A nonsense correlation

4 Draw a scatter diagram and draw on a line of best fit for the following set of data.

Candidate	A	B	C	D	E	F
Maths Paper 1 result	96	84	75	93	85	90
Maths Paper 2 result	79	93	73	85	82	94

5 **a** Plot the following additional data on the scatter diagram you drew in question **4** and draw on a new line of best fit.

Candidate	G
Maths Paper 1 result	85
Maths Paper 2 result	33

b Can you offer some examples for the cause of the outlier data?

S8: The product moment correlation coefficient (pmcc)

Learning objectives

You will learn how to:

- Understand that the strength of correlation is given by the pmcc.
- Understand that pmcc always has a value in the range from –1 to +1.
- Appreciate the significance of a positive, zero or negative value of pmcc in terms of correlation of data.
- Use a calculator to calculate the pmcc from raw data.

Introduction

A more accurate method of assessing the strength of correlation is to find the **correlation coefficient**. There are a number of different correlation coefficients but you are going to concentrate on Pearson's product moment correlation coefficient (abbreviated to r), developed by the English mathematician Karl Pearson in the 1890s.

Correlation coefficient: A number used to assess the strength of a correlation.

Mathematics in the real world

Here is another of Tyler Vigen's spurious correlations. On Tyler's website you can see the strong, negative correlation between people who literally worked themselves to death in the US and physical copies of video games sold in the UK. The correlation coefficient is –0.9342.

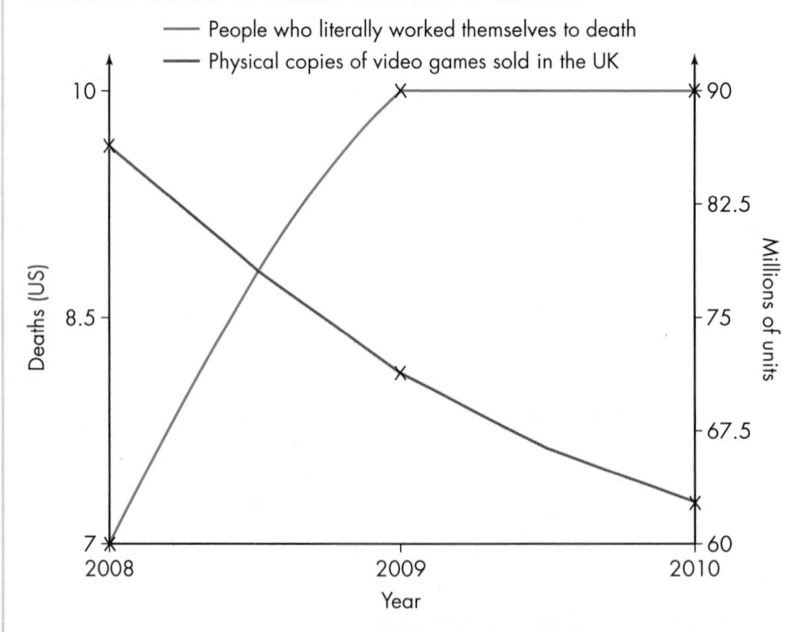

You should, quite rightly, be thinking that correlation does not imply causation. A correlation does not necessarily mean that the more people who work themselves to death in the US the fewer video games will be sold in the UK. The fact that the correlation coefficient has been calculated does not make the case for the correlation any stronger. But you can see here that $r = –0.9342$. So how was the figure calculated?

Discussion

Choose one of the following and discuss it with your peers.

1. Imagine that you have been asked by Sebastian Coe, chairman of the London 2012 Olympics Organising Committee, to draw a scatter diagram to illustrate the relationship between the number of countries participating and the number of medals won since 1896 (when the modern Olympic Games started).

   Talk about the strength of the correlation that you think you will see.

2. Imagine you have been asked by the UK Met Office to draw a scatter diagram to illustrate the relationship between sea level and world temperature since the Battle of Hastings in 1066.

   Talk about the strength of the correlation that you think you will see.

3. Imagine you are a statistician at the Office for National Statistics (ONS) and you have been asked to draw a scatter diagram to illustrate the relationship between daily screen time (on any device) and A-level results.

   Talk about the strength of the correlation that you think you will see.

Some pmcc theory

Calculations of correlation coefficients use the differences between data points on a scatter diagram and the mean point $(\bar{x}, \bar{y})$. If you select a plotted point with coordinates (x_i, y_i) and then measure the distances to the two means then

- the horizontal distance will be given by $x_i - \bar{x}$
- the vertical distance will be given by $y_i - \bar{y}$.

You can find the pmcc using your calculator. Calculators use the following formula:

$$r = \frac{S_{xy}}{\sqrt{S_{xx}S_{yy}}}$$

$$= \frac{\Sigma xy - \frac{\Sigma x \Sigma y}{n}}{\sqrt{\left(\Sigma x^2 - \frac{(\Sigma x)^2}{n}\right)\left(\Sigma y^2 - \frac{(\Sigma y)^2}{n}\right)}}$$

where S_{xy}, S_{xx} and S_{yy} are measures of sums and differences of mean and individual x and y values. (You don't need to remember the formulae).

Interpreting the pmcc

The pmcc (r) tells you the strength of a correlation. It ranges from –1 to 1 and can have any value in this range.

- If $r = -1$ then this is a perfect negative linear correlation. For negative values of r, the closer it is to –1, the stronger the correlation; the closer it is to 0, the weaker the correlation.
- If $r = 0$ then there is no linear correlation. However, this doesn't mean another type of correlation might not exist.
- If $r = 1$ then this is a perfect positive linear correlation. For positive values of r, the closer it is to 1, the stronger the correlation; the closer it is to 0, the weaker the correlation.

The diagram shows a range of scatter diagrams and their correlation coefficients.

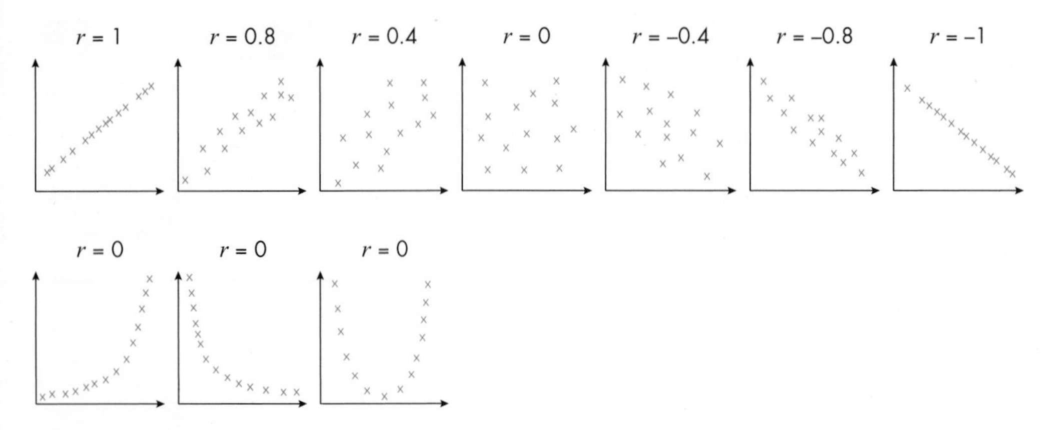

At the ends of the top row of scatter diagrams, the tightly packed plots forming a straight line have r values of 1 and –1, meaning the plots show perfect positive correlation and perfect negative correlation respectively. As the scatter diagrams become less tightly packed then the value of r moves towards 0. The scatter diagram where $r = 0.8$ is a stronger positive correlation than the scatter diagram where $r = 0.4$. The scatter diagrams with $r = 0.8$ and $r = -0.8$ have the same strength of correlation, but one correlation is positive and the other is negative.

In the bottom row, there are no linear correlations, so $r = 0$ in all cases. However, you can see patterns so these are non-linear correlations.

Calculating the pmcc

You now need to become very good friends with your calculator. You need to find the instruction booklet or the relevant video online and learn how to calculate the pmcc using your calculator. You should make sure you know how to find the pmcc in this way. This usually involves simply entering all the corresponding x and y data values and then selecting the necessary option to calculate the pmcc.

Use your calculator find the pmcc for the following set of data.

Candidate	A	B	C	D	E	F
Maths Paper 1 result	96	84	75	93	85	90
Maths Paper 2 result	79	93	73	85	82	94

You should get $r = 0.325\ 576$. (If you don't get this answer, try again and make sure you have carried out the calculation correctly.) So what does this mean? It means that there is a weak, positive correlation between the maths paper 1 results and the maths paper 2 results.

Outside of the examination room you can also use MS Excel to do the calculation. Enter the data into a spreadsheet with one column for the x values and another column for the corresponding y values. Then use the function PEARSON and follow the onscreen instructions.

Exercise 7B

1 Two variables have a positive linear correlation.

What is the range of values for the product moment correlation coefficient for this kind of relationship?

2 Write an accurate mathematical description for each of the following values of the product moment correlation coefficient.

a 0.98

b –0.36

c 1.00

d 0.02

3 Calculate the correlation coefficient and comment on the type and degree of the correlation, if any, for the following set of data.

Candidate	A	B	C	D	E	F	G
Maths Paper 1 result	96	84	75	93	85	90	72
Maths Paper 2 result	79	93	73	85	82	94	74

4 Which of the following values shows the strongest correlation and which shows the weakest?

0.70 –0.05 0.5 –0.85

5 Calculate the correlation coefficient and comment on the type and degree of the correlation, if any, for the following set of data.

Candidate	A	B	C	D	E	F	G
Maths Paper 1 result	96	84	75	93	85	90	72
Art Paper 1 result	25	32	50	78	12	45	48

S9: Regression lines

Learning objectives

You will learn how to:

- Understand the concept of a regression line.
- Use a calculator to calculate the equation of the regression line from raw data.
- Plot a regression line from its equation.
- Use interpolation with regression lines to make predictions.
- Understand the potential problems of extrapolation.

Introduction

Drawing a line of best fit by eye is not a very accurate method of finding the best fit. A more accurate way to find the line of best fit is to find the linear equation of the line.

The general form of an equation of a straight line is:

$$y = mx + c$$

where

- m is the gradient (the steepness of the line)
- c is the y-intercept (where the line crosses the y-axis).

For this analysis, you use the same equation but with different letters:

$$y = a + bx$$

where

- b is the gradient
- a is the y-intercept.

If b is positive then there is a positive correlation.

If b is negative then there is a negative correlation.

You call a line of best fit a 'regression line' and it has the equation $y = a + bx$.

Regression line: A line of best fit for a given set of values, using the equation of a straight line $y = a + bx$.

Mathematics in the real world

In recent years the economies of the Eurozone (countries that use the Euro as their currency) have been in turmoil. The scatter diagram below shows the relationship between percentage change in nominal gross domestic product (x-axis) and percentage change in the ratio of debt to gross domestic product (y-axis).

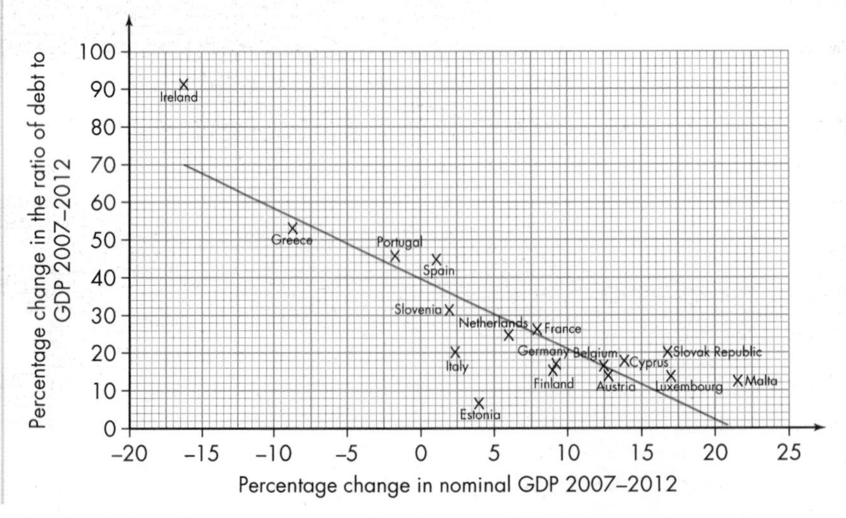

Data source: Austerity Is the Only Deal-Breaker, Yanis Varoufakis

What does this mean?

The x-axis (percentage change in nominal gross domestic product) is a measure of how much the country is earning.

- A positive percentage means that the country is doing well selling its products and services and its economy is growing.
- A negative percentage change means that the country is not doing so well selling its products and services and its economy is shrinking.

The y-axis (percentage change in the ratio of debt to gross domestic product) is a measure of the amount of debt the country has as a proportion of the gross domestic product.

- A high percentage means that the country has high debt levels and indicates the economy is not performing well.
- A low percentage means that the country has low debt levels and indicates the economy is performing well.

The regression line is showing a linear relationship between percentage change in nominal gross domestic product and percentage change in the ratio of debt to gross domestic product.

Discussion

Choose one of the following and discuss it with your peers.

Imagine that you are an economist and have been asked to make some predictions about the economic performance of certain European countries.

1. Using the scatter diagram above, talk about the countries experiencing most difficulty and why.
2. Using the scatter diagram above, talk about the countries experiencing least difficulty and why.
3. Using the scatter diagram above and, even though the UK is not in the Eurozone, talk about the UK's percentage change in the ratio of debt to gross domestic product if the percentage change in nominal gross domestic product is +2.6% .

Finding the equation of the regression line

The equation of the regression line of y on x is:

$$y = a + bx$$

where $b = \frac{S_{xy}}{S_{xx}}$ and $a = \bar{y} - b\bar{x}$.

(You don't need to remember these formulae.)

To find the equation of the regression line, you again need to become very good friends with your calculator. You need to find the instruction booklet or the relevant video online and learn how to calculate the values of a and b using your calculator. This usually involves simply entering all the corresponding x and y data values and then selecting the necessary options to calculate a and b.

Use your calculator find the values of a and b and hence state the equation of the regression line for the following set of data.

Candidate	A	B	C	D	E	F
Maths Paper 1 result	96	84	75	93	85	90
Maths Paper 2 result	79	93	73	85	82	94

You should get $a = 53.616\ 97$ and $b = 0.352\ 385$, so the equation of the regression line is:

$$y = 53.616\ 97 + 0.352\ 385x$$

where

- x is the maths paper 1 result
- y is the maths paper 2 result.

Outside the examination room you can also use MS Excel to do the calculation. Enter the data into a spreadsheet with one column for the x values and another column for the corresponding y values. Then use the function INTERCEPT to calculate a and SLOPE to calculate b and follow the onscreen instructions.

You can now use the equation of the regression line and the mean point to draw the regression line on the scatter diagram.

- From the equation, you know the y-intercept is at (0, 53.616 97).
- You can calculate the coordinates of the mean point to be (87.2, 84.3).

Drawing a line between these two points give you the regression line (line of best fit).

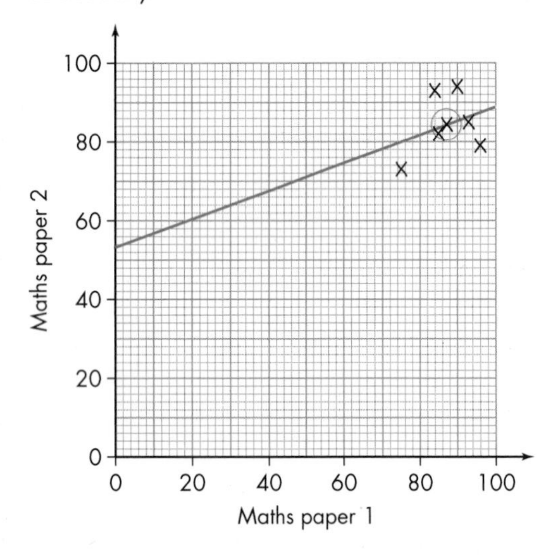

Interpolation and extrapolation

Once you have drawn a line of best fit, either by eye or more accurately by plotting the equation of the regression line, you can start using the line of best fit to make predictions.

- A prediction within the range of data is **interpolation**.
- A prediction outside the range of data is **extrapolation**.

Interpolation: A prediction of a value within the range of the data.

Extrapolation: A prediction of a value outside the range of the data.

Extrapolation comes with a health warning. For example, the physical law described by Hooke's law relates the force F needed to extend a spring to the distance x of the extension.

$$\text{Force} = -kx$$

where $-k$ is a constant relating to the stiffness of the spring.

The relationship is linear but Hooke's law is only applicable before the spring reaches its elastic limit. After that, Hooke's law no longer applies and the relationship is not linear.

If you had data within the elastic limit and made predictions within your data range then this would be interpolation. However if you didn't know about the elastic limit and started making predictions outside your data range then you could be completely wrong.

Using the data for the two maths papers previously plotted, you can predict the result in maths paper 2 when you know the result from maths paper 1 is 45%.

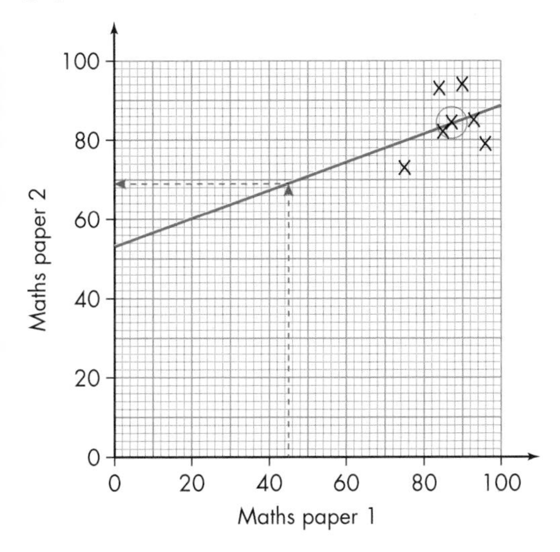

Using the regression line, a maths paper 1 result of 45% gives you a predicted result of 69.5% on paper 2.

You could also have used the equation of the regression line.

$$y = 53.616\,97 + (0.352\,385 \times 45)$$

$$= 69.474\,295$$

This is an example of extrapolation. The value of 45 is outside the data range, and consequently the prediction should not be relied upon.

Exercise 7C

1 Find the equation of the regression line for the set of data shown in the table.

Candidate	A	B	C	D	E	F	G
Maths Paper 1 result	96	84	75	93	85	90	72
Maths Paper 2 result	79	93	73	85	82	94	74

What do the values of a and b tell you in relation to the base data?

2 Using the equation of the regression line in question **1**, predict a maths paper 2 result if the maths paper 1 result is 78.

Comment on the prediction.

3 Using the equation of the regression line in question **1**, predict a maths paper 2 result if the maths paper 1 result is 33.

Comment on the prediction.

4 Find the equation of the regression line for the set of data shown in the table.

Candidate	A	B	C	D	E	F	G
Maths Paper 1 result	96	84	75	93	85	90	72
Art Paper 1 result	25	32	50	78	12	45	48

What do the values of a and b tell you in relation to the base data?

5 Plot the scatter diagram for the data in question **4** and draw on the regression line.

Using the regression line, predict an art paper 1 result if the maths paper 1 result is 76.

Comment on the prediction.

6 Ed Sheeran has a house in Suffolk not far from the market town of Framlingham.

You are going to investigate whether his popularity has affected the number of visitors to Framlingham Castle.

The following table details the number of visitors to Framlingham Castle and Ed Sheeran's record sales for the past five years.

Year	Number of visitors to Framlingham castle	Ed Sheeran record sales
2011	72 800*	0
2012	75 500*	0
2013	83 435	0
2014	92 403	572 000*
2015	100 000*	1 428 000

*estimated

a Plot the scatter diagram for this data.

b Calculate the correlation coefficient.

c Find the equation of the regression line and give an interpretation of the y-intercept and gradient.

d Do you think Ed Sheeran's popularity has resulted in more people visiting Framlingham castle?

Comment on using your graph to predict how many visitors Framlingham castle can expect in 2016 if Ed Sheeran releases another album in 2016 and it sells 3 million copies.

Case study

Could you be the manager of a Premiership football club? One of the key attributes of being a good manager in the 21st century is an ability to understand and interpret statistics! Have a look at the following charts and results. What are they telling you?

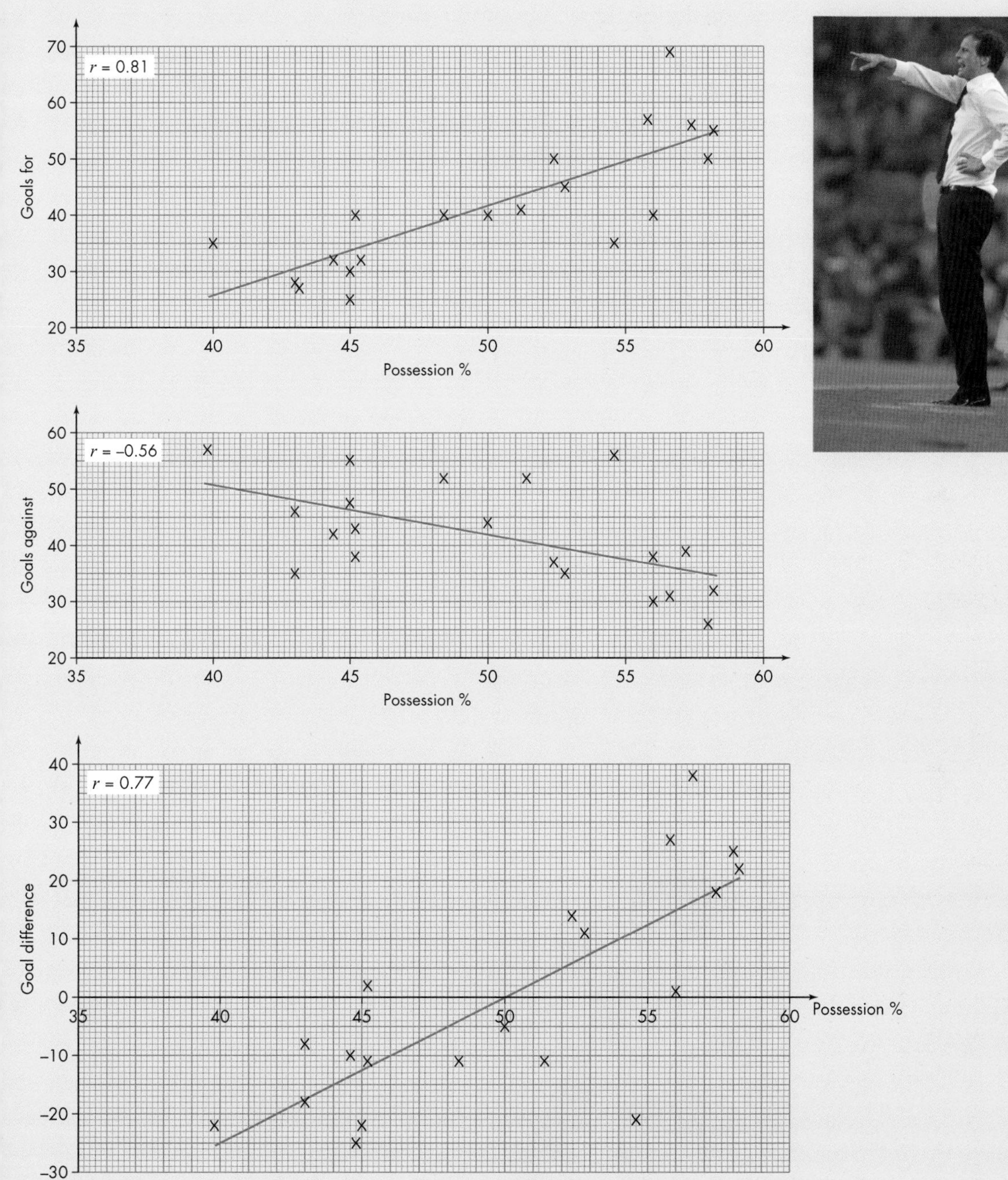

Project work

You are mathematically perfect! Many parts of human bodies are in the same proportion, for example, arm length and forearm length. If you take measurements from a group of people and divide arm length by forearm length and then take an average of the results you will see that the result is close to 1.618. This number is called the golden ratio. Enter "golden ratio" into your search engine for lots more information.

Ask the head of your year group to provide you with list of the names of all the students in your year group. For each student measure and record:

- the length of one arm from the top of the shoulder to the end of the middle finger
- the length of the same arm from the elbow to the end of the middle finger.

Plot the results on a scatter diagram, describe the correlation, calculate the pmcc and find the equation of the regression line.

Draw the regression line on the scatter diagram and make some predictions.

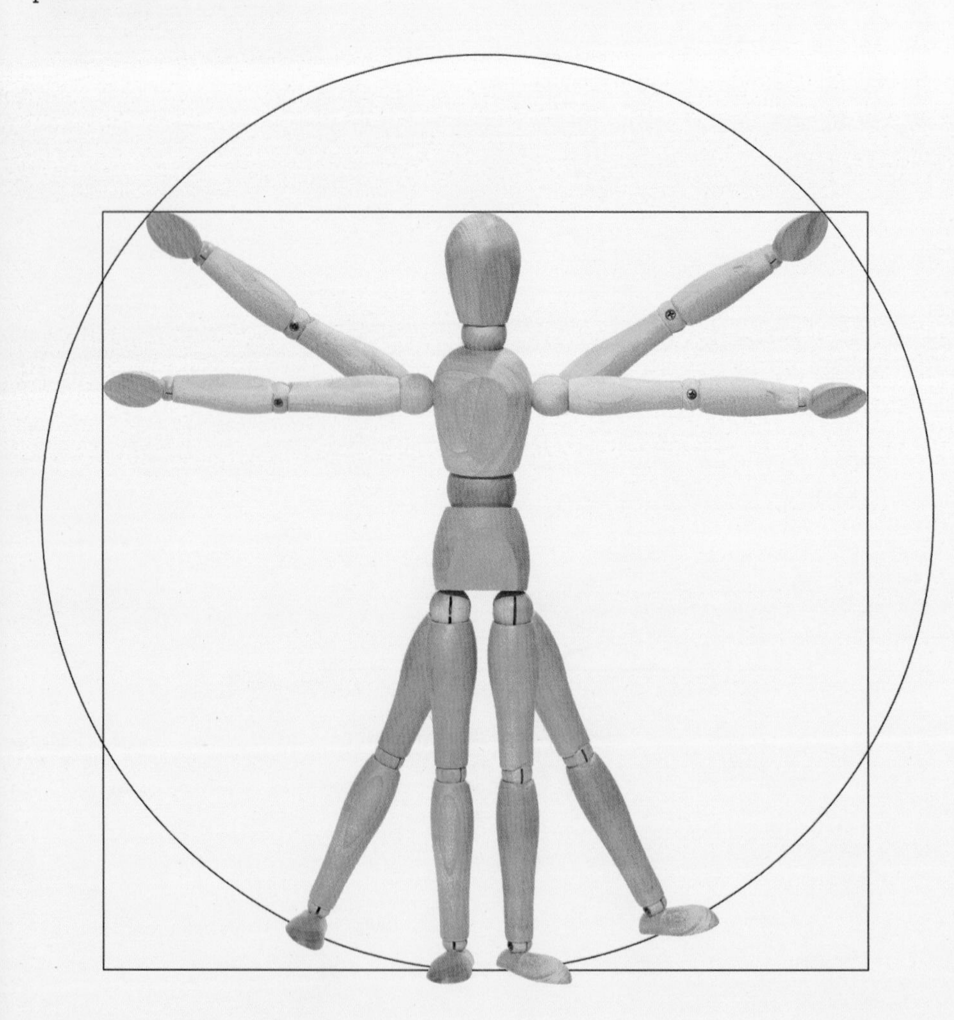

Check your progress

How confident are you feeling in your level of knowledge? What do you need to practise more?

Spec reference	Learning objective	▶	▶▶	▶▶▶
S7.1	Recognise when pairs of data are uncorrelated, correlated, strongly correlated, positively correlated and negatively correlated			
S7.2	Appreciate that correlation does not necessarily imply causation			
S9.1	Plot data pairs on scatter diagrams and draw, by eye, a line of best fit through the mean point			
S7.3	Understand the idea of an outlier; identify and understand outliers and make decisions whether or not to include them when drawing a line of best fit			
S8.1	Understand that the strength of correlation is given by the pmcc			
S8.2	Understand that pmcc always has a value in the range from −1 to +1			
S8.3	Appreciate the significance of a positive, zero or negative value of pmcc in terms of correlation of data			
S10.1	Use a calculator to calculate the pmcc from raw data			
S9.2	Understand the concept of a regression line			
S10.1	Use a calculator to calculate the equation of the regression line from raw data			
S9.3	Plot a regression line from its equation			
S9.4	Use interpolation with regression lines to make predictions			
S9.5	Understand the potential problems of extrapolation			

8 Critical path analysis

Like most people, mathematicians like all sorts of things but they are particularly partial to numbers. So here are a few numbers for you. But what are they describing?

- 900 acres of the Vale of Avalon used for the festival site.
- 200 000 revellers at the 2014 festival.
- 15 000 hand-painted signs that have been used for festivals so far.
- 5000 long-drop toilets on site.
- 1000 compost toilets.
- 2 years that compost from these toilets can fertilise Worthy Farm.
- 800 market stalls on the festival site.
- 400 water taps serve the festival.
- 150 years that Worthy Farm has been in the Eavis family.

You should have worked out that the figures are for the annual Glastonbury festival, one of the biggest music festivals in the world. So what has Glastonbury got to do with mathematics? The Glastonbury festival doesn't just happen. It requires careful planning (1000 compost toilets don't just magically appear) and if this didn't happen then it couldn't go ahead.

In preparation for Glastonbury each year, many jobs have to be done. Some of these jobs will start even before the end of the current festival, while some jobs can be left until a few days before the next festival starts. There will also be jobs that can't start until other jobs have finished or nearly finished. The jobs that are dependent on other jobs and determine the length of the preparation time are known as the critical path.

Working out what all the activities are for Glastonbury, and when they need to be done, is known as project management. Project management also involves allocating activities (or tasks) to the people who need to do them, and involves keeping control of the costs. For large projects, project management is an essential part of the project, and good project management can make the difference between projects succeeding or failing.

Critical path analysis is part of a project manager's tool kit to help them deliver their project on time, within budget and to the expected level of quality. (It's no use having 1000 environmentally-friendly compost toilets if there's no toilet roll!)

R1: Compound projects

Learning objectives

You will learn how to:

- Represent compound projects by activity networks using activity-on-node representation.

What is project management?

More specifically, what is a project? It's a temporary endeavour undertaken to create a unique product, service or result.

A project is temporary in that it has a defined beginning and end in time, and therefore defined scope and resources.

And a project is unique in that it is not a routine operation, but a specific set of operations designed to accomplish a singular goal. So a project team often includes people who don't usually work together – sometimes from different organizations and across multiple geographies.

The development of software for an improved business process, the construction of a building or bridge, the relief effort after a natural disaster, the expansion of sales into a new geographic market – all are projects.

And all must be expertly managed to deliver the on-time, on-budget results, learning and integration that organizations need.

Project management, then, is the application of knowledge, skills, tools, and techniques to project activities to meet the project requirements.

The Project Management Institute (pmi.org)

One of the key project management processes is planning. The planning of a project involves a number of different strands with key elements being:

- the definition of the activities that need to be performed
- the dependencies between them
- how long they are going to take.

A number of different models can be used to represent a project. You are going to look at activity networks, specifically activity-on-node representations.

Mathematics in the real world

In 2011 the government scrapped an NHS computer scheme project which was said to have a forecast cost of £11.4 billion, with an actual spend of £6.46 billion at the end of March 2011. At the time it was the biggest civilian IT project in the world.

The aim of the project was to develop a national IT system for the NHS. There are many reasons why complex projects fail but a report written by the Major Projects Authority concluded that:

'There can be no confidence that the programme has delivered or can be delivered as originally conceived.'

'The project has not delivered in line with the original intent as targets on dates, functionality, usage and levels of benefit have been delayed and reduced.'

Discussion

Choose one of the following and discuss it with your peers.

1. Imagine that you have been asked by the Major Projects Authority to identify a big public spending project completed in the last ten years, for example the London Olympics.

   Talk about when the project started, how long it took, some of the key activities and dependencies between key activities.

2. Imagine you have been asked by the organiser of Glastonbury to help with the preparation for the next Glastonbury festival.

   Talk about as many activities as possible that you will need to include in the project plan and the dependencies between them.

3. Imagine you are hoping to buy and move into a new house.

   Talk about as many activities as possible that you will need to include in the project plan and the dependencies between them.

Activity networks

Most activities can be written in terms of activity networks. An obvious example is school or college homework. The table lists all the activities involved from start to finish, and shows their dependencies and durations (how long it takes to do). This is sometimes called scoping the project.

Activity		Immediate predecessor	Duration (in days)
A	Set homework	-	1.0
B	Write down homework	A	1.0
C	Research homework	B	3.0
D	Review class notes	B	1.0
E	Cross-reference textbook	B	1.0
F	Write up homework	C, D, E	3.0
G	Check homework	F	1.0
H	Hand in homework	G	1.0
I	Mark homework	H	5.0

Not everyone tackles homework in the same way. You might include additional activities in this list whereas others would delete some. You will notice that not all of the above activities will be completed by the same person. For example, the activity 'set homework' would be done by the teacher, whereas the activity 'write down homework' would be done by the student.

Resourcing of projects (allocating the activities to people) is not a concern in this chapter but it is important that you include all the activities within a project regardless of who performs them. The diagram shows the activity network for the project homework.

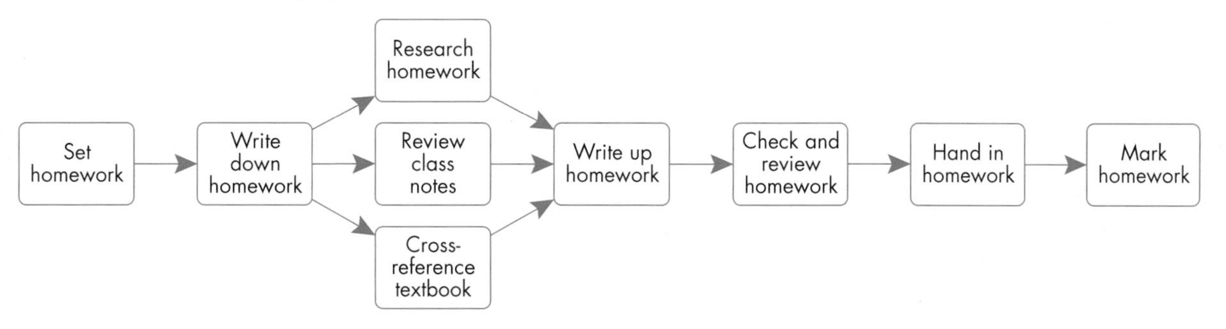

The activity network should be read from left to right. The boxes in this network represent activities and the arrows indicate dependencies between activities. For example, the activity 'write down homework' cannot start until the predecessor activity 'set homework' is complete. Likewise the successor activity 'review class notes' cannot start until the activity 'write down homework' is complete.

The activity network can also be drawn to show the durations of activities, as shown in the diagram.

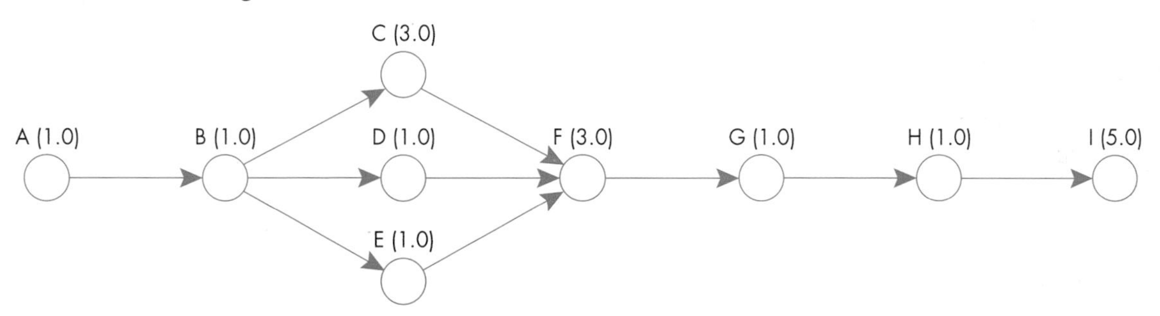

The nodes (circles) in this network represent activities and the arcs (arrows) indicate dependencies between activities. Each node is labelled with the identifier of the activity and the duration of the activity.

For example A(1.0) is the node for activity A 'Set homework', which has a duration of 1 day. This type of diagram is known as an **activity-on-node network**.

Activity-on-node network:
A diagrammatic representation of a set of activities and their durations, showing the dependencies between activities.

Exercise 8A

1 A new mathematics textbook is to be written. The work involved has been divided into a number of activities, as shown in the table below.

The minimum duration needed to complete each activity is also shown.

Activity	Immediate predecessor	Duration (in weeks)
A	-	2.0
B	A	3.0
C	A	1.0
D	B	4.0
E	C	3.0
F	D, E	2.0
G	F	2.0
H	G	1.0
I	H	2.0

Construct an activity network for the project.

2 You are going to go Interrailing in Europe over the summer with a group of friends. (If you are not sure what Interrailing is, ask your teacher or google 'Interrailing'.)

a Write a list of all the activities involved in preparing for the trip.

b Make a table of your activities and include dependencies and durations. (Use the table format shown in question **1**.)

c Construct an activity network for your Interrailing in Europe preparation.

d Swap your Interrailing in Europe preparation work with a fellow student and review the accuracy of their activity network. Discuss any differences between your networks.

R2: Critical activities

Learning objectives

You will learn how to:

- Use early time and late time algorithms to identify critical activities and find the critical path(s).

Introduction

A crucial part of project planning is the identification of the longest path (or paths) from the start of the project to its end. These are the **critical paths** of a project, and you use critical path algorithms to analyse the project. (An algorithm is simply a process.)

Critical path: The longest path (or paths) through a project from the start to the end of the project.

Mathematics in the real world

A mass burial site suspected of containing 30 victims of The Great Plague of 1665 has been unearthed at Crossrail's Liverpool Street site in the City of London.

The discovery was found during excavation of the Bedlam burial ground at Crossrail's Liverpool Street site, which will allow construction of the eastern entrance of the new station.

Jay Carver, Crossrail Lead Archaeologist said: "The construction of Crossrail gives us a rare opportunity to study previously inaccessible areas of London and learn about the lives and deaths of 16th and 17th Century Londoners.

"This mass burial, so different to the other individual burials found in the Bedlam cemetery, is very likely a reaction to a catastrophic event. Only closer analysis will tell if this is a plague pit from The Great Plague in 1665 but we hope this gruesome but exciting find will tell us more about the one of London's most notorious killers."

From www.crossrail.co.uk/

What have the deaths of thousands of people hundreds of years ago got to do with critical activities?

Crossrail is Europe's largest infrastructure project. Stretching from Reading and Heathrow to the west of London, across to Shenfield and Abbey Wood to the east of London, the new railway covers over 100 km of track including 21 km of new twin-bore rail tunnels and ten new stations.

What was the biggest engineering challenge that you have encountered since starting the project and what steps have you taken to overcome this?

Bill Tucker, Crossrail central section delivery director:

... In total, approximately 30 design consultant companies have been engaged in some aspect of the Crossrail design for either ourselves or our contractors. The Crossrail Technical Assurance Plan (TAP) outlines how we manage and accept designs from both our FDCs and contractors. The TAP specifies a gated acceptance process, which incorporates single-discipline and inter-discipline reviews and gains the concurrence of our infrastructure maintainers, London Underground and Rail for London.

Achieving the acceptance gates for our contractors' detailed design is important to maintaining our critical path programme. Managing the contractors' design in a manner that considers the complexity of these interfaces is a top priority of our project teams and chief engineer's group every day.

From www.theengineer.co.uk/

There will be numerous activities that are on the critical path for the Crossrail project. What will be the impact of finding a plague pit?

Discussion

Choose one of the following and discuss it with your peers.

1. Imagine that you have been asked by the Major Projects Authority to identify a big, public spending project completed in the last ten years.

   Identify and talk about activities that you think might be on the critical path and why.

2. Imagine you have been asked by the organiser of Glastonbury to help with the preparation for the next Glastonbury festival.

   Identify and talk about activities that you think might be on the critical path and why.

3. Imagine you are hoping to buy and move into a new house.

   Identify and talk about activities that you think might not be on the critical path and why.

Forward scan

A forward scan shows the earliest possible start time for each activity. The earliest time that an activity can start requires all the predecessor activities to be complete. To do this you work from the start of the project to the end of the project (from left to right) in the activity network.

Forward scan: A forward scan shows the earliest possible start time for each activity.

The activity network for the homework project can be amended to include the durations for each activity. The diagram shows this, with the durations shown in the middle box of the second row of each node.

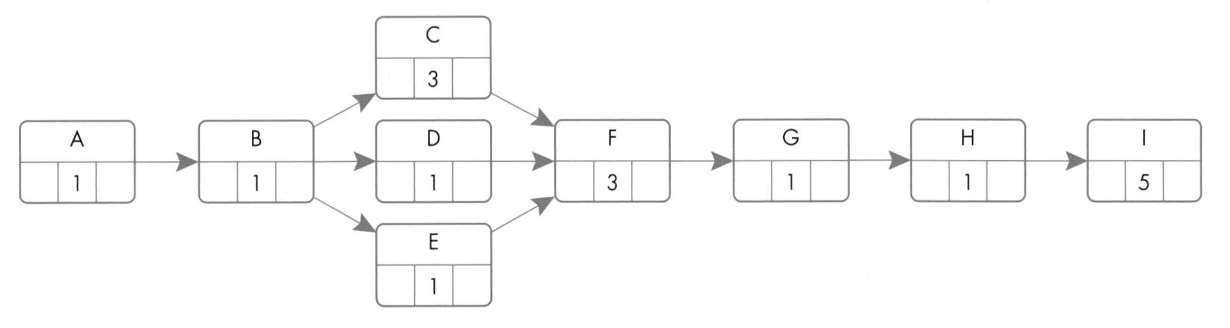

By following the network from left to right, you can find the earliest possible time for each activity.

Activity A

Activity A is at the start of the project and consequently has no predecessors. The earliest time that Activity A can start is 0.

Activity B

Activity B is dependent on the completion of activity A which has a duration of 1 day. The earliest time that Activity B can start is:

Activity A earliest time + Activity A duration = 0 + 1 = 1

The table shows the complete process from the start.

Activity	Duration	Predecessors	Predecessor earliest start	Predecessor duration	Activity earliest start
A	1	-	-	-	0
B	1	A	0	1	0 + 1 = 1
C	3	B	1	1	1 + 1 = 2
D	1	B	1	1	1 + 1 = 2
E	1	B	1	1	1 + 1 = 2
F	3	C	2	3	2 + 3 = 5
		D	2	1	2 + 1 = 3
		E	2	1	2 + 1 = 3
G	1	F	5	3	5 + 3 = 8
H	1	G	8	1	8 + 1 = 9
I	5	H	9	1	9 + 1 = 10

Note that Activity F is dependent on the completion of activities C, D and E. Activity F cannot start until the longest of C, D or E is complete. So the earliest time that activity F can start is 5.

The activity network can now be amended to show the earliest start for each activity, shown in the first box of the second row of each node.

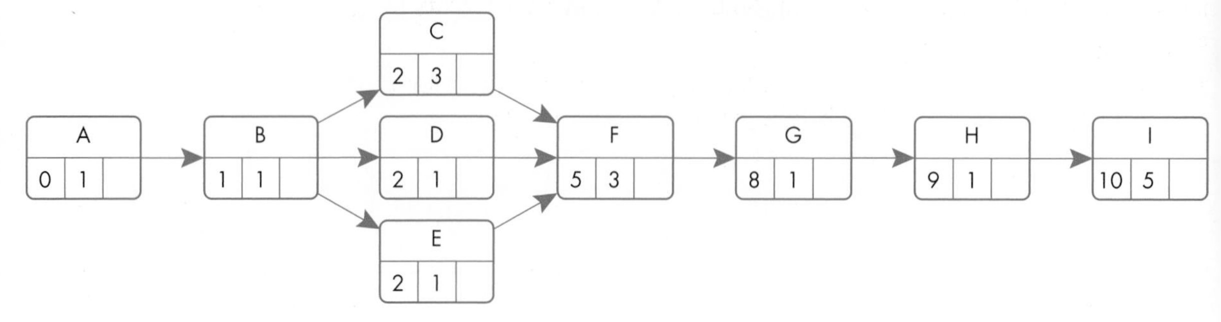

If you add the duration of the final activity, activity I (5 days), to its earliest start time (10 days) you can calculate the length of the project.

Length of project = Activity I earliest start time + Activity I duration

= 10 + 5 = 15

This is the length of the longest critical path.

Backward scan

You also need to find the latest time that an activity can be completed without extending the length of the critical path. To do this you perform a backward scan, working from the end of the project back to the start of the project (from right to left) in the activity network.

Backward scan: A backward scan shows the latest finish time for each activity without extending the length of the critical path.

By following the network from right to left, you can find the latest possible finish time for each activity without extending the length of the critical path.

Activity I

Activity I is at the end of the project and so the latest time is the end of the project which is 15.

Activity H

Activity H has a successor, activity I, which has a duration of 5 days. The latest time that Activity H can finish is:

Activity I latest time – Activity I duration = 15 – 5 = 10

The table shows the complete process from the end.

Activity	Duration	Successors	Successor latest time	Successor duration	Activity latest time
I	1	-	-	-	15
H	5	I	15	5	15 – 5 = 10
G	1	H	10	1	10 – 1 = 9
F	1	G	9	1	9 – 1 = 8
E	3	F	8	3	8 – 3 = 5
D	1	F	8	3	8 – 3 = 5
C	1	F	8	3	8 – 3 = 5
B	3	C	5	3	5 – 3 = 2
		D		1	5 – 1 = 4
		E		1	5 – 1 = 4
A	1	B	2	1	2 – 1 = 1

Note that Activity B is succeeded by activities C, D and E. The latest time that activity B can finish without extending the length of the critical path is the smallest of 5 – 3, 5 – 1, and 5 – 1. So the latest time that activity B can finish is 2.

The activity network can now be amended to show the latest start for each activity, shown in the third box of the second row of each node.

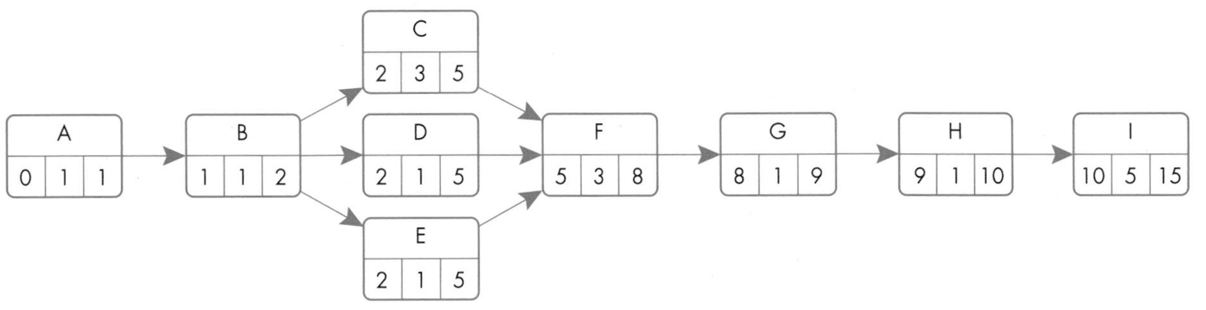

So the **critical activities** are those that are on the critical path. In the example these are:

Critical activities: activities where the latest finish minus earliest start minus duration = 0.

A Set homework

B Write down homework

C Research homework

F Write up homework

G Check homework

H Hand in homework

I Mark homework

If you examine activities D and E, which are not critical activities, you can see that as long as they are completed by day 5, the length of the critical path will not be extended. They can each start as early as day 2 but as they only each have a duration of 1 day they could start as late as day 4.

Exercise 8B

1 In Exercise 8A question **1** you drew an activity network.

Redraw this network with the two-row boxes for each node.

- **a** Do forward and backward scans to complete the new activity network with the earliest and latest times.
- **b** List the activities that are on the critical path.

  What is the length of the critical path?
- **c** What could you change to take an activity off the critical path?

  List the activities that are now on the critical path.

  What is the length of the new critical path?

2 For the Interrailing activity network that you constructed in Exercise 8A question **2**, do forward and backward scans to determine the earliest and latest times.

What is the length of the critical path?

R3: Gantt charts

Learning objectives

You will learn how to:

- Use Gantt charts (cascade diagrams) to represent project activities.

Introduction

Activity networks are one way of representing a project in visual terms. Another useful way to visualise a project is by using a graphical representation known as a **Gantt chart** (cascade diagram).

Gantt chart:
A visualisation of a project using a graphical representation which allocates activities and durations to give a project schedule.

What is a Gantt chart?

A Gantt chart is a visual representation of a project. All the activities that comprise a project are shown on the Gantt chart, usually as bars, extending from left to right. The bars tend to cascade down the page with the earliest starting activity at the top of the Gantt chart and the latest finishing activity at the bottom. Time is the horizontal variable. Gantt charts are a useful and powerful way of representing a project because they show:

a) all the activities

b) the start and end of each activity

c) the duration of each activity

d) any overlap between activities

e) the entire timescale of the project.

As a project progresses the Gantt chart will be updated and the impact of variances can be assessed.

Gantt chart history

In the early 20th century Henry Gantt, an American engineer and management consultant, developed scientific management techniques which included what we know today as the Gantt chart.

A Polish engineer, Karol Adamieck, had developed a similar chart in the 1890s to help show production schedules but as he didn't publish his work in a language that was widely spoken, Gantt was able to make the charts accessible to a much wider audience.

Consequently the charts are called Gantt charts rather than Adamiecki charts. There are various software packages that now incorporate Gantt charts and this allows them to be generated and easily updated for use in complex projects.

Henry Gantt

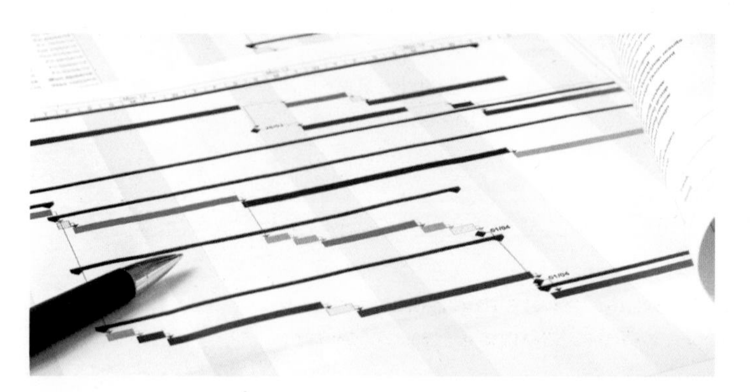

Mathematics in the real world

Large sports events, such as the football and rugby World Cups and the Olympics, require careful planning and preparation. Planning such an event covers a huge range of activities, and the overall planning task is usually broken into a range of sub-activities. For example, one of the key components of a successful rugby World Cup is the quality and robustness of the playing surface. The extract below is from a project plan to develop a rugby pitch for a world class tournament.

	August Week commencing					September Week commencing				October Week commencing				
	1	8	15	22	29	5	12	19	26	3	10	17	24	31
Contractor starts														
Prepare site														
Set up site														
Spray pitch														
Install primary drainage														
Install secondary drainage														
Sand amelioration														
Prepare seed bed														
Seeding														
Grow grass and upkeep														

What does this Gantt chart tell you? Can you see all the activities? Do you think you could identify dependencies from this chart? Can you identify the critical activities? Can you see the timescale for the project?

Discussion

Imagine that you have been asked to write an article about the differences between activity networks and Gantt charts.

Talk with your peers about both the elements that are the same and those that are different. What are the advantages and disadvantages of the two different representations?

Time analysis

Let's look at the homework project again, specifically the durations and earliest and latest times.

Activity		Duration (in days)	Earliest time	Latest time
A	Set homework	1.0	0	1
B	Write down homework	1.0	1	2
C	Research homework	3.0	2	5
D	Review class notes	1.0	2	5
E	Cross reference textbook	1.0	2	5
F	Write up homework	3.0	5	8
G	Check homework	1.0	8	9
H	Hand in homework	1.0	9	10
I	Mark homework	5.0	10	15

For each activity you need to find the difference between the earliest time and the latest time and compare this to the duration to determine if there is any **float**. Float is the amount of time an activity can be delayed without impacting other activities.

Float: The amount of time an activity can be delayed without impacting other activities.

Activity D:

latest time – earliest time = 5 – 2 = 3

So there is a window of 3 days in which activity D can be completed. The duration of activity D is only 1 day so it can float within this 3-day window. You can calculate the float for activity D as follows.

Float = Latest time – Earliest time – Duration

= 3 – 1 = 2

The table shows the calculated float for all of the homework activities.

Activity		Duration (in days)	Earliest time	Latest time	Float
A	Set homework	1.0	0	1	0
B	Write down homework	1.0	1	2	0
C	Research homework	3.0	2	5	0
D	Review class notes	1.0	2	5	2
E	Cross reference textbook	1.0	2	5	2
F	Write up homework	3.0	5	8	0
G	Check homework	1.0	8	9	0
H	Hand in homework	1.0	9	10	0
I	Mark homework	5.0	10	15	0

What do you notice?

Critical activities have float = 0.

Non-critical activities have float > 0.

You can now draw the Gantt chart (cascade diagram) for the homework project.

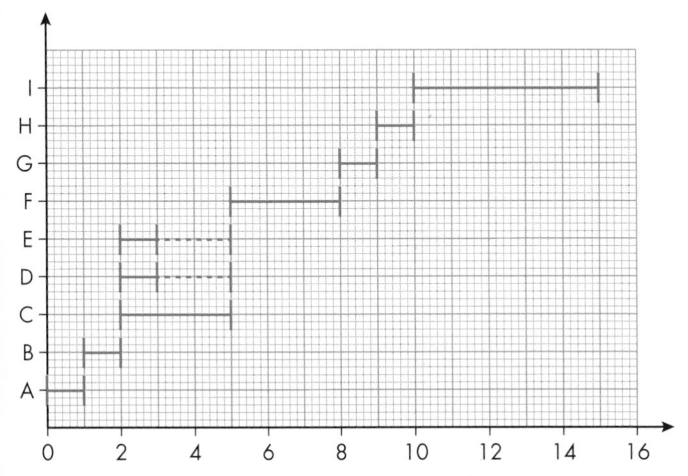

The critical activities are drawn with solid lines. They have zero float. For each of the other activities the dashed line indicates the window in which the activity can float.

Exercise 8C

1 In Exercise 8B question **1** you drew an activity network with earliest and latest times.

- **a** Tabulate this information and calculate the float for each activity.
- **b** Draw the Gantt chart (cascade diagram).
- **c** Do the activities with zero float correspond with the critical activities?

2 In Exercise 8B question **2** you calculated the earliest and latest times for your Interrailing project.

- **a** Tabulate this information and calculate the float for each activity.
- **b** Draw the Gantt chart (cascade diagram).
- **c** Do the activities with non-zero float correspond with the non-critical activities?

3 You and a group of friends decide to go to the next Glastonbury festival.

- **a** Make a list of all the things that you will need to do before you actually depart for the festival, for example, buy tickets, organise transport, buy a tent, etc.
- **b** Turn your list into a table of activities. Include the predecessor(s) of each activity and the durations.
- **c** Draw the activity network with the two rows of boxes for each node.

  Do forward and backward scans to complete the activity network with the earliest and latest times.
- **d** List the activities that are on the critical path.

  What is the length of the critical path?
- **e** What could you change to take an activity off the critical path?

  List the activities that are now on the critical path.

  What is the length of the new critical path?
- **f** Tabulate the duration and forward and back scan information and calculate the float for each activity.
- **g** Draw the Gantt chart (cascade diagram).
- **h** Are you going to the next Glastonbury festival after all this planning?

Case study

In 2013 the Construction Programme for the London 2012 Olympic and Paralympic Games won an award for the Programme of the Year 2012 given by the Association of Project Management (APM).

Here is the overview of the project.

Category: Programme of the Year 2012

Winner: The Construction Programme for the London 2012 Olympic and Paralympic Games, Olympic Delivery Authority

Overview

London's successful bid for the 2012 Olympic and Paralympic Games created the need for a major regeneration and construction programme to provide the venues and infrastructure needed to stage the Games. The programme of construction was extensive, technically and politically challenging, and up against a fixed deadline of the Opening Ceremony of the Games in July 2012. Turning the vision of the Olympic bid into the reality of roads, bridges and stadia was the job of the Olympic Delivery Authority (ODA), a new publicly funded body established by an Act of Parliament in April 2006.

You can read more from the case study on the APM website.

Why do you think this programme was given this award?

Project work

In February 2014 part of the railway track and sea wall was swept away at Dawlish in Devon during severe winter storms. By April 2014 the track had been repaired by a 300-strong Network Rail team at a cost of £35m.

By researching the project to rebuild the track, determine as many of the project activities as possible, their durations and dependencies. By finding the earliest and latest times, calculate the critical path and construct an activity network.

Check your progress

How confident are you feeling in your level of knowledge? What do you need to practise more?

Spec reference	Learning objective			
R1.1 R1.2	Represent compound projects by activity networks using activity-on-node representation			
R2.1	Use early time and late time algorithms to identify critical activities and find the critical path(s)			
R3.1	Use Gantt charts (cascade diagrams) to represent project activities			

9 Expectation

The National Lottery celebrated its 20th birthday in November 2014. By looking at the number of times that the different lottery numbers have been drawn, you might think you could develop a strategy to win the jackpot.

Out of 1758 draws up to November 2014, the seven numbers to have been drawn the most are 23, 25, 31, 33, 38, 43 and 44. 38 is the most frequently drawn number having been picked 290 times. The number that has been drawn the fewest times is 13 which has only been drawn 203 times. Of course, for a lot of people, 13 is an unlucky number. But if you are Chinese, then 4 is an unlucky number and that isn't at the bottom of the list. And if you are Japanese, then 9 is an unlucky number and that isn't at the bottom of the number of draws list either. So how can you use this knowledge to help you win? Should you buy a lottery ticket with the numbers 23, 25, 31, 33, 38, 43 because these are the luckiest numbers? Or has luck got nothing to do with it?

R4: Probability

Learning objectives

You will learn how to:

- Understand that uncertain outcomes can be modelled as random events with estimated probabilities.
- Apply ideas of randomness, fairness and equally likely events to calculate expected outcomes.

Introduction

It's a cold, dark Saturday morning in the middle of the winter. You have been working hard studying all week and are consequently very tired but you are still up early to go to your Saturday job. All you want to do is bury your head under the covers and sleep until noon. You can't be bothered to turn the light on. You open one of your drawers and pull out a T-shirt. You have a dozen T-shirts of different colours. You don't care what colour it is as long as it is clean. What is the **probability** that it will be black to match the sky outside? What is the probability that it will be white to match the frost on the window? What is the probability it won't be either of these colours?

Probability: the likelihood of an outcome or event.

Mathematics in the real world

Probability is all about the chance or likelihood of something happening. You will more than likely encounter and think about probability frequently in everyday life and not even realise that you are considering a mathematical concept. Which of the following questions have crossed your mind?

- What is the chance of me going to university?
- What is the chance of that gorgeous person in Year 13 saying yes when I ask them out?
- What is the chance of a tsunami hitting the coast of Britain?
- What is the chance of me getting a ticket to Glastonbury next year?
- What is the chance of my football team either staying in or getting into the Premiership?
- What is the chance of my parents getting divorced?
- What is the chance of me getting married?
- What is the chance of me becoming a millionaire by the age of 30?

Discussion

Choose one of the following and discuss it with your peers.

1. Imagine that it is Christmas Day and you have just taken the seal off a family-size tin of chocolates.

   Talk about the likelihood that the first chocolate you take out of the tin will be your favourite.

2. Imagine that you have just buttered a slice of toast for breakfast. In your haste you drop your toast on the floor.

   Talk about the likelihood the toast will land butter-side down.

3. Imagine that your name, along with the names of the rest of Year 12, has been put into a hat to go to watch one of the X-Factor auditions.

   Talk about the likelihood that your name will be selected.

Experimental probability

Experimental probability is a measure of the number of times a particular event happens in an experiment or a survey.

When a drawing pin is dropped, it can land point up or point down.

In an experiment, a drawing pin is dropped 50 times and the results are recorded. The drawing pin lands point up 33 times and point down 17 times. Each time the drawing pin lands is a random event.

You can use the results to calculate experimental probability.

$$\text{experimental probability} = \frac{\text{number of times an event happens}}{\text{total number of trials}}$$

The experimental probability that a single drop results in the drawing pin landing point up is 33 out of 50 trials:

$$\text{P(point up)} = \frac{33}{50}$$

P(point up) means the probability of the event the drawing pin will land point up.

Similarly the probability of the drawing pin landing point down is 17 out of 50 trials:

$$\text{P(point down)} = \frac{17}{50}$$

Experiment: an action where the outcome is uncertain.

Event: a single result of an experiment.

Random event: an event with a probability of occurrence determined by some probability distribution.

Experimental probability: the ratio of the number of times an event occurs to the total number of trials.

Point up and point down are the only two possible options. Consequently they are said to be an exhaustive set because they encompass the entire range of possible outcomes. The sum of all the probabilities for an exhaustive set is 1.

$$\text{P(point up)} + \text{P(point down)} = \frac{33}{50} + \frac{17}{50} = 1$$

Exhaustive set: the entire range of possible outcomes.

The probability of an outcome not happening is always equal to 1 minus the probability of it happening.

$$\text{P(outcome does not happen)} = 1 - \text{P(outcome happens)}$$

In the drawing pin experiment, the probability of the drawing pin landing point up is equal to 1 minus the probability of it landing point down:

$$\begin{aligned}\text{P(point up)} &= 1 - \text{P(point down)}\\ &= 1 - \frac{17}{50}\\ &= \frac{33}{50}\end{aligned}$$

If the pin is dropped 1000 times, the experimental probability can be used to work out an estimate for the number of times the drawing pin will land pin up.

$$\begin{aligned}\text{Number of times} &= \text{P(point up)} \times 1000\\ &= \frac{33}{50} \times 1000\\ &= 660\end{aligned}$$

The estimated number of 'pin up' outcomes in 1000 trials is 660.

Here is another experiment. Many of you will have played the rock–paper–scissors game at school. If not, the rules of the game are simple and straightforward: after an agreed count, for example, one, two, three, you put out your hand in a fist, to represent a rock, flat to represent paper, or with two fingers stuck out, to represent scissors. The winner is determined as follows:

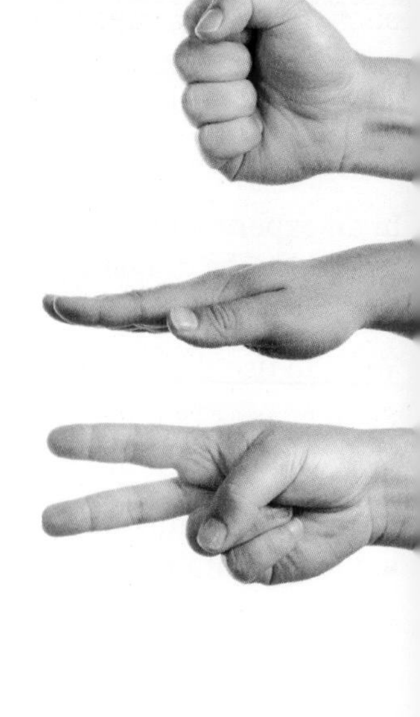

- Rock beats scissors. (Rock smashes scissors.)
- Paper beats rock. (Paper covers rock.)
- Scissors beat paper. (Scissors cut paper.)

If both players make the same shape then the game is a draw. You play a pre-agreed number of games, for example, best out of twenty, and the winner is then decided.

Imagine you play the game 20 times with a neighbour and you win 5 times, lose 7 times and draw 8 times. What is the experimental probability you will win the next game?

The experimental probability that you win is 5 out 20 games:

$$P(\text{win}) = \frac{5}{20} = \frac{1}{4}$$

Similarly the experimental probability that you lose is 7 out of 20 games:

$$P(\text{lose}) = \frac{7}{20}$$

And the experimental probability that you draw is 8 out of 20 games:

$$P(\text{draw}) = \frac{8}{20} = \frac{2}{5}$$

This is another example of an exhaustive set of outcomes: win, lose and draw. The sum of the probabilities of these outcomes should be 1. You can check:

$$P(\text{win}) + P(\text{lose}) + P(\text{draw}) = \frac{5}{20} + \frac{7}{20} + \frac{8}{20} = 1$$

If you then decided to play the game a further 100 times how many times would you, for example, expect to draw? Well $\frac{2}{5}$ of the games resulted in a draw

so $\frac{2}{5}$ of 100 gives:

$$\frac{2}{5} \times 100 = 40$$

The estimated number of draws in 100 games is 40.

But what if you played the game 10 000 times? How many times would expect to win, lose and draw?

Theoretical probability

Theoretical probability predicts the likelihood of an event happening, if all outcomes are equally likely.

The **theoretical probability** of an event is the number of ways that the event can occur, divided by the total number of outcomes.

You can use theoretical probability to describe how likely the different outcomes of an experiment are. It is calculated using this formula.

$$\text{theoretical probability} = \frac{\text{number of times an event can happen}}{\text{total number of possible outcomes}}$$

The possible outcomes of the rock–paper–scissors game for Player A are shown in the **sample space diagram** below.

A **sample space diagram** shows all the possible outcomes for an experiment.

		Player A		
		rock	**paper**	**scissors**
Player B	**rock**	draw	win	lose
	paper	lose	draw	win
	scissors	win	lose	draw

So in a game of rock–paper–scissors there are nine possible outcomes.

- Three wins for player A. (Three losses for player B.)
- Three losses for player A. (Three wins for player B.)
- Three draws.

If the players make their choice randomly then the probabilities are:

$$P(\text{win}) = \frac{3}{9} = \frac{1}{3}$$

$$P(\text{lose}) = \frac{3}{9} = \frac{1}{3}$$

$$P(\text{draw}) = \frac{3}{9} = \frac{1}{3}$$

As rock–paper–scissors is a **fair game**, how many times would you expect to win, lose and draw if you play the game 10 000 times?

Fair game: a game in which each player is equally likely to win.

The earlier experiment would suggest:

$$P(\text{win}) = \frac{1}{4} \times 10\,000 = 2500$$

$$P(\text{lose}) = \frac{7}{20} \times 10\,000 = 3500$$

$$P(\text{draw}) = \frac{2}{5} \times 10\,000 = 4000$$

But the experiment only included 20 games. This is not very many. Over an increasing number of games you would expect the experimental probability to tend towards the theoretical probability:

$$P(\text{win}) = \frac{1}{3} \times 10\,000 = 3333\tfrac{1}{3}$$

$$P(\text{lose}) = \frac{1}{3} \times 10\,000 = 3333\tfrac{1}{3}$$

$$P(\text{draw}) = \frac{1}{3} \times 10\,000 = 3333\tfrac{1}{3}$$

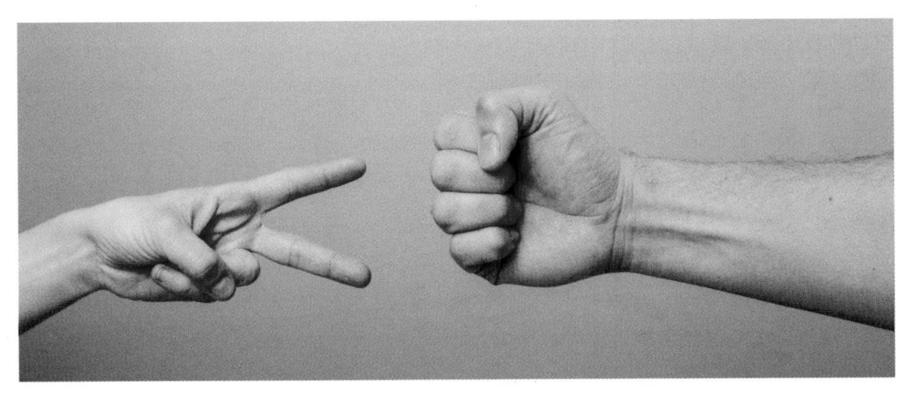

Exercise 9A

1 Drop a drawing pin on your desk 60 times and record whether it lands point up or point down.

Work out the experimental probability of the drawing pin landing point up and point down.

Comment on whether you think this experiment is fair.

If you were to drop the drawing pin 1000 times, estimate how many times you would expect the drawing pin to land point up.

2 The probability that you will have at least six emails when you log in to your school email account is 0.36.

What is the probability that you will have fewer than six emails?

3 Play the rock–paper–scissors game with a classmate 100 times and record your results.

How do your results compare to the experimental results earlier in this chapter?

How close are your results to the theoretical probabilities for this game?

4 What is the theoretical probability of a draw when playing the rock–paper–scissors game?

What is the theoretical probability that the rock–paper–scissors game does not result in a draw?

5 A computer game has five different levels.

The probability of a player getting to a particular level is as follows:

P(level 1) = 0.46

P(level 2) = 0.33

P(level 3) = 0.12

P(level 4) = 0.06

P(level 5) = ?

What is the probability that the player will reach level 5?

6 Two students roll two dice backwards and forwards between them.

One student says "If we add the two numbers that are on top of each die, I think the number that will come up most frequently is 6."

The other student disagrees. She says "I think the number that will come up most frequently is 7."

Construct a sample space diagram and calculate the probability for each possible sum.

Who is right?

7 If you and a classmate were to roll two dice, as described in question **6**,

a what is the probability of getting an even number?

b what is the probability of getting a multiple of 5?

c what is the probability of not getting a multiple of 5?

8 If you and a classmate were to roll two dice as described in question **6** 600 times,

a how many times would you expect to get a sum of 10?

b how many times would you expect not to get a sum of 10?

R5: Diagrammatic representations

Learning objectives

You will learn how to:

- Understand and apply Venn diagrams and simple tree diagrams.

Introduction

John Venn was an English mathematician born in Hull. He is best known for Venn diagrams, which give a pictorial representation of the relationships between sets.

Some artists have used the Venn diagram model creatively by overlapping concepts and categories. This Venn diagram illustrates the overlap between science and art. The image is from the Hubble telescope and is very beautiful. However, it could only be generated using some very accurate mathematical transformations of raw numbers from the telescope.

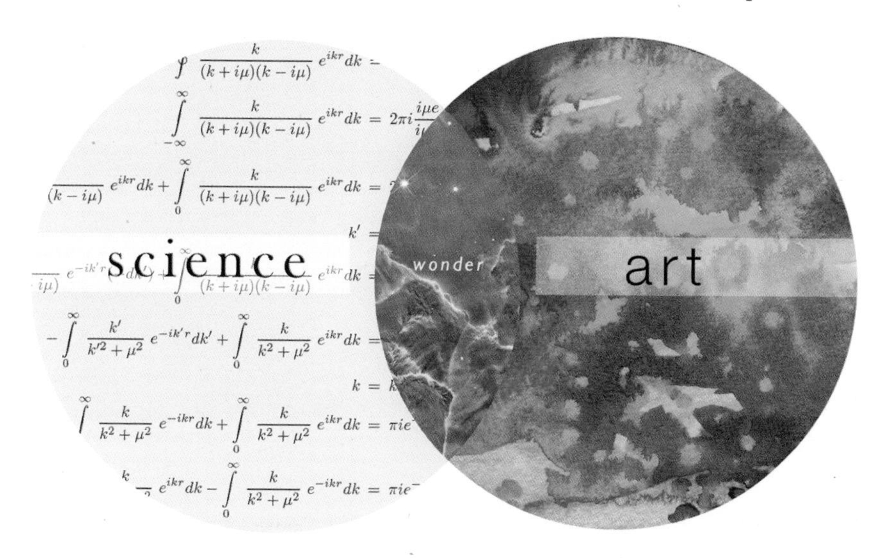

Venn diagrams are used to summarise information and determine probabilities from this information. Mathematical notation (a set of symbols) is used to write different probabilities.

Probabilities can also be represented and calculated using tree diagrams.

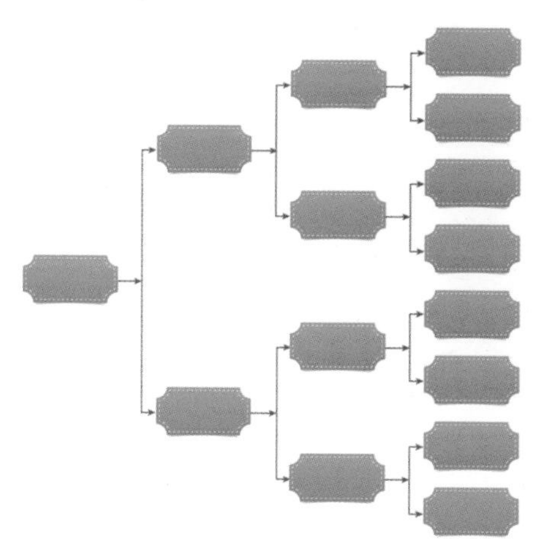

Mathematics in the real world

Venn diagrams are used in many areas of life to categorise or group items, as well as to compare and contrast different items. Here are some examples.

Venn diagrams are used a lot in modern marketing analysis.

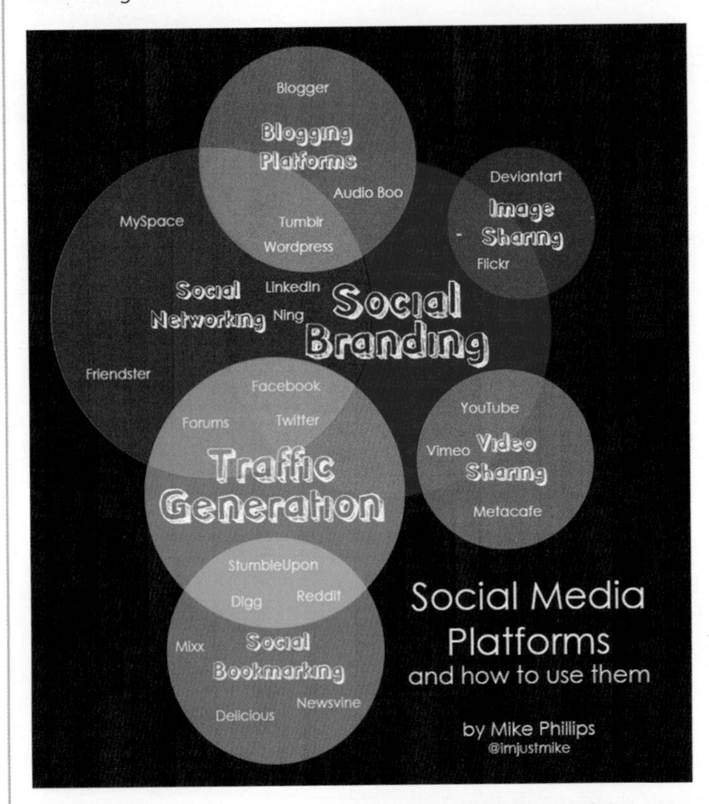

Here is another creative use of the Venn diagram model.

To find many varied, real-life examples of uses of Venn diagrams in everyday life just type 'Venn diagrams in everyday life' into your search engine and look at the images.

Discussion

Choose one of the following, discuss it with your peers and then try drawing the associated Venn diagram.

1. Imagine you have to represent film and TV stars on a Venn diagram.

   Talk about examples of stars who have only appeared in films, others who have only appeared on TV and finally stars who have appeared in films and been in TV shows.

2. Imagine you have to represent foodstuffs on a Venn diagram.

   Talk about examples of hamburger contents that are not vegetables, vegetables that will never be found in hamburgers and finally vegetables that are in hamburgers.

3. Imagine that you have to represent Daniel Craig's film appearances on a Venn diagram.

   Talk about examples of Bond films that Daniel Craig has starred in, examples of Bond films that Daniel Craig hasn't starred in and films that Daniel Craig has starred in but are not Bond films.

Venn diagrams

A group of 60 students are discussing summer music festivals. Of these students:

- 11 have tickets for Glastonbury
- 19 have tickets for the V festival
- 6 students have tickets for both festivals.

There are 36 students who currently don't have tickets for either summer music festival.

This information is summarised on the Venn diagram.

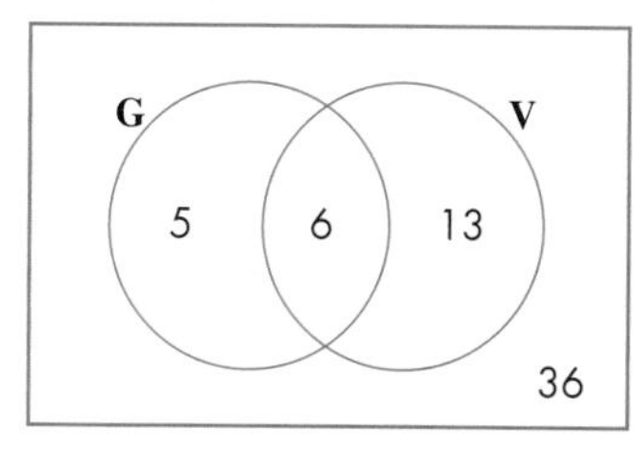

The rectangle represents the **universal set**. It is the sample space of all the possible outcomes.

The left-hand circle, labelled G, represents the 11 students who have tickets for Glastonbury.

The right-hand circle, labelled V, represents the 19 students who have tickets for the V festival.

The overlap of the two circles represents the 6 students who have tickets for both Glastonbury and the V festival.

The 36 who don't have tickets to either festival are outside the circles but inside the rectangle.

The **universal set** is the collection of all things in a particular context. All other sets within the universal set are subsets.

A **subset** is a group of things within the defined universal set.

Venn diagrams can be used to calculate probabilities. The probability P(G) that a student has a ticket to Glastonbury is:

$$P(G) = \frac{5+6}{60} = \frac{11}{60}$$

The probability P(V) that a student has a ticket to the V festival is:

$$P(V) = \frac{6+13}{60} = \frac{19}{60}$$

Students who have tickets for *both* Glastonbury *and* the V festival are shown in the overlap of the two circles. In a Venn diagram this overlap is the **intersection** of the two sets, and is written as $G \cap V$.

$$P(G \cap V) = \frac{6}{60}$$

Students who have tickets for *either* Glastonbury *or* the V festival *or* both are shown in the two circles. In a Venn diagram, this is the **union** of the two sets and is written as $G \cup V$.

$$P(G \cup V) = \frac{5+6+13}{60} = \frac{24}{60}$$

Students who don't have tickets to Glastonbury are shown as everyone outside the circle labelled G, written as G′. You can find the probability of this in one of two ways:

$$P(G') = \frac{13+36}{60} = \frac{49}{60}$$

The top line of the fraction is the sum of the students who have tickets to the V festival, but not Glastonbury, and the students who aren't going to either festival.

But you know that $P(G) = \frac{11}{60}$, and you know that the sum of all probabilities in a sample space is 1, so you could work out P(G′) as follows:

$$P(G') = 1 - P(G) = 1 - \frac{11}{60} = \frac{49}{60}$$

This is known as **complementary probability**.

Complementary probability: If A′ represents the event not A then P(A′) = 1 − P(A).

Tree diagrams

Tree diagrams are used to show sequences of events, and in particular they are used to calculate the probabilities of combinations of events.

A tin contains 12 coloured buttons. Four buttons are blue and eight are red. If one button is taken out and then replaced in the tin, and then a second button is taken out, what is the probability that both buttons are red? What is the probability they are two different colours? What is the probability that at least one of the buttons is blue?

A tree diagram can help understand the information given and answer the questions. First you need some notation. Let B stand for the event 'took a blue button' and R stand for the event 'took a red button'.

The probability of taking a blue button, P(B), is:

$$P(B) = \frac{4}{12} = \frac{1}{3}$$

The probability of taking a red button, P(R), is:

$$P(R) = \frac{8}{12} = \frac{2}{3}$$

These events can be represented on a tree diagram.

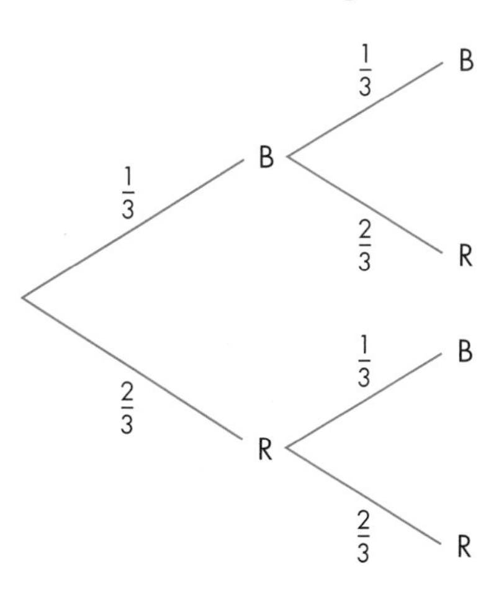

The probabilities have been written along the branches of the tree.

The probabilities on pairs of branches sum to 1:

$$\frac{1}{3} + \frac{2}{3} = 1.$$

Groups of branches from the same source should always sum to one on a tree diagram.

If a third button is taken from the tin, you could add another set of branches to the tree diagram.

So, what is the probability that both buttons taken are red? You are looking for the probability of a red *and* a red, i.e. P(RR). To find this you go along the relevant branches of the tree diagram and multiply the probabilities:

$$P(RR) = \frac{2}{3} \times \frac{2}{3} = \frac{4}{9}$$

What is the probability two different coloured buttons are taken?

The possible combinations are:

- red then blue (RB)
- blue then red (BR).

The probability of 'red then blue' is:

$$P(RB) = \frac{2}{3} \times \frac{1}{3} = \frac{2}{9}$$

The probability of 'blue then red' is:

$$P(BR) = \frac{1}{3} \times \frac{2}{3} = \frac{2}{9}$$

The probability two different coloured buttons are taken is:

$$P(RB) + P(BR) = \frac{2}{9} + \frac{2}{9} = \frac{4}{9}$$

What is the probability that at least one blue button is taken?

The possible combinations are:

- red then blue (RB)
- blue then red (BR)
- blue then blue (BB).

The probabilities P(RB) and P(BR) have been calculated above. The probability P(BB) is:

$$P(BB) = \frac{1}{3} \times \frac{1}{3} = \frac{1}{9}$$

The sum of the probabilities is:

$$P(RB) + P(BR) + P(BB) = \frac{2}{9} + \frac{2}{9} + \frac{1}{9} = \frac{5}{9}$$

These probabilities can be found using the tree diagram.

As you go along the branches of a tree diagram you multiply.

As you go down the tree diagram you add between branches.

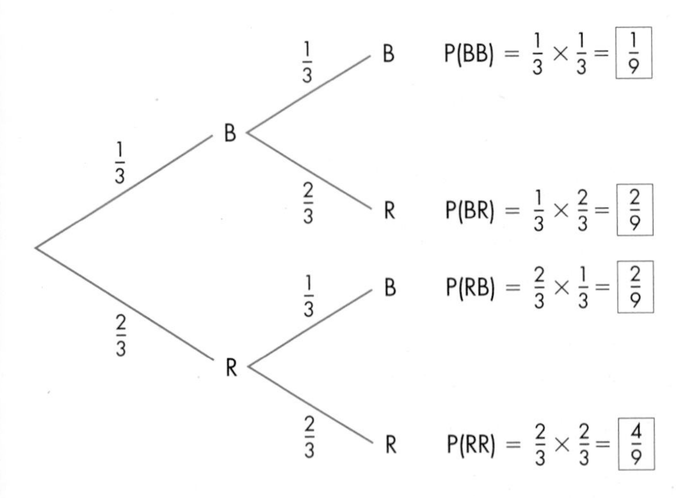

Exercise 9B

1 For each section of this question copy following Venn diagram.

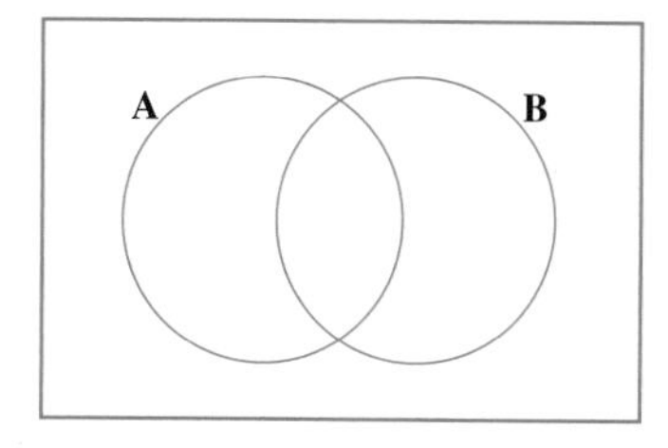

Shade the part of the diagram that represents:

a A

b B

c $A \cap B$

d $A \cup B$

e A'

f B'

g $A' \cap B'$

h $A' \cup B'$

2 Study the following Venn diagram which contains some films in which Nicole Kidman and Tom Cruise have starred.

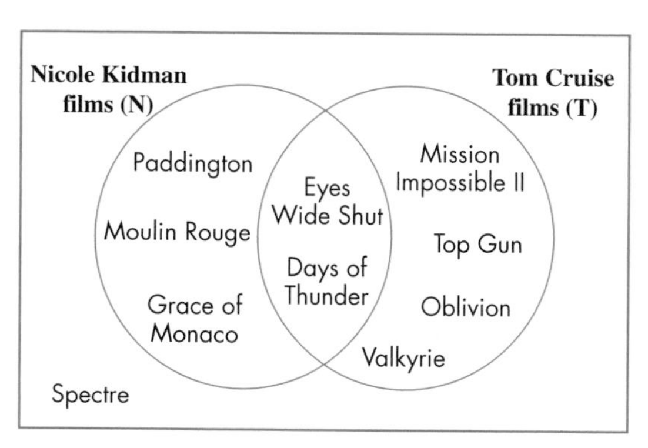

a List all the films in N.

b List all the films in T.

c List all the films in $N \cap T$.
Are these just Nicole Kidman films, just Tom Cruise films, Nicole Kidman and Tom Cruise films, Nicole Kidman or Tom Cruise films or none of these?

d List all the films in $N \cup T$.
Are these just Nicole Kidman films, just Tom Cruise films, Nicole Kidman and Tom Cruise films, Nicole Kidman or Tom Cruise films or none of these?

e List all the films in $N' \cap T'$.
Are these just Nicole Kidman films, just Tom Cruise films, Nicole Kidman and Tom Cruise films, Nicole Kidman or Tom Cruise films or none of these?

3 Use the Venn diagram in question **2** which contains some films in which Nicole Kidman and Tom Cruise have starred.

a Find the probability that a randomly selected film will star Nicole Kidman, P(N).

b Find the probability that a randomly selected film will star Tom Cruise, P(T).

c Find the probability that a randomly selected film will star Nicole Kidman and Tom Cruise, $P(N \cap T)$.

d Find the probability that a randomly selected film will star Nicole Kidman or Tom Cruise, $P(N \cup T)$.

e Find the probability that a randomly selected film will not star Nicole Kidman, $P(N')$

f Find the probability that a randomly selected film will not star Nicole Kidman and will not star Tom Cruise, $P(N' \cap T')$.

g Find the probability that a film starring Tom Cruise also stars Nicole Kidman.

4 Two children are playing with a bag which contains ten marbles. Three marbles are yellow and the rest are green.

One child pulls out a marble and puts it back. The other child then pulls out a marble and puts it back.

a Illustrate this game on a tree diagram.

b Determine the probability that both marbles are yellow.

c Determine the probability that the two marbles are different colours.

d Determine the probability that neither of the marbles is green.

e Determine the probability that both marbles are the same colour.

5 Three children are playing with a bag which contains ten marbles. Three marbles are yellow and the rest are green.

One child pulls out a marble then puts it back. Then the second child pulls out a marble and puts it back. Then the third child pulls out a marble and puts in back.

a Illustrate this game on a tree diagram.

(Hint: your tree diagram will need two branches for the first child, which will be followed by branches for the second child which will be followed by branches for the third child).

b Determine the probability that all three marbles are yellow.

c Determine the probability that the three marbles are different colours.

d Determine the probability that two of the marbles are the same colour and the other is different.

e Determine the probability that all three marbles are the same colour.

6 Two children are playing with a bag which contains ten marbles. Three marbles are yellow, six marbles are red and the rest are green.

One child pulls out a marble and then puts it back. Then the other child pulls out a marble and puts it back.

a Illustrate this game on a tree diagram.

(Hint: your tree diagram will need three branches for each child, one for each colour).

b Determine the probability that both marbles are yellow.

c Determine the probability that the two marbles are different colours.

d Determine the probability that neither of the marbles is green.

e Determine the probability that both marbles are the same colour.

R6: Combined events

Learning objectives

You will learn how to:

- Calculate the probability of combined events: both A and B, neither A nor B and either A or B (or both).
- Understand the difference between independent and dependent events.

Introduction

Does the amount of time spent doing revision affect examination results? The simple answer is yes. You can say that examination results are dependent on the amount of time spent doing revision. Does the number of showers a student takes during Year 10 affect their GCSE examination results? Unlikely. So you can say that a student's GCSE examination results are independent of the number of showers they took in Year 10.

Probabilities and their associated calculations are affected by dependence and independence.

Dependent events: events that are affected by previous events

Independent events: events that are not affected by other events

Mathematics in the real world

Dependent events could be:

- Parking illegally and getting a parking ticket. You will only get a parking ticket if you have parked illegally.
- Eating too much food and feeling full. You will only feel full if you eat too much food.
- Buying a National lottery ticket and winning the lottery. You will only win the lottery if you buy a ticket in the first place.

Independent events could be:

- Parking illegally and watching your favourite DVD.
- Eating too much food and it being sunny.
- Buying a lottery ticket and the school bus being late.

Discussion

Choose one of the following and discuss it with your peers.

1. Imagine that you have been asked to do a presentation to Year 7 students about why you need a GCSE in maths.

   Talk about things that are dependent on having a GCSE in maths.

2. Imagine that you have been asked to do a presentation to Year 10 students about why you need a good night's sleep.

   Talk about things that are dependent on having a good night's sleep.

3. Imagine that you have been asked to do a presentation to Year 11 students about good nutrition.

   Talk about things that are dependent on eating the right food.

Dependent events

When you looked at tree diagrams in the previous section, you considered a tin containing 12 coloured buttons. Four buttons were blue and eight were red. One button was taken out and replaced, and then another button was taken out and replaced.

This is called sampling with replacement. The selection of a second button is not affected by the first selection and so the events are independent.

If the first button is not put back in the tin, this would be sampling without replacement. The selection of the first button would affect the probability of the colour of the second button, and so the events are dependent.

How would the tree diagram change?

As before, let B stand for the event 'took a blue button' and R stand for the event 'took a red button'.

The probability of taking a blue button, P(B), the first time is:

$$P(B) = \frac{4}{12} = \frac{1}{3}$$

The probability of taking a red button, P(R), the first time is:

$$P(R) = \frac{8}{12} = \frac{2}{3}$$

If a blue button is taken out the first time, there are now 11 buttons left, eight of which are red, so the probability of a red button being taken out next is:

$$P(R) = \frac{8}{11}$$

If a blue button is taken out the first time, there are now 11 buttons left, three of which are blue, so the probability of a blue button being taken out the second time

$$P(B) = \frac{3}{11}$$

If a red button is taken out the first time, there are now 11 buttons left, seven of which are red, so the probability of a red button being taken out the second time is:

$$P(R) = \frac{7}{11}$$

If a red button is taken out the first time, there are now 11 buttons left, four of which are blue, so the probability of a blue button being taken out the second time is:

$$P(B) = \frac{4}{11}$$

These events can be represented on a tree diagram as shown below.

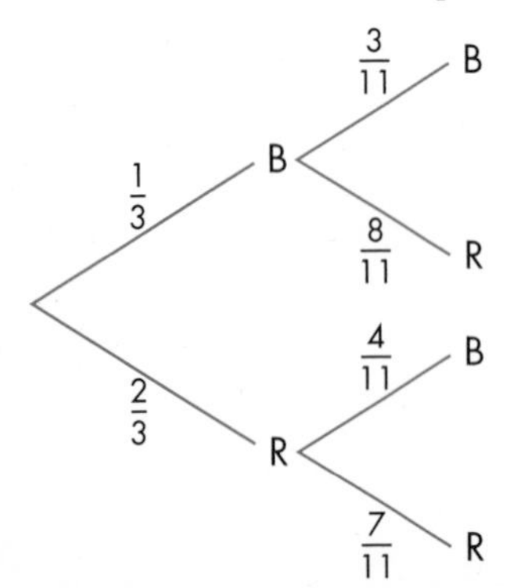

The probabilities have been written along the branches of the tree. Notice that the probabilities on each pair of branches always sum to 1.

- In the first pair of branches the sum is $\frac{1}{3} + \frac{2}{3} = 1$.
- In the second (top right) pair of branches the sum is $\frac{3}{11} + \frac{8}{11} = 1$.
- In the third (bottom right) pair of branches the sum is $\frac{4}{11} + \frac{7}{11} = 1$.

Groups of branches from the same source should always sum to one on a tree diagram. You can use this property to check your tree diagrams.

What is the probability that two red buttons will be taken? This is the probability for a red button first and then another red button, shown as P(RR).

$$P(RR) = \frac{2}{3} \times \frac{7}{11} = \frac{14}{33}$$

What is the probability that two different coloured buttons will be taken? The different combinations are:

- red then blue (RB)
- blue then red (BR).

The probability is the sum P(RB) + P(BR).

$$P(RB) = \frac{2}{3} \times \frac{4}{11} = \frac{8}{33}$$

$$P(BR) = \frac{1}{3} \times \frac{8}{11} = \frac{8}{33}$$

$$P(RB) + P(BR) = \frac{8}{33} + \frac{8}{33} = \frac{16}{33}$$

What is the probability that at least one blue button will be taken?

The different combinations are:

- red then blue (RB)
- blue then red (BR)
- blue then blue (BB).

The probability is the sum P(RB) + P(BR) + P(BB).

P(RB) and P(BR) have been calculated above.

$$P(BB) = \frac{1}{3} \times \frac{3}{11} = \frac{3}{33}$$

$$P(RB) + P(BR) + P(BB) = \frac{8}{33} + \frac{8}{33} + \frac{3}{33} = \frac{19}{33}$$

Probabilities can also be expressed as decimal fractions and tree diagrams can be used to show probabilities for combinations of different types of events.

Consider a student who works out that the amount of time he spends playing games on his PlayStation is related to his maths homework grades.

- The probability he will spend more time on his homework than playing on his PlayStation is 0.7.
- If he spends more time on his homework than playing on his PlayStation, the probability his maths homework is graded A is 0.9.
- If spends more time on his PlayStation than doing his homework, the probability his maths homework is not graded A is 0.8.

This is another example of a dependent event. His maths homework grade is dependent on how much time he spent on his homework compared to how much time he spent playing on his PlayStation.

What is the probability he spends more time on his homework than playing on his PlayStation and his maths homework is graded A?

What is the probability he spends more time on his PlayStation than his homework and his maths homework is graded A?

What is the probability his maths homework is graded A?

This sounds complicated, but can be made simple using a tree diagram.

Let H stand for the event 'doing more homework'.

Let A stand for the event 'awarded maths homework grade A'.

These events can be represented on a tree diagram, as shown below.

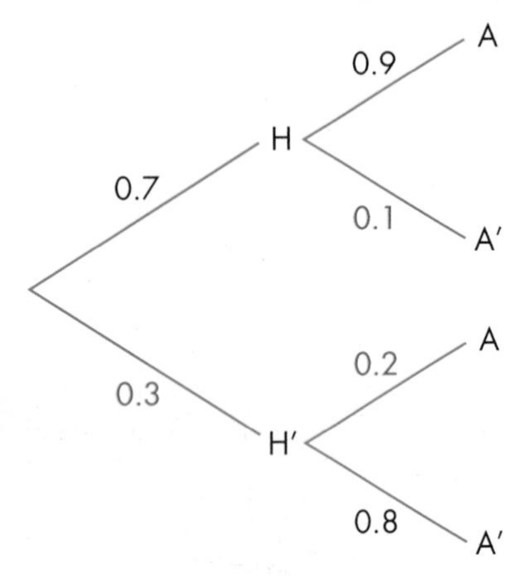

The probability he will spend more time on his homework than playing his PlayStation is 0.7, so the probability he spends more time on his PlayStation that his homework is:

$1 - 0.7 = 0.3$ (represented by H′)

If he spends more time on his homework than playing on his PlayStation, the probability his maths homework is graded A is 0.9, so the probability his maths homework is not graded A is:

$1 - 0.9 = 0.1$ (represented by A′)

If spends more time on his PlayStation than doing his homework, the probability his maths homework is not graded A is 0.8, so the probability his homework is graded A is:

$1 - 0.8 = 0.2$ (represented by A)

All of the probabilities have been written along the branches of the tree diagram. All the pairs of probabilities sum to 1.

What is the probability he spends more time on his homework than playing on his PlayStation and his maths homework is graded A?

The probability is given by P(HA).

$P(HA) = 0.7 \times 0.9 = 0.63$

What is the probability he spends more time on his PlayStation than his homework and his maths homework is graded A?

The probability is given by P(H′A):

$P(H'A) = 0.3 \times 0.2 = 0.06$

What is the probability his maths homework is graded A?

The possible combinations for this are:

- he spends more time doing homework than playing on his PlayStation and gets an A
- he spends more time playing on his PlayStation than doing homework but still gets an A.

In both circumstances he will get an A but the probabilities for the combinations are different. Read along the branches for H and H′:

$$P(A) = P(HA) + P(H'A)$$
$$= (0.7 \times 0.9) + (0.3 \times 0.2)$$
$$= 0.63 + 0.06 = 0.69$$

Independent events

In a group of 18 sixth formers:

- 6 are on Facebook
- 3 are on Twitter
- 1 student is on both Facebook and Twitter
- 4 are not on Facebook or Twitter but they do have call-only mobile phones
- 6 students are not on Facebook or Twitter and do not have call-only mobile phones.

For this group of students are the events 'being on Facebook' and 'being on Twitter' independent?

The information can be shown on a Venn diagram.

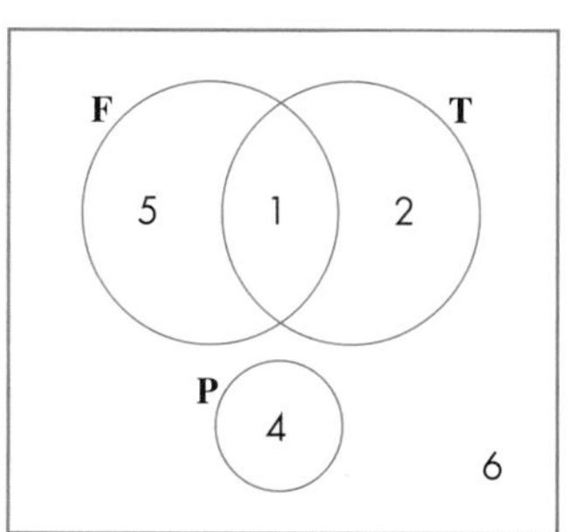

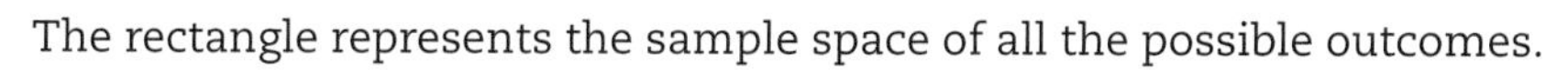

The rectangle represents the sample space of all the possible outcomes.

The left-hand circle (F) represents the students who are on Facebook.

The right-hand circle (T) represents the students who are on Twitter.

The overlap of the two circles represents students who are on both Facebook and Twitter.

The circle at the bottom (P) represents the students who have call-only mobile phones.

If the events 'being on Facebook' and 'being on Twitter' are independent, then the following must be true.

$$P(F \cap T) = P(F) \times P(T)$$

You know from the Venn diagram that the intersection of (F) and (T) has 1 student, so

$$P(F \cap T) = \frac{1}{18}$$

You can calculate P(F) and P(T) from the information given above the Venn diagram.

$P(F) = \frac{6}{18}$ and $P(T) = \frac{3}{18}$

So:

$P(F \cap T) = \frac{6}{18} \times \frac{3}{18} = \frac{1}{18}$

So you can say that, for this group of students, the events 'being on Facebook' and 'being on Twitter' are independent.

If two events A and B are **independent** then $P(A \cap B) = P(A) \times P(B)$.

Exercise 9C

1 You have 12 T-shirts of different colours in a drawer: six black, four blue and two white.

You take one T-shirt and put it to one side. You then take another T-shirt from the drawer.

What is the probability that you select:

a two white T-shirts

b one black T-shirt and one blue T-shirt.

c two T-shirts of the same colour.

2 You throw two dice.

Is the probability of throwing a six on each die dependent or independent?

Explain and justify your answer.

3 The probability of your mum letting you go out shopping next Saturday is 0.3. ($P(S) = 0.3$).

The probability of the school football team winning on Saturday afternoon is 0.4. ($P(F) = 0.4$).

Given that these two events are independent, find:

a $P(S \cap F)$

b $P(S \cup F)$

c $P(S' \cap F')$

4 You buy two raffle tickets for a raffle of ten prizes.

Is the probability of winning a prize with one of the raffle tickets dependent on winning a prize with the other raffle ticket?

Explain and justify your answer.

5 Two children are playing with a bag of marbles which contains ten marbles. Three marbles are yellow and the rest are green.

One child pulls out a marble and doesn't put it back. Then the other child pulls out a marble.

a Calculate the probability that both marbles are yellow.

b Calculate the probability that the two marbles are different colours.

c Calculate the probability that neither of the marbles is green.

d Calculate the probability that both marbles are the same colour.

6 You buy a lottery ticket for the UK National Lottery.

What is the probability of you winning with this one ticket? (Give your answer to 3 significant figures.)

R7: Expected value

Learning objectives

You will learn how to:

- Calculate the expected value of quantities such as financial loss or gain.

Introduction

There are many circumstances in everyday life when people weigh up the pros and cons of doing something. For example, you might decide, based on a favourable weather forecast, not to carry an umbrealla. For most people in the area this is good decision but for the odd person there may be a localised shower which makes it a bad decision. Insurance companies are continually making such decisions. They decide whether or not they are going to insure an individual or company based on an element of risk. The lower the risk for the insurance company the lower the cost of the insurance and vice versa. They hope that they make good decisions so that at the end of the financial year the insurance company makes a profit. This section is about similar scenarios. Is the reward worth the risk?

Mathematics in the real world

Suppose you are studying for a qualification where the final mark consists of 50% from coursework and 25% from each of two examinations. If you score 75/100 on your coursework and 69/100 and 83/100 on your examinations, how would the overall mark be calculated?

This calculation uses a weighted average which takes into consideration the different values of each element of the qualification. The overall mark would be calculated as follows:

Overall mark = $(0.5 \times 75) + (0.25 \times 69) + (0.25 \times 83) = 75.5$

If you replace the weightings with probabilities, instead of calculating a weighted average you calculate the expected value.

Discussion

Choose one of the following and discuss it with your peers.

1. Imagine that you have been asked to choose between two different mathematics qualifications.

   The first qualification is weighted as follows: 50% coursework, two examinations each of which are 25% of the overall mark.

   The second qualification is weighted as follows: three examinations each of which are $\frac{1}{3}$ of the overall mark.

   Talk about reasons why you might choose one of these qualifications instead of the other.

2. Imagine that you will be asked by your maths teacher to do extra homework if, when two dice are rolled, the sum of the top faces is 6, 7 or 8.

   Talk about how likely it is that you will be asked to do extra homework.

3. Imagine that you will get a prize if, when three coins are tossed, they all show heads.

   Talk about how likely it is that you will get a prize.

Calculating expected value

You have two £1 coins. You decide that when you toss the coins, you will spend the coins that land heads-up. How much can you expect to spend?

The possible combinations that can occur are:

- HH
- HT
- TH
- TT.

If the coins show HH, then you will spend £2. This combination can only occur once out of the four possible outcomes so the probability is $\frac{1}{4}$ or 0.25.

If the coins show HT or TH, then you will spend £1. This combination can occur twice out of the four possible outcomes so the probability is $\frac{2}{4}$ or 0.5.

If the coins show TT, then you won't be spending any money. This combination can only occur once out of the four possible outcomes so the probability is $\frac{1}{4}$ or 0.25.

This is shown in the table.

Outcome of throws	HH	HT or TH	TT
Amount to spend	£2	£1	£0
Probability	0.25	0.5	0.25

You calculate the expected value, $E(X)$, by multiplying each amount by its probability:

$$E(X) = (2 \times 0.25) + (1 \times 0.5) + (0 \times 0.25) = £1$$

So you can expect to spend £1.

You can say that the amount you can spend is a random variable as it is the value obtained from an experiment. A random variable must be a numerical value.

The set of all the possible amounts you can spend and their associated probabilities are known as a probability distribution.

The expected value of X is calculated as follows:

$$E(X) = \Sigma x \, P(X = x)$$

This means that you multiply each value, x, by its associated probability, $P(X = x)$, and then sum all of the products (multiplications).

A **random variable** is the value obtained from an experiment. A random variable must be a numerical value.

Probability distribution: the set of all possible values a random variable can have with their associated probabilities.

The **expected value** of X is also known as the mean of X and is written as $E(X)$.

Example 1

You develop a game to use at a school fundraising event.

The game involves one spin of a four-sided spinner numbered 1 to 4.

- If the side the spinner rests on is prime (2 or 3), the player wins the value in £s on the spinner.
- If the side the spinner rests on is not prime (1 or 4), the player gives the value in £s on the spinner to the stall-holder.

Is the game guaranteed to make a profit?

Outcome of spinner	1	2	3	4
Stall-holder win	£1	−£2	−£3	£4
Probability	0.25	0.25	0.25	0.25

$$E(X) = (1 \times 0.25) + (-2 \times 0.25) + (-3 \times 0.25) + (4 \times 0.25) = 0$$

The expected value is 0, so the game is not going to make a profit. Instead it will just break even with no profit and no loss.

Example 2

What would happen if the rules of the game in Example 1 are changed as follows?

- If the side the spinner rests on is odd (1 or 3), the player wins the value in £s on the spinner.
- If the side the spinner rests on is even (2 or 4), the player gives the value in £s on the spinner to the stall-holder.

Outcome of spinner	1	2	3	4
Stall-holder win	−£1	£2	−£3	£4
Probability	0.25	0.25	0.25	0.25

$$E(X) = (-1 \times 0.25) + (2 \times 0.25) + (-3 \times 0.25) + (4 \times 0.25) = 0.5$$

The expected value this time is 0.5, or 50p, so the game will make a profit.

Exercise 9D

1 Find the expected value of the random variable with outcomes and associated probability distribution as shown in the following table.

Outcome	1	2	3	4
Probability	$\frac{1}{3}$	$\frac{1}{6}$	$\frac{1}{6}$	$\frac{1}{3}$

2 Find the expected value of the random variable with outcomes and associated probability distribution as shown in the following table.

Outcome	−2	0	1
Probability	0.4	0.1	0.5

3 Two fair dice are rolled and the sum of their scores is recorded.

a Find the probability distribution for the sum of the scores.

b Calculate $E(X)$.

4 You have three £1 coins. You decide that when you toss the coins, you will spend the coins that land heads-up.

How much can you expect to spend?

5 Using the probability distribution in question **3**, invent a fair game, so that $E(X) = 0$.

Show that $E(X) = 0$.

6 Using the probability distribution in question **3**, invent a biased (unfair) game, so that $E(X) \neq 0$.

Show that $E(X) \neq 0$.

Case study

The Monty Hall Problem

There is a classic probability problem that academics have argued about. Can you work it out?

A friend of yours challenges you to win a car. They put you in front of three doors. Behind one door is a car and behind each of the other two is a goat.

You choose a door and your friend, who knows what's behind the other two doors, opens one of them to reveal a goat.

Your friend then asks you whether you want to stick with your first choice or swap to the other unopened door.

What do you do and why?

If you don't understand the problem or you are not sure whether to swap, use the internet to help you.

Project work

What is the probability that in a random group of people at least one pair will share the same birthday?

Ask your Head of Year 12 to give you a list of Year 12 birthdays. Randomly pick different sized groups from this list and see how many shared birthday pairs you find for each group.

What is the probability that in a random group of three people at least one pair will share the same birthday? What about five people? What about 12 people? What about 30 people?

Check your progress

How confident are you feeling in your level of knowledge? What do you need to practise more?

Spec reference	Learning objective	▶▷▷	▶▶▷	▶▶▶
R4.1	Understand that uncertain outcomes can be modelled as random events with estimated probabilities			
R4.2	Apply ideas of randomness, fairness and equally likely events to calculate expected outcomes			
R5.1	Understand and apply Venn diagrams and simple tree diagrams			
R6.1	Calculate the probability of combined events: both A and B, neither A nor B and either A or B (or both)			
R6.1	Understand the difference between independent and dependent events			
R7.1	Calculate the expected value of quantities such as financial loss or gain			

10 Cost benefit analysis

Cost benefit analysis is an approach to estimating strengths and weaknesses of various requirements for a business or any activity that involves some sort of risk. These requirements can be based on many things, such as health and safety, time considerations, employee satisfaction and finance.

Cost benefit analysis has two purposes. One is to determine whether a project is a sound decision or investment. The other is to provide a basis for comparing projects. It was the French engineer and economist, Jules Dupuit (1804–1866) who is credited with the creation of cost benefit analysis. This came about from an article on flood management he wrote in 1848.

Cost benefit analysis for transport investment started in the UK in 1960 with the M1 project and was later applied to London Underground's Victoria line. Since 2011 it has been the cornerstone of transport appraisal in the UK and is used by the Department for Transport.

Guns versus butter

Although you will be dealing with cost benefit analysis on a small scale in this chapter, cost benefit analysis also plays a part in the macroeconomics used by governments. The 'guns versus butter' model demonstrates the relation between what a nation invests in defence and in civilian goods. It is a simplification of national spending and the trade-off between military and consumer spending and can be a useful predictor of election success. One theory of this 'guns versus butter' model comes from the USA at the outbreak of World War I when the USA imported sodium nitrate from Chile for both its use in gunpowder and in chemical fertiliser in farming. The USA decided to manufacture it themselves and the National Defense Act of 1916 directed 'the Secretary of Agriculture to manufacture nitrates for fertilizers in peace and munitions in war'. The news media at the time presented this as 'Guns and Butter'.

This phrase was used in Germany in 1936 when Joseph Goebbels stated (in translation) 'We can do without butter, but, despite all our love of peace, not without arms. One cannot shoot with butter but with guns.'

In 1976 the British Prime Minister, Margaret Thatcher, made a similar remark comparing the Soviets putting guns before butter but the UK putting guns below nearly everything else.

Austerity versus floods?

During the winter of 2015 there were many floods in England, causing havoc in the north, especially in Cumbria and Yorkshire. York in particular had been devastated by floods in 2000 and 2012, so people were concerned that perhaps more should have been done to prevent the area from flooding.

Discussion

Discuss the following with your peers.

How do you think the government should spend money in your area? In other areas? Is it right to spend money equally across the whole of the UK or should some areas receive more? Does your area suffer from floods? Is it enough to take temporary measures to try and avoid flood damage or should more expensive long-term solutions be sought? What considerations need to be taken into account when allocating funds?

R8: Living with uncertainty

Learning objectives

You will learn how to:

- Understand that many decisions have to be made when outcomes cannot be predicted with certainty.

How random are you? – An experiment

Being human always means that you introduce some bias, no matter how hard you try to avoid it, so you cannot replicate **random events** that arise from inanimate sources.

Random events: Events which follow no predictable pattern and whose outcome cannot be predicted.

To see this, try the following experiment.

- Imagine that you toss a coin 50 times.
- Write down a sequence of 50 possible tosses using H for heads and T for tails.
- Now do a frequency table of the lengths of runs of heads and tails: a run starts and finishes when the sequence changes from H to T or vice versa.

A sequence of twenty tosses could look like this:

HHTHTHHHTTHHHHTHHTTT.

This can be separated into runs as follows:

HH T H T HHH TT HHHH T HH TTT.

The corresponding frequency table would be as follows:

Length of run	1	2	3	4
Frequency	4	3	2	1

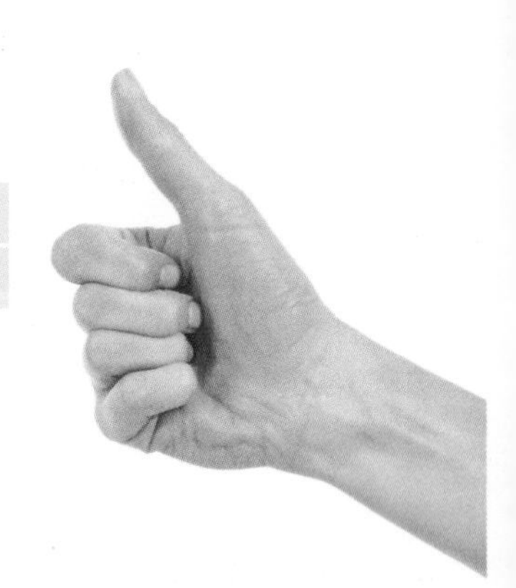

This shows there are four occurrences of a run of one side, H or T, three occurrences of a run of two, HH or TT, and so on.

Now toss a coin 50 times and record the results in the same fashion.

Make a frequency table and compare the two tables.

What do you notice? Any surprises? Discuss what you found with a partner.

Discussion

Discuss the following with your peers.

1. Take one of the quotes below and discuss the opinion expressed. Do you agree or disagree with it? Why?
   - Uncertainty and expectation are the joys of life.
     Security is an insipid thing.
     William Congreve
   - Poverty with security is better than plenty in the midst of fear and uncertainty.
     Aesop
   - The power of the lawyer is in the uncertainty of the law.
     Jeremy Bentham

- Medicine is a science of uncertainty and an art of probability.
  William Osler
- There is no such uncertainty as a sure thing.
  Robert Burns

2. Uncertainty is all around you. To some people, that is what makes life interesting. To others it is a constant worry. The quotes above come from people in many different areas.

   When has uncertainty played a major part in your life? How did you deal with it?

Dealing with uncertainty

One way of making a decision is to list the pros and cons of the situation and balance them out.

Here is an imaginary list made by a student who is trying to decide whether to buy an iPhone.

Pros	Cons
Popular	Expensive – can I afford it?
Lots of apps	Do I need them all?
Excellent camera	I have a digital camera already
Compatible with other Apple products	Not compatible with other digital hardware
Ease of use	Need to recharge the battery often

Not all the pros and cons are equally important, so it is possible to apply **weighting** to the list items.

Weighting: Weighted values are adjusted to reflect their relative importance.

This can be done by moving the dividing line one way or the other depending on what is important to them.

Pros	Cons
Popular	Expensive – can I afford it?
Lots of apps	Do I need them all?
Excellent camera	I have a digital camera already
Compatible with other Apple products	Not compatible with other digital hardware
Ease of use	Need to recharge the battery often

Now the overall picture shows that the cons outweigh the pros, so the student decides not to buy an iPhone.

Big decisions cause stress in your life – it might be what university to attend, what career to follow, who to take to the prom, what car or bike to buy, what flat to rent, etc.

Some other strategies that can help in dealing with such uncertainty are listed below.

- Advice from a friend:

  This strategy involves thinking about the advice you would give a friend if they asked you for advice about the situation that is causing you uncertainty. It helps because it tends to put you at a distance from the decision so you view it more objectively.
- KISS

  (Keep it simple, silly): Do not try to take on board too much information – it is easy to get overwhelmed by information and miss the main points because of small inconsequential detail.
- Use a spreadsheet:

  This keeps track of all the information – categorise it simply under pros, cons, qualities and include your ranking.

People who make decisions can be classified as satisfiers or maximisers. Satisfiers are generally happier because they spend less time and energy making a decision and are often not anxious about whether they made the right choice.

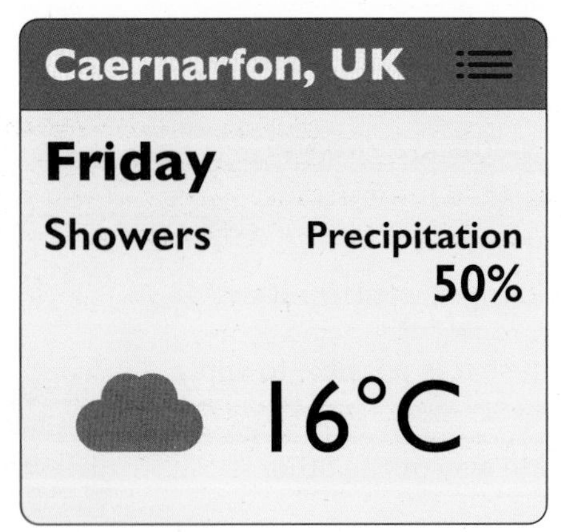

Often you cannot predict with certainty. For example the weather forecast often uses probability when predicting rain. However by considering the probabilities of various events you can make informed decisions.

Exercise 10A

For this exercise use a different strategy for each situation.

Choose from pros and cons, advice for a friend or using a spreadsheet and remember to KISS!

1. Should I go to university?
2. Should I change my mobile phone?
3. Should I move out of home and get a flat?

R9: Control measures

Learning objectives

You will learn how to:

- Understand that the actions that can be taken to reduce or prevent specific risks may have their own costs.

A **control measure** is any measure taken to eliminate or reduce risk. Risk might be injury or bodily harm to humans or other living creatures, loss or damage to property or goods, disruption in the production of goods, etc.

Control measure: any measure taken to eliminate or reduce risk.

In order to effectively implement a control measure, the likelihood and potential severity of a risk must first be assessed. After assessment, the most effective way of eliminating or reducing the risk needs to be considered. This process might involve personal judgements or consultations with other people or organisations.

You may have heard announcements at bus stations, railway stations or airports about not leaving any 'unattended luggage' or reporting 'suspicious packages' to staff. These announcements are examples of simple control measures. There can be physical control measures as well, such as fire extinguishers placed at strategic points in a building, metal detectors at airports or software devices in cars to measure exhaust emissions. If materials for control are not readily available, it should be considered whether or not they can be procured or suitable actions taken to implement controls.

A risk must be evaluated for its necessity. For example, there is always a risk in crossing the road, yet if you need to get to the other side of the road to get home you consider that risk so small that it is insignificant compared to the need to cross the road. If, however, you were going to cross the road to avoid someone with a questionnaire you would consider that not to be necessary because the risk need not be taken. Prevention of injury to humans and other living creatures is paramount and you see many safety measures taken from a simple 'STOP' shouted by a parent to a toddler who may be about to walk into the road to safety barriers on motorways and ultimately to the heroes of our world who intervene to tackle rogue people shooting innocents. If the risk is due to faulty equipment that will be eliminated once the equipment is repaired or replaced.

There are specific control measures for dangerous industries, such as those using hazardous chemicals. Some control measures used with hazardous chemicals involve the use of safety masks, gloves and other protective equipment designed to shield the body. These control measures must be implemented every day. In other industries, control measures are usually only implemented when a risk presents itself and only last while the risk is active.

Control measures can also be described in terms of mitigation. **Mitigation** describes measures or actions which will reduce the impact of an incident, so for example, in an area prone to flooding, mitigation measures can be put in place to reduce the probability of flooding (such as improved drainage) and to respond as effectively as possible if a flood does happen (such as having rescue boats available and adequate shelter and food for anyone evacuated from their homes as a result of flooding).

Mitigation: measures to control a hazard and to reduce the damage in the event of the hazard happening.

The Health and Safety Executive publish documents to raise awareness of control measures that should be undertaken in various situations. Here is an extract from 'Controlling fire and explosion risks in the workplace'.

Prioritise your control measures as follows:

- reduce the quantity of dangerous substances to a minimum;
- avoid or minimise releases of dangerous substances;
- control releases of dangerous substances at source;
- prevent the formation of an explosive atmosphere, including by ventilation;
- collect, contain and remove any releases to a safe place;
- avoid ignition sources;
- avoid adverse conditions (such as exceeding pressure/temperature limits) that could lead to danger;
- keep incompatible substances apart.

Of course, these control measures have a cost. It might be the cost of having to employ more people so that work shifts are not too long (which makes people tired and prone to error). It could be the cost of having to buy more storage space.

Discussion

Discuss the following with your peers.

Think about control measures in your school or college. What precautions are taken? Why are they important? Can you think about any that are not in place that would benefit your environment?

Discuss the costs involved in these and what else might have a cost when implementing control measures.

Exercise 10B

For each situation, first identify the risks in the situation and categorise the type of risk (injury or bodily harm to humans or other living creatures, loss or damage to property or goods, disruption in the production of goods, etc.), then list at least four control measures you think should be in place. Prioritise them so that the most important control measure appears first.

1. In your school or college chemistry laboratory
2. In the operating theatre of a large hospital
3. In a prison visiting area
4. In an aeroplane
5. In a fish market

Performing a cost benefit analysis

Cost benefit analysis is a way of calculating and comparing the costs of a project (or a decision, or a piece of work) with the benefits, or gains, of that project. In business, cost benefit analysis can be used to work out if the costs of a particular project (such as developing a new product) are appropriate to the estimated income from the project. This can help when making decisions about business investments. Cost benefit analysis can also be used to compare the relative effectiveness of different projects.

Cost benefit analysis: Cost benefit analysis is used to compare the costs of carrying out an action or a project with the benefits delivered by the action or project.

When deciding on a course of action, cost benefit analysis can be used to:

- decide whether to undertake a project
- decide which of several projects to undertake
- clarify the aims and outcomes of the project
- develop ways of measuring the success (or otherwise) of a project
- calculate what is needed (finance, time, resources, political agreement) to deliver the project.

Cost benefit analysis can be applied to most actions which involve spending money. Without realising it, you probably do a cost benefit analysis every time you decide whether to spend your own money. For example, is it worth spending £5 on a download – what is the benefit you get? It is quite hard to put a financial value on this kind of benefit.

Other benefits are more easily calculated. A factory owner might want to know if it is worth spending money on a new piece of equipment. To do this, he needs to know the cost of the equipment and the benefits he will see. The benefits might be an increase in the number of items made, so he makes more profit, or a reduction in running costs compared with his old equipment. In many cases, there will more than one benefit. In these cases, all the benefits should be calculated individually to find the overall benefit.

Cost benefit analysis also has to consider options such as:

- the cost of not doing the project
- the cost of the project if it fails.

Calculating the costs and benefits

When carrying out a cost benefit analysis, the costs should include:

- initial costs, such as spending (sometimes known as capital investment for large spending programmes)
- implementation costs – the cost of making the project happen
- ongoing costs, such as staff costs, when the project is up and running.

The benefits should include:

- direct benefits, such as improved output or productivity
- indirect benefits, for example, a new piece of machinery might be cleaner, so the atmosphere in the factory is better.

Analysis is most easily done when costs and benefits can be expressed in financial terms, as amounts of money. However, some costs and benefits are hard to express as amounts of money. These tend to be costs and benefits of environmental and health impacts. How do you put a value on clean air in a city, for example?

Car insurance

Insurance is part and parcel of everyday life. Like it or not, there are certain insurances that are compulsory. If you drive a car or motorbike on the road, it is a criminal offence not to be insured. This is to protect both you and anyone else involved in a car accident from financial loss, whether it is your fault or theirs. The cost of car insurance depends on many factors: the age of the person to be insured and the make and model of car or motorbike to name just two factors. There are three types of car (or motorbike) insurance.

Third party only

Third party insurance is the minimum level of cover required by law.

The cover provided by this type of policy is very basic. It only agrees to pay for damage caused to other people's property or compensation or costs related to injuries they sustain, if the accident is the fault of the covered driver. Third party insurance is limited, as it would not cover damage to your car or yourself, so is better suited to a motorist who drives an old car with low value and who is not bothered if the car is damaged.

Third party insurance policies are cheapest because the cover is very limited. However, in recent years many more young and newly-qualified drivers have opted for this type of policy because those who have just passed their test can find the cost of car insurance extremely expensive. Since younger drivers are statistically more likely to crash – and therefore to make a claim – insurers have put up the cost of this type of cover.

Third party, fire and theft

Third party, fire and theft insurance offers more protection than third party only. It offers the same level of cover as that offered by third party only policies, but also offers protection against loss or damage if your own car is burnt or stolen. Third party, fire and theft insurance is cheaper than fully comprehensive cover, so may be more suitable for cheaper vehicles.

Fully comprehensive

Fully comprehensive insurance offers the greatest level of cover. It includes cover for damage to your own vehicle as well as any damage suffered by others from a range of causes, including accident, fire and theft. Fully comprehensive insurance is usually quite expensive.

Comprehensive insurance can come with a number of extras, such as breakdown cover, legal expenses cover and courtesy cars, and cover to drive another car. These add-ons are often not quoted as part of the initial insurance premium and may not be available with every policy.

Discussion

Discuss the following with your peers.

What factors do you think play a part in the cost of car or motorbike insurance? How do you think insurance companies decide on the amount you have to pay (the premium)? Do you think it fair that there is supposed to be no difference in premiums between male and female drivers' insurance?

Other insurances

You can pay to be insured against almost any sort of risk. Some of the most common types of insurance that people have are listed below.

- Life insurance – This pays out a specified figure to named beneficiaries on the death of the insured.
- Personal accident insurance – This will compensate you if an event causes you disability, injury or death.
- Medical and health insurance – This covers your earnings if you fall sick or get injured to the extent that you cannot work and earn as before. It also covers the cost of medical care.

- Home insurance – This cover is to insure your home against loss or damage as a result of fire, electrical fault, plumbing malfunction, flood, etc. It is unlikely you will be allowed a mortgage without taking out home insurance.
- Travel insurance – When travelling, this cover ensures you are compensated for any loss, damage, injury, sickness or inconvenience that comes up as a result. It may cover personal accidents, hijackings, travel delays and more.
- Pet insurance – This covers vet's bills if your pet gets ill. It might also cover you against a claim if your pet attacks someone or causes damage.
- Portable electronic device insurance – This covers portable devices such as mobile phones, laptops or tablets. The cover ensures replacement or repair of such devices if they are stolen, lost or damaged.
- Professional liability insurance – This protects professionals such as doctors and architects if their clients bring negligence claims against them.
- Mortgage insurance – This cover comes to the aid of a lender if a homebuyer defaults.

Discussion

Discuss the following with your peers.

1. Discuss what insurance you have and why you have it. What insurance do you think everyone over 16 should have? Do you think your views on insurance will change in the next five years? In 20 years?
2. There are some wacky insurances that you can take out, such as alien abduction cover. Talk about some others that you might have heard of and discuss whether they are worthwhile.

Is insurance worth it?

You can get free National Health Service (NHS) treatment if you are lawfully entitled to be in the UK and usually live here. (There are some others who are eligible as well.) However, some people take out private medical insurance as this pays all – or some – of your medical bills if you are treated privately. It gives you a choice in the level of care you get and how and when it is provided.

You don't have to take out private medical insurance – but if you don't want to use the NHS, you might find it hard to pay for private treatment without insurance, especially for serious conditions. There are some things (such as organ transplants and chronic illnesses) that are not covered by some policies, so you should always consider the cost, exclusions and benefits of such insurance.

Exercise 10C

1 Investigate NHS dental charges and the cost of private dental care.

Is it worth taking out a dental plan?

2 Investigate the cost of private medical insurance.

Are you eligible?

Is it worth the cost?

3 Some stores offer breakdown insurance on various items.

Here are some kettles, their cost and the cost of a three-year breakdown insurance.

Based on the cost of the insurance, which kettle do you think is most likely to break down?

How often does your kettle break down? Is it worth insuring a kettle?

- Kettle A, cost £4.99, insurance £0.99
- Kettle B, cost £19.99, insurance £5.99
- Kettle C, cost £99.99, insurance £27.99

Is it worth it?

One way of bringing car insurance down is to have a tracker device in your car. This records its location, how carefully you drive and how you could improve your driving. By showing that you are a careful driver it can ensure lower premiums for car insurance and, if your car is stolen, it allows the police to find out where it is. A telematics box will record anything a car does whether it is the insured person behind the wheel or not. The five key areas it looks at are: cornering, swerving, braking, speed and acceleration.

Discussion

Discuss the following with your peers.

Do you think this big brother technology restricts drivers too much? What could be the benefits? What about the disadvantages? Do you think your friend should have such a device in their car? Why?

R10: Risk analysis

Learning objectives

You will learn how to:

- Use probabilities to calculate expected values of costs and benefits of decisions.
- Understand that calculating an expected value is an important part of such decision making.

Expected value

The **expected value** is the anticipated value for a given risk or investment. It is calculated by multiplying each of the possible outcomes by the likelihood that each outcome will occur, and summing all of those values.

Everyone has to carry out some sort of **risk analysis** at various points. You do it every time you cross the road by asking yourself whether it is safe to cross, especially if you see a car approaching. You wonder whether you can get across before it reaches you and in your mind you make that decision. Most of the time you will be correct.

Responsible organisations do risk analyses as part of their routine activities. Depending on the nature of the organisation, the risk analysis may be carried out annually or every few years. The case study at the end of this chapter shows one example of this.

Expected value: the anticipated value for a given risk or investment.

Risk analysis: assessment of potential risks and management of those risks.

Is it worth it?

To help businesses make decisions the probabilities of various events happening and their costs are taken into account before decisions are made. As a simple example, take the case of the kettle that costs £4.99 and the three-year breakdown insurance that costs £0.99. Is this worth taking out?

You are in a better position if you know the probability of a kettle breaking down in that period. To determine whether it is worth taking out the insurance you need to allocate a probability to the kettle breaking down. For the purposes of this example, you can allocate a probability of 0.01 – so there is a 1% chance that the kettle will fail within the thee-year period of the insurance.

If you take out the insurance, the total cost of kettle is the cost plus the insurance.

total cost with insurance = £4.99 + £0.99 = £5.98

If you do not take out the insurance the expected cost of the kettle is £4.99 plus the expected replacement cost.

The expected replacement cost is calculated by multiplying the probability of the kettle breaking down by the cost of the new kettle.

total cost without insurance = £4.99 + 0.01 × £4.99
= £4.99 + £0.05 = £5.04

So the total cost without insurance is less than the total cost with insurance, so in this case it does not seem worth taking out the insurance.

However, with more expensive items, it may become more worthwhile.

Control measures have a cost, whether they are compulsory (like car insurance) or not.

Example 1

An author has been given a fee in advance of £5000 on the understanding that £3000 has to be repaid if the book is not written by a certain deadline.

The author estimates the probability of a delay is 0.3 due to possibly having to take care of a sick relative.

As a control measure the author can employ a care attendant at a cost of £1000, but this has to be booked and paid for in advance.

What assumptions might the author make?

What will be the expected penalty if no control measure is taken?

One assumption the author might make is that the penalty for delay will be implemented and that there will be no mitigating circumstances. Another is that the author might not get future work if the deadline is not met.

Expected penalty = loss of advance × probability of delay

= £3000 × 0.3 = £900

Cost of control measure = £1000

There is little difference between the expected penalty and the cost of the control measure.

Discussion

Discuss the following with your peers.

Should the author take the control measure? Remember, the loss could be £3000 but it might be £0. Is it worth a gamble? What about the author's peace of mind and reputation?

Example 2

When a car is serviced at a garage a courtesy car is often provided for the driver to use while their car is out of use. This is covered by the garage insurance but the driver has to pay the first £500 of any claim on the insurance.

The driver has the option of paying £10 to reduce this amount to just £50.

If the probability of the driver having an accident while driving the courtesy car is 0.1, should the driver pay the £10?

What would the probability of an accident be if the penalty and cost of the control measure were the same?

Assumptions:

- The driver cannot afford to pay £500 if there is an accident.
- The driver meets the conditions of the insurance policy.

Expected penalty = loss of £500 × probability of accident

= £500 × 0.1 = £50

Cost of control measure = £10 + loss of £50 × probability of accident

= £10 + £50 × 0.1 = £15

The cost of the control measure is less than the expected penalty, so it is probably worth paying the extra.

To work out when it is worth paying for the control measure, you can calculate the probability which will give equal costs to the penalty and the control measure.

The probability of an accident is x.

Expected penalty = loss of £500 × probability of accident

= £500 × x = £500x

Cost of control measure = £10 + loss of £50 × probability of accident

= £10 + £50 × x = £(10 + 50x)

If these are equal, you can equate them and ignore the units.

$$500x = 10 + 50x$$
$$450x = 10$$
$$x = \frac{10}{450}$$
$$x = 0.022 \text{ to 3 d.p.}$$

So there is no benefit in paying the £10 if the probability of an accident is less than 0.022.

Exercise 10D

1 A cupcake baker has a monthly order worth £2000.

There is a penalty of £500 if any order is burnt.

The thermostat on the baker's oven is old and the probability of the oven overheating is 0.2.

Should the baker replace the thermostat at a cost of £80?

What if the cost was £120?

What assumptions might the baker make?

2 Your laptop is getting old and the probability of it crashing and losing all your work is 0.1.

You reckon your work is worth £3000.

The cost of a new laptop is £400. This is the control measure.

a What is your expected loss?

Should you take the control measure?

What assumptions are you making?

You decide to take this control measure.

The probability of a new laptop crashing is 0.001.

An extended warranty is available at a cost of £50 (another control measure).

b What is your expected loss if you do not take this new control measure?

Should you take it?

What assumptions are you making?

3 Your new mobile phone costs £360.

   The probability you leave it on the bus is 0.1.

   The probability someone will hand it in to lost property and you will recover it is 0.4.

   The cost of insurance to cover the loss of the phone is £30.

   Should you pay for this control measure?

   List any assumptions that you make.

4 A young driver finds that third party only insurance is £2400 and fully comprehensive insurance is £3000.

   The value of the insured car is stated as £3000.

   The probability the driver will write off the car in the first year is 0.07.

   What type of insurance should the young driver buy?

   List any assumptions that you make.

5 A carpet firm has a £50 000 contract to provide carpets for a new student residence.

   If it is not carpeted in time then the penalty is £20 000.

   The probability of any one of the two carpet fitters being sick and delaying the job is 0.05.

   Should the project manager pay £800 as a control measure to a firm that will supply carpet fitters if the need arises?

   List any assumptions that you make.

6 Think about some insurances or control measures that you have.

   What do you estimate are the probabilities of loss or damage in each case?

   Calculate the expected values of loss if you don't have these controls.

   List any assumptions that you make.

   Are the control measures worth it?

Discussion

Discuss the following with your peers.

Should you always go for minimising the maximum loss?

Think about reputations or other consequences if things go wrong.

Two (or more) control measures

With large businesses it may be necessary to have more than one control measure in place. As well as machinery malfunctioning, employees might be ill and deliveries delayed by roadworks, for example. These can all be taken into account when doing a risk analysis to determine the most economical course of action. However the most economical course might not always be the best course of action and factors such as reliability, reputation and feedback on internet websites should always play an important factor in all decisions. Taking a moral and ethical stance rather than a purely economic view can be far more rewarding in the long run.

In the questions that follow the example you will be asked to list any assumptions you make. There is no absolute right and wrong in these cases, so it is expected that you will discuss any assumptions you make with your peers and teachers.

Example 3

A vet faces a £80 000 compensation claim for delay if a horse being treated does not recover after a period of two weeks. The recovery will be delayed if either or both of two critical medicines are not effective.

- The probability of medicine A being effective is 0.8.
- The cost of using a more effective medicine than A that will definitely work is £10 000.
- The probability of medicine B being effective is 0.65.
- The cost of using a more effective medicine than B that will definitely work is £30 000.

a Work out what the expected compensation claim would be if no control measures are taken.

State any assumptions that you make.

b Which control measures, if any, would you recommend the vet to use?

Justify your recommendation.

---

a Assumptions:

- The two medicines are independent of each other.
- The horse will not become injured or harmed in any other way.

The probability of a successful treatment is found by multiplying the probabilities.

P(success) = $0.8 \times 0.65 = 0.52$

So P(delay in recovery) = $1 - 0.52 = 0.48$

Expected compensation due = £80 000 × 0.48 = £38 400

b If a more effective medicine than medicine A is the only control measure used, the total cost will be the cost of that medicine plus the expected penalty.

Expected cost of use of control measure for A
= £10 000 + (1 − 0.65) × £80 000 = £38 000

If a more effective medicine than medicine B is the only control measure used, the total cost will be the cost of that medicine plus the expected penalty.

Expected cost of use of control measure for B
= £30 000 + (1 − 0.8) × £80 000 = £46 000

If both control measures are used the total cost will just be the cost of the two medicines as there will be no expected penalty.

Cost of use of both control measures
= £30 000 + £10 000 = £40 000

In this case the recommendation is to use both control measures.

Using the control measure for B only is pointless as it is the most expensive by far. There is little difference in cost between using the control measure for A and using both control measures, so on balance it would probably be best to use both control measures: it is only £2000 more than using the control measure for A alone and this would guarantee no delay, so the vet's reputation would remain intact.

Exercise 10E

1 A courier firm faces a £2000 penalty if 4000 magazines are not all delivered on time to subscribers.

The magazines will be delayed if one or both of the following events happen.

- A – a courier falls ill.
- B – a delivery van breaks down.

The probability of event A happening is 0.2.

The probability of event B happening is 0.1.

The control measures available to the courier firm are to reserve a driver at a cost of £300 or reserve a delivery van at a cost of £150. The firm can reserve both at a discount price of £425.

a Work out what the penalty would be if no control measures are taken.
State any assumptions that you make.

b Which control measures, if any, would you recommend the firm to take?
Justify your recommendation.

2 An IT manager looks after the computers in a university.

An organisation requires the use of the PC suite for a conference and is charged £50 000. They insist that a penalty of £30 000 is paid if any of the facilities malfunction.

This penalty is paid if

- any computer fails (with probability 0.15)
- the internet goes down (with probability 0.2).

The control measures and their cost are:

- provision of back-up computers: £2000
- internet dongles: £3000

a Work out what the penalty would be if no control measures are taken.
State any assumptions that you make.

b Which control measures, if any, would you recommend the IT manager to take?
Justify your recommendation and state the expected profit based on it.

3 A video game developer sells a new game called *Cor: moths!* to a company for £500 000.

However if the game is pirated or has a bug, the developer has to pay back a penalty of £300 000.

- The probability the game is pirated is 0.7.
- The probability the game has a bug is 0.2.

The control measures and their cost are:

- Provision of an embedded code to stop pirating the game at a cost of £30 000.
- A thorough test for bugs at a cost of £10 000.

a Work out what the expected penalty would be if no control measures are taken.

State any assumptions that you make.

b Which control measures, if any, would you recommend the developer to take?

Justify your recommendation and state the expected profit based on it.

4 An entrepreneur decides to invest £60 000 in buying a container full of mobile phones that should sell for a total of £100 000.

However there is a probability of 0.1 that they will get damaged and become unsalable on route from Asia and the probability of theft is 0.15.

There are two control measures the entrepreneur can take.

- Pay £10 000 for them to be bubble-wrapped so no damage occurs.
- Pay a team of security guards £20 000 to ensure they are not stolen.

a Work out what the expected profit would be if no control measures are taken.

State any assumptions that you make.

b Which control measures, if any, would you recommend the entrepreneur to take?

Justify your recommendation and state the expected profit based on it.

5 Use a spreadsheet for this question.

A removal firm has to pay a penalty of £1000 to a family that moves house if the removal is delayed or any goods are broken in transit.

The probability that the removal is delayed is twice the probability that any goods are broken.

There are two control measures the removal firm can take.

- A: Employ extra people to ensure no delay. This will cost £200.
- B: Pay a specialist packer £150 to guarantee no breakages.

a Set up a spreadsheet with column headings as shown.

	A	B	C	D	E	F	G
1	P(goods broken)	P(removal delayed)	P(penalty if no control measures taken)	Expected penalty	Expected cost if control measure A taken	Expected cost if control measure B taken	Expected cost if both control measures taken
2							

Pick a value for P(goods broken) such that 0.01 < P(goods broken) < 0.3 and use the spreadsheet to work out what the expected profit would be if no control measures are taken.

State any assumptions that you make.

b Use your spreadsheet to investigate the penalty and costs when the probabilities change.

c Calculate the probabilities when the expected cost if control measure A is taken is the same as the expected cost if control measure B is taken?

d Calculate the probabilities when the expected penalty with no control measures is the same as the expected cost if both control measures are taken?

Case study

Here is an example of a risk analysis of a moderately sized charitable organisation.

Note that the probabilities are estimated (some places prefer to just put high, medium or low rather than to allocate numbers). The penalties are not listed since they depend on too many unquantifiable variables, but the impact is stated instead. The dates are the dates when that risk was last examined.

The word 'mitigation' is used here rather than 'control measure'.

Risk ID	Date	Impact	Probability	Details	Mitigation	Mitigation status
1	4/9/15	High	0.8	Loss of key personnel (including turnover and illness), or inability to replace staff	Annual staff reviews Contingency plans HR policies	Ongoing
2	4/9/15	High	0.4	Failure of computer systems or loss of files (including resource files)	Back-up of server held off-site Resources held by department chairs	Backed up daily
3	17/12/14	High	<0.1	Organisation sued (by students, schools, clients or staff)	Insurance policies: £2m cover for professional indemnity £10m cover for public liability £20m cover for employers' liability	Reviewed annually
4	9/10/14	Medium	0.2	Poor customer service arising from inadequate working practices	Reviewed as complaints arise	Ongoing
5	10/12/12	High	<0.001	Disaster makes premises unusable	£10m building insurance in place	Reviewed annually
6	3/1/15	Medium	<0.1	Website hacked or goes down	Service providers come highly recommended Loss of resources in the short term would affect our reputation	Ongoing
7	5/4/15	Low	<0.001	Failure to meet legal obligations	Long-term relationship with our solicitors who look at and advise on all contracts	Ongoing

The risk analysis has identified the impact and the probabilities for a range of risks. With a comprehensive risk analysis, this organisation can feel confident that it has taken all the precautions necessary to minimise the risk. The costs to the organisation are the insurances, solicitors and time involved in looking after the human resource side.

Project work

Project work – 1: hand games

Odds or Evens is a game for two players. One is called 'odds', the other 'evens'. Each player holds one hand out in front of them and on the count of 'one – two – three – shoot', shows one or two fingers. If the total number of fingers shown is odd, 'odds' wins. If the total is even, 'evens' wins.

Morra is a hand game that dates back to the ancient Romans and Greeks. The most popular version consists of all players (two, three or more) throwing out a single hand, each showing zero to five fingers, and calling out loud their guess at what the sum of all the fingers shown will be. All players who guess the sum earn one point. The first player to reach three points wins the game.

This project can be a short one by analysing just the Odds or Evens game, explaining the outcomes and why there is never a tie.

It can be a medium one by dealing with both Odds or Evens and Morra.

It can be a long one by extending the medium one to include research into other hand games.

Project work – 2: insurance

Find out the cost of insuring your belongings at home and at university.

Think about whether it is cheaper to take a risk and not insure or pay an insurance premium and feel safer.

Check your progress

How confident are you feeling in your level of knowledge? What do you need to practise more?

Spec reference	Learning objective			
R8.1	Understand that many decisions have to be made when outcomes cannot be predicted with certainty			
R9.1	Understand that the actions that can be taken to reduce or prevent specific risks may have their own costs			
R10.1	Use probabilities to calculate expected values of costs and benefits of decisions			
R10. 2	Understand that calculating an expected value is an important part of such decision making			

11 Graphical methods

In October 2011 the population of the world passed 7 billion (7 000 000 000) people. In 2015 China announced that it planned to relax its 'one child per couple' policy. China has an ageing population and the younger generation will have to take an increasing burden of providing support and care to the elderly. All couples will now be allowed to have two children. What impact will this have on overall world population growth? The cartoon here shows how world population growth is accelerating. But this isn't just a cartoon, it's a graph. A graph is a visual representation showing the relationship between variables. What sort of a relationship is this graph showing you?

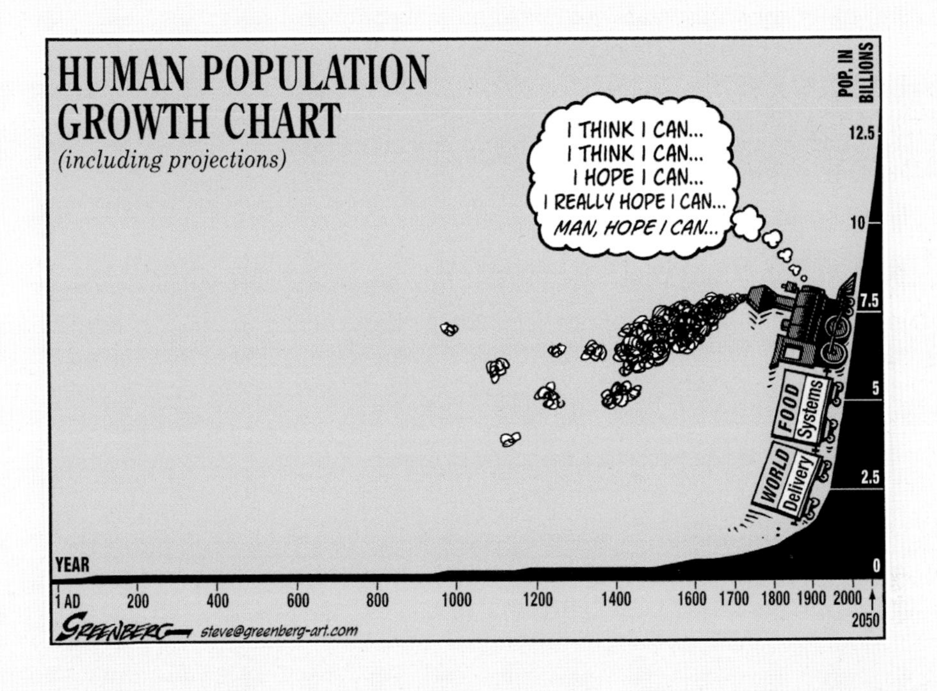

G1: Graphs of functions

Learning objectives

You will learn how to:

- Sketch and plot curves defined by simple equations.
- Know the shapes of the graphs of linear, quadratic, cubic and exponential functions.

Introduction

A student has finished her A-levels and is going to university. She has a mobile phone, which she currently only uses for texting, on a pay-as-you-go tariff. The charge for sending a text is 12p. She thinks that when she is away at university she will send fewer texts and make more calls. Her mobile phone provider charges 30p per minute for a call. How could the relationship between cost and duration of calls be presented so that the student can make a decision about staying with pay-as-you-go or buying a bundle? What sort of mathematical relationship links the cost and duration of calls in this instance?

Mathematics in the real world

You and many of your friends will celebrate your 18th birthdays during Year 13. An 18th birthday is a good excuse for a party. If you're invited to an 18th birthday party, you might have to use a taxi to get home. Different taxi firms will have different charges, based on the distance travelled and any initial call-out rate.

The graph shows a graphical representation of the fares for two different taxi companies. Understanding the graph can help you decide which taxi company to use.

Discussion

Choose one of the following and discuss it with your peers.

1. Imagine that you are going to be sponsored to run the London marathon. You want to draw a graph showing how much money you will raise if you are paid by the mile.

   Talk about what this graph will look like and the relationship between miles covered and money raised.

2. Imagine that when you were 11 years old you were told that you would get £2 per week pocket money. The next year, your pocket money was doubled, and your pocket money will keep doubling every year until you are 18.

   Talk about what this graph will look like and the relationship between your age and the amount of pocket money you will receive per week.

3. Imagine that you throw a ball up into the air and then catch it.

   Talk about what shape graph the path of the ball will make.

Linear graphs

A **linear graph** is a straight line graph. The graph of the taxi fares in the Mathematics in the real world box shows two straight line, or **linear graphs**. This means that there is a linear relationship between the miles covered and the cost of the taxi: the cost increases in direct proportion to the number of miles.

Linear graph: a straight line graph.

Another taxi firm has a different charge rate. It charges a flat rate of £6 for the initial hire and then £1.50 per mile. What would the graph for this taxi company look like?

Every linear graph can be represented with an equation.

Let y be the total cost of a journey and x be the number of miles.

The total cost is:

y = initial hire (£6.00) + (rate per mile × number of miles)

= initial hire (£6.00) + (£1.50 × number of miles)

So the equation for the cost is:

$y = 6 + 1.5x$

You can draw the graph of this in two ways.

- Make a table of values and plot the line.
- As you know that this will be a straight line graph, you can work out the axes intercepts and then plot the line.

Make a table of values

A table of values will give sets of pairs of coordinates which can be plotted on the axes.

The lowest value x can have is zero. (You might call out the taxi and then change your mind but you would still be charged the flat rate of £6.00 even though you didn't go anywhere!) You don't plan to travel in a taxi for more than 10 miles. You can now make a table of values as follows.

x	0	1	2	3	4	5	6	7	8	9	10
y											

To complete the table you need to substitute each value of x into the equation.

When $x = 0$, $C = 6 + (1.5 \times 0)$ so $y = 6$

When $x = 1$, $C = 6 + (1.5 \times 1)$ so $y = 7.5$

When $x = 2$, $C = 6 + (1.5 \times 2)$ so $y = 9$

And so on until you have completed the table.

x	0	1	2	3	4	5	6	7	8	9	10
y	6	7.5	9	10.5	12	13.5	15	16.5	18	19.5	21

You then plot corresponding values of x and y as pairs of coordinates: (0, 6), (1, 7.5), (2, 9), etc.

The plotted points are shown in the graph and are joined with a straight line.

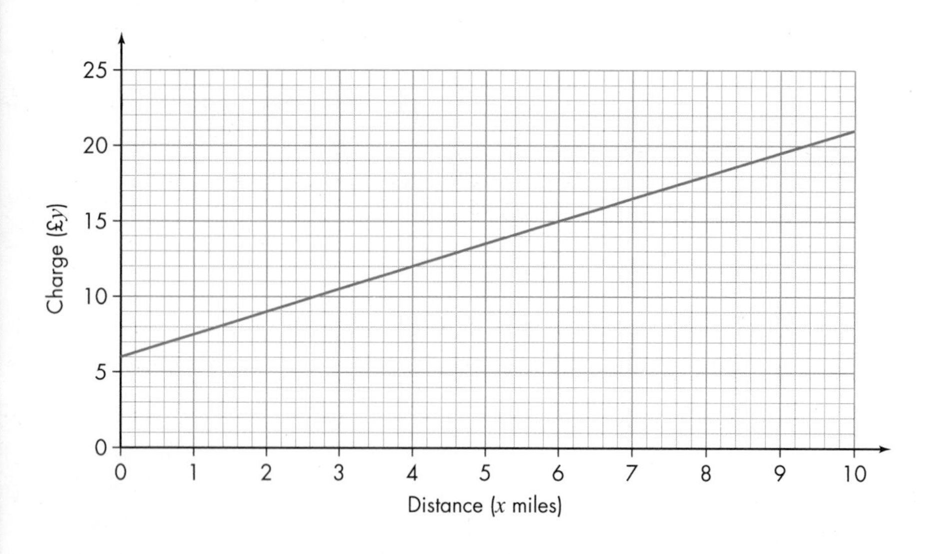

Work out the axes intercepts

An **axes intercept** is the point where the graph meets or crosses one of the coordinate axes. In this example this means when the graph crosses the x-axis (horizontal) or the y-axis (vertical).

Axes intercept: where a graph meets or crosses one of the coordinate axes.

When the graph crosses the y-axis, $x = 0$.

Substituting $x = 0$ into the equation gives:

$$y = 6 + (1.5 \times 0)$$

So $y = 6$

The graph will intercept the y-axis at (0, 6).

When the graph crosses the x-axis, $y = 0$.

Substituting $y = 0$ into the equation gives:

$$0 = 6 + 1.5x$$

Rearranging gives

$$1.5x = -6$$

So $x = -4$

Although you cannot have a negative number of miles, the equation is still valid to find that the point where the graph intercepts the x-axis is (–4, 0).

Joining two pairs of coordinates will always result in a straight line and you might have made a mistake. When you draw a straight line graph, you should always plot a minimum of three pairs of coordinates.

If you substitute $x = 2$ into the equation, then $y = 6 + (1.5 \times 2) = 9$. So also plot (2, 9).

After joining up the points using a ruler, the graph looks as shown.

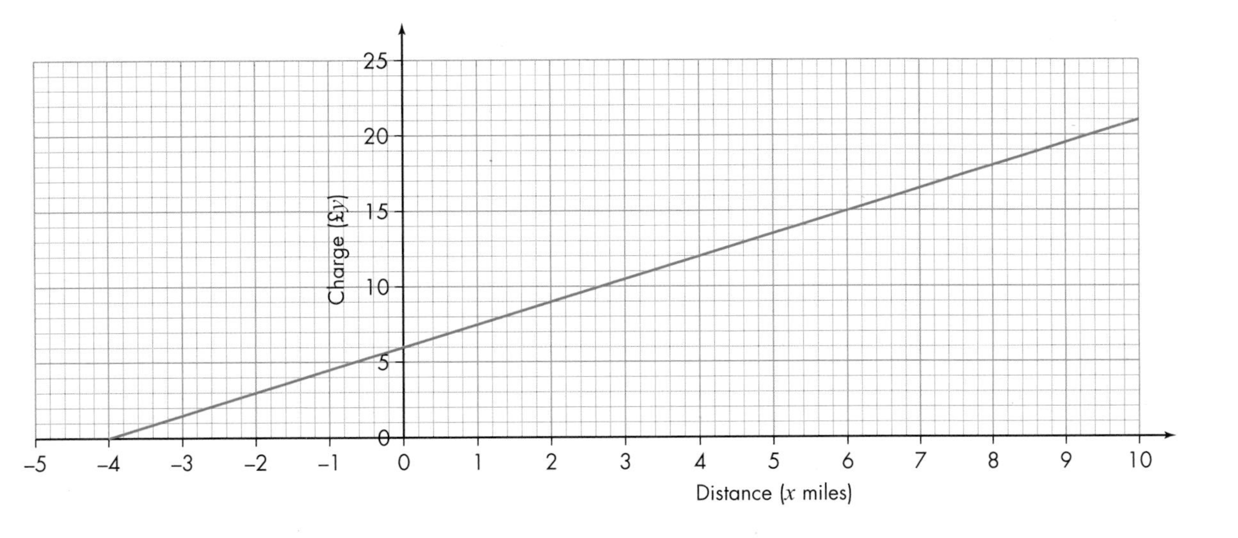

The two different methods give the same outcome, although using the axes intercept method you have to use negative values of x when in reality you can't have a negative number of miles.

Exercise 11A

1 Your friends are sponsoring you to run the London Marathon (26.2 miles).

You want to draw a graph showing how much money you will raise if you are paid by the mile.

A friend has agreed to pay you 50p per whole mile (hint: convert 50p to £) and will give you £10 for just starting the race.

Form an equation for this information, make a table and plot the graph for this equation.

2 Complete a table of values and subsequently draw the graph of $y = 3x - 2$ for $-5 \leqslant x \leqslant 5$.

3 A plumber charges his clients a £50 call-out fee then £30 per hour thereafter.

Another plumber doesn't have a call-out fee but charges £50 per hour.

Form an equation for each plumber's charges, complete a table of values for each equation and on the same axes draw the two graphs.

Which plumber would you employ to fix a dripping tap and which would you employ to install a new bathroom? Justify your answers.

4 Complete a table of values and subsequently draw the graph of $y = \frac{4 - x}{2}$ for $-5 \leqslant x \leqslant 5$.

Compare this graph to the one you drew in question **2**. What do you notice?

5 A mathematics student has completed the table of values below for the equation $y = 3 - 2x$.

Use this table of values to draw the graph of $y = 3 - 2x$.

Can you spot what has gone wrong?

Correct the student's mistakes by redoing the table of values and plotting the graph again.

x	–4	–3	–2	–1	0	1	2	3	4
y	–5	–3	–1	1	3	1	–1	–1	–5

Quadratic graphs

A **quadratic graph** is a graph of a **quadratic function**. A quadratic function always has a term in x^2. You can write a general quadratic function as:

$y = ax^2 + bx + c$

where a, b and c can take any value except a cannot be 0.

Here are some examples of quadratic functions.

Quadratic function: a function with a term in x^2 and the general form $y = ax^2 + bx + c$

Quadratic graph: a graph of a quadratic function.

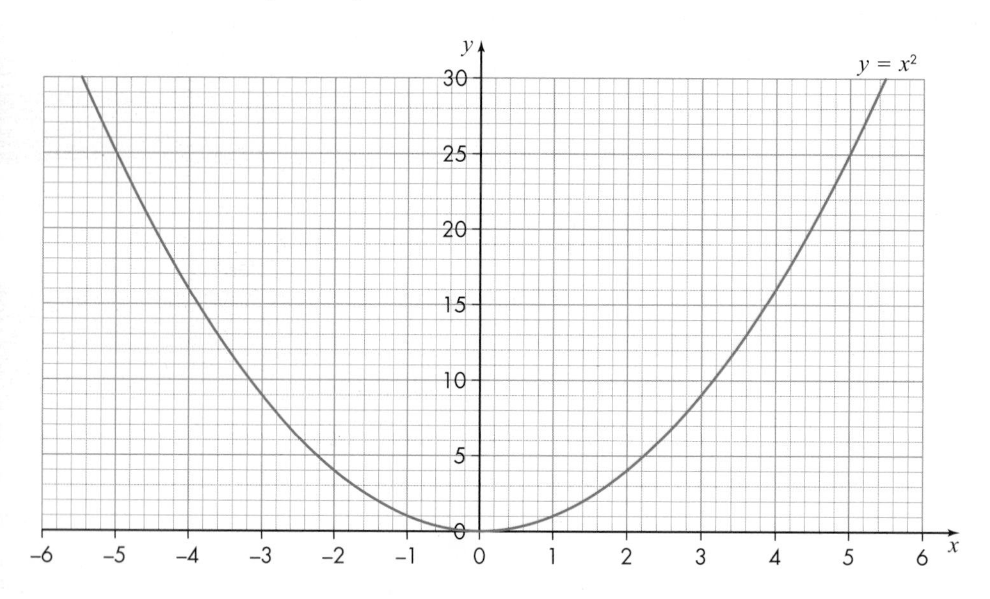

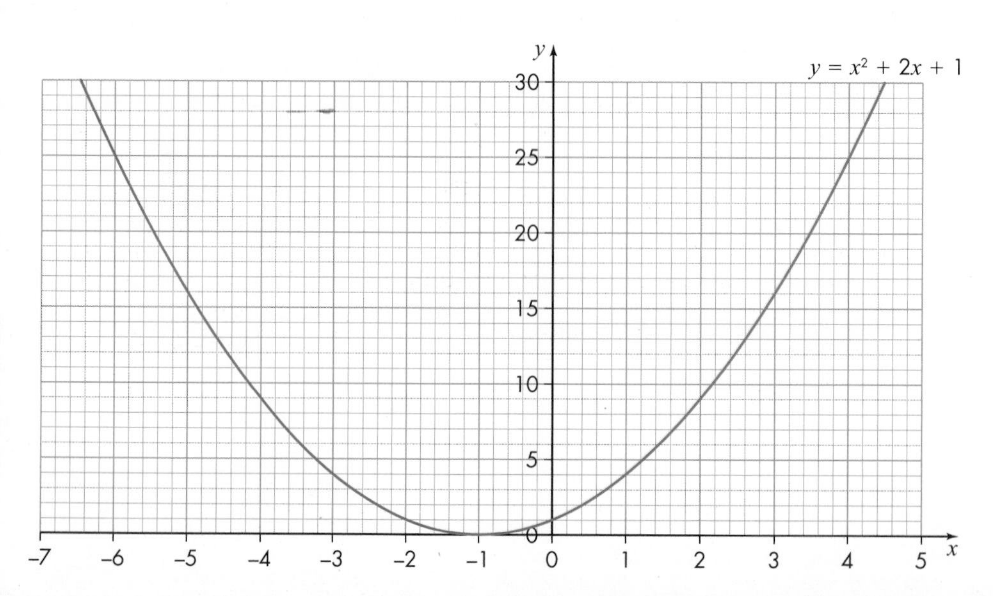

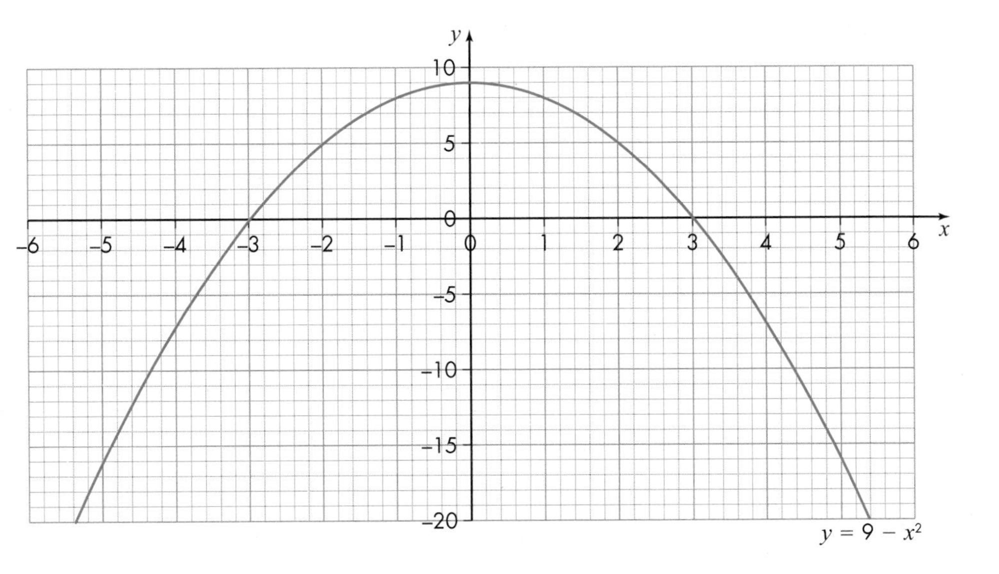

Quadratic graphs have one **turning point**. This is either at the top of the arch or the bottom of the arch (if it is upside down). In the case of the graph of $y = x^2$ the coordinates of the turning point are (0, 0). This turning point is a **minimum point** where the gradient changes from negative to positive as the curve is travelled from left to right. This is explored in more detail in Chapter 12.

Turning point: a point on a curve is where the gradient is zero.

Maximum point: a turning point where the gradient changes from positive to negative as the curve is travelled from left to right.

Minimum point: a turning point where the gradient changes from negative to positive as the curve is travelled from left to right.

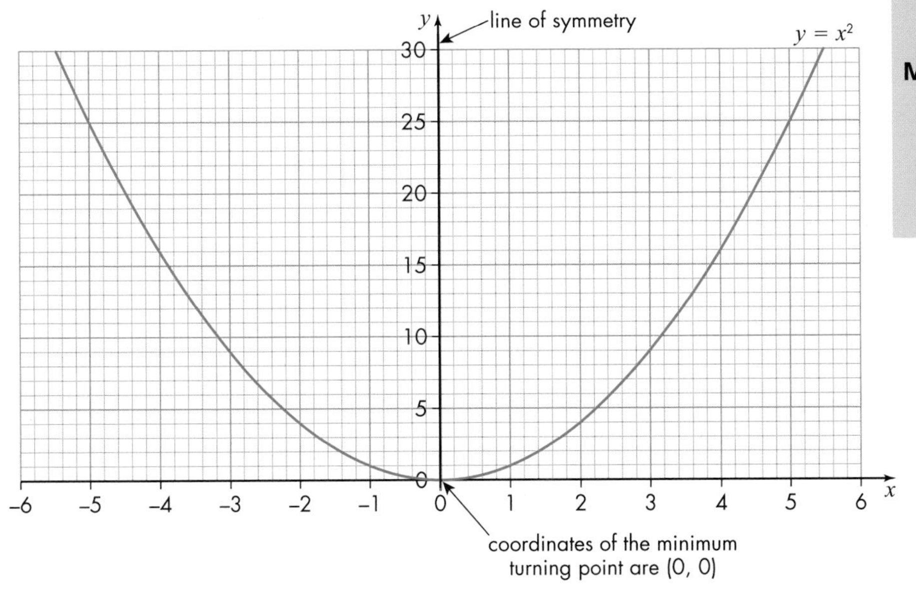

The shape of a quadratic graph is always a **parabola**, which is a special curve shaped like an arch (it can be upside down). A parabola has a line of symmetry through the turning point. So in the case of the graph of $y = x^2$ the line of symmetry goes vertically through the turning point at (0, 0) and is actually the y-axis.

Parabola: a special curve shaped like an arch (it can be upside down) with a line of symmetry that goes vertically through the turning point.

When a ball is thrown into the air, the path it follows is a parabola. The path could be modelled as a quadratic graph such as:

$$y = -x^2 + 4x + 5$$

To draw the graph of this function you make a table of values. The table will give pairs of coordinates, which can be plotted to give the quadratic graph.

When finding values of quadratic functions it helps to split the quadratic function into parts and work out each bit separately and then add them up at the end.

x	−1	0	1	2	3	4	5
$-x^2$	−1	0	−1	−4	−9	−16	−25
$+4x$	$(4 \times -1) = -4$	$(4 \times 0) = 0$	$(4 \times 1) = 4$	$(4 \times 2) = 8$	$(4 \times 3) = 12$	$(4 \times 4) = 16$	$(4 \times 5) = 20$
$+5$	+5	+5	+5	+5	+5	+5	+5
y	0	5	8	9	8	5	0

You then use corresponding values of x and y as pairs of coordinates:

(−1, 0), (0, 5), (1, 8), (2, 9), (3, 8), (4, 5), (5, 0).

You plot the points and join them with a smooth curve.

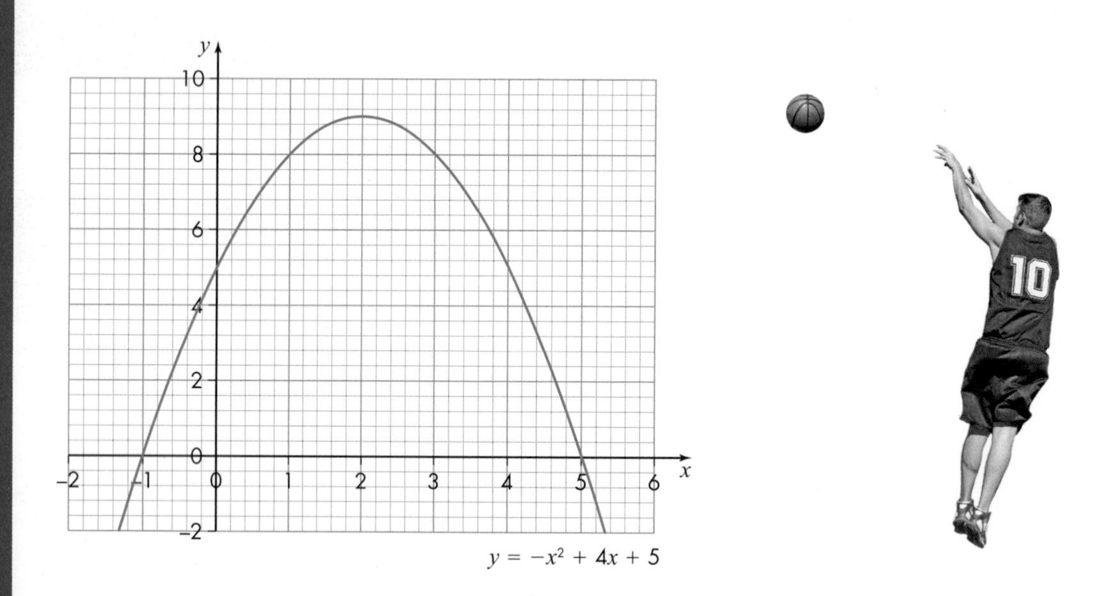

Cubic graphs

A **cubic graph** is a graph of a **cubic function**. A cubic function always has a term in x^3. You can write a general cubic function as:

$$y = ax^3 + bx^2 + cx + d$$

where a, b, c and d can take any value except a cannot be 0.

Cubic function: a function with a term in x^3 and the general form $y = ax^3 + bx^2 + cx + d$

Cubic graph: a graph of a cubic function.

Here are some examples of cubic functions.

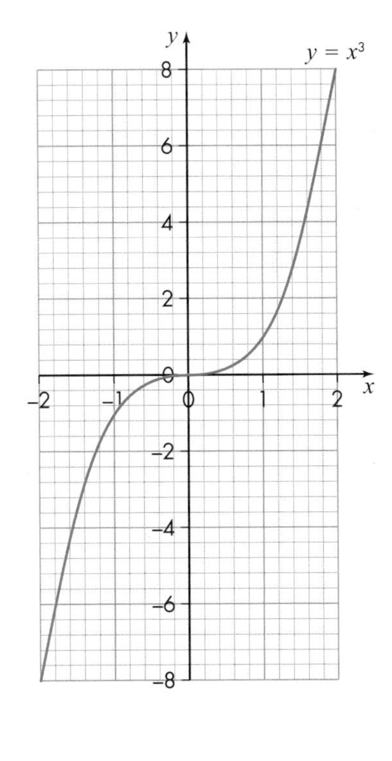

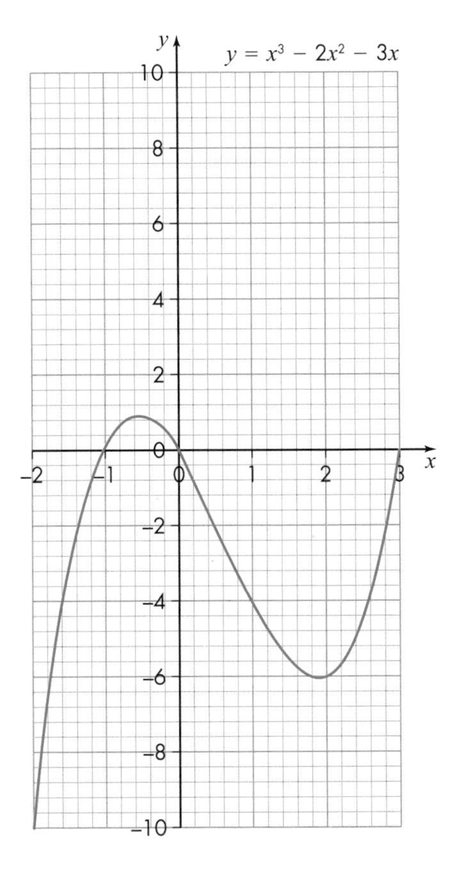

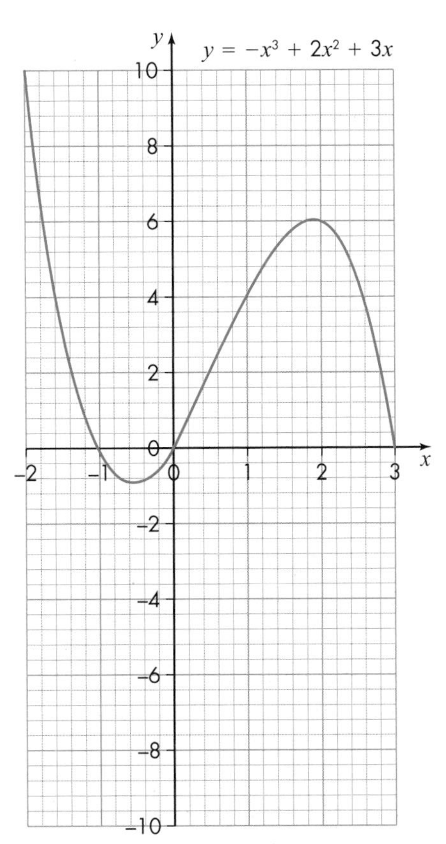

You can draw a cubic graph in the same way you draw a quadratic graph, by making a table of values. For example, if the equation of the curve is:

$y = x^3 - 8$,

the table below shows the coordinate pairs.

x	−3	−2	−1	0	1	2	3
x^3	−27	−8	−1	0	1	8	27
−8	−8	−8	−8	−8	−8	−8	−8
y	−35	−16	−9	−8	−7	0	19

You then use corresponding values of x and y as pairs of coordinates:

(−3, −35), (−2, −16), (−1, −9), (0, −8), (1, −7), (2, 0), (3, 19).

You plot the points and join them with a smooth curve.

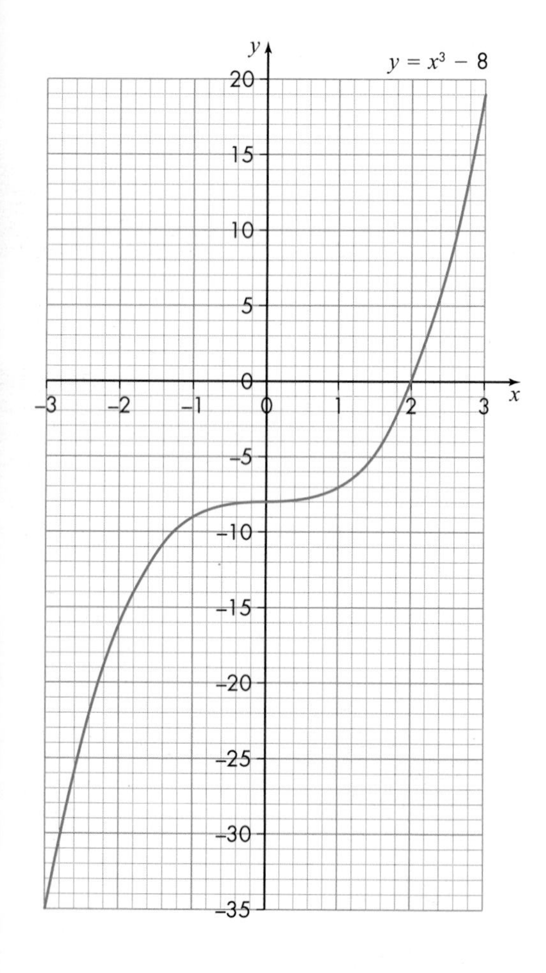

You will notice that at (0, −8) the curve changes from being concave to convex. This is known as a **point of inflexion**.

If you were to go through the process above to plot the graph of $y = x^3 + 6x^2$, the graph would look like this.

Point of inflexion: a point on a curve at which the curve changes from being concave (curving downward) to convex (curving upward), or vice versa.

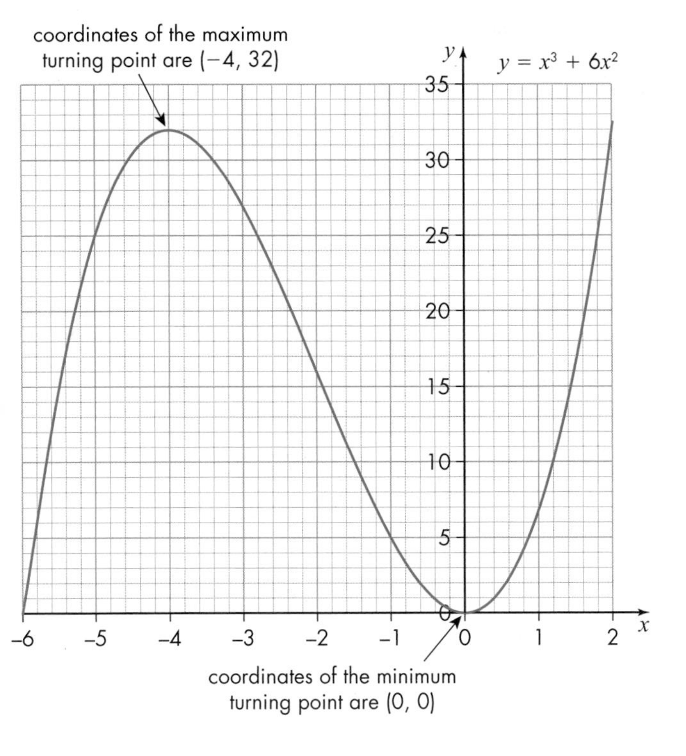

This graph has two turning points: a maximum at the top of the first arch at (–4, 32) and a minimum at the bottom of the second arch at (0, 0).

Exercise 11B

1 Complete a table of values and subsequently draw the graph of $y = x^2 - 5x + 6$ for $-1 \leqslant x \leqslant 6$.

2 A footballer kicks a ball from the ground into the air. The path the ball follows is given by the equation $y = -x^2 + 5x - 6$.

By making a table of values for $-1 \leqslant x \leqslant 6$ and drawing a graph, plot the path of the ball.

What is the maximum height (assume metres) the ball reaches?

Compare your graph to the one you drew in question **1**. What do you notice?

3 Complete a table of values and subsequently draw the graph of $y = x^3$ for $-5 \leqslant x \leqslant 5$.

What are the coordinates of the point of inflexion?

4 Complete a table of values and subsequently draw the graph of $y = x^3 - x$ for $-5 \leqslant x \leqslant 5$.

5 Complete a table of values and, on the same pair of axes as question **4**, draw the graph of $y = x - x^3$ for $-5 \leqslant x \leqslant 5$.

What do you notice?

Exponential graphs

An **exponential graph** is a graph of an **exponential function**. An exponential function is one where x, or whatever the input variable is, appears as an exponent (a power).

Exponential function: a function where the input variable appears as an exponent.

Exponent: another name for a power or index.

Exponential graph: a graph of an exponential function.

Here are some examples of exponential functions.

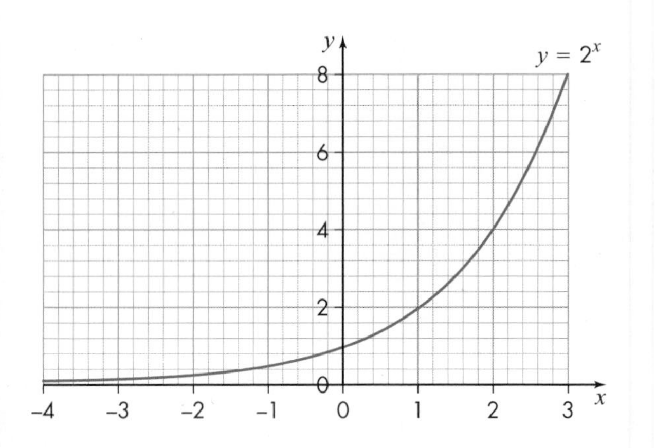

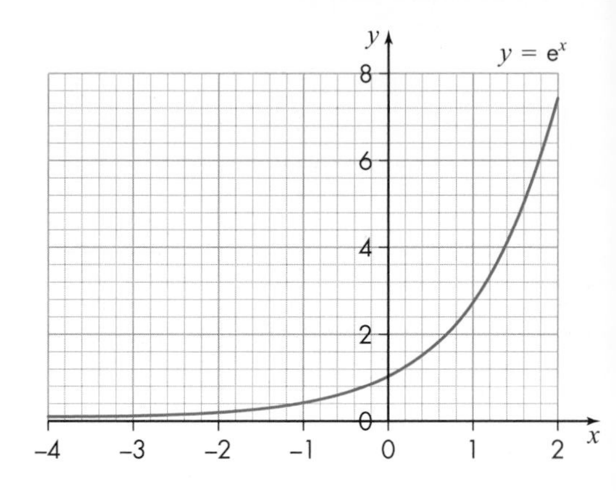

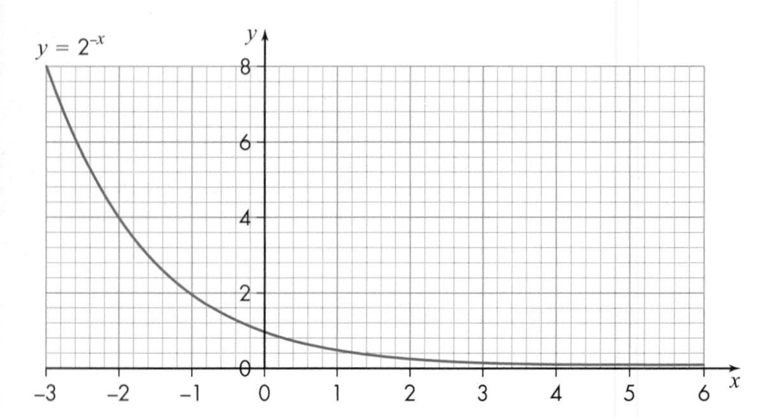

The pocket money discussion at the beginning of this chapter is an example of an exponential function. You can use a table to calculate the weekly pocket money as you grow up.

Age	Weekly pocket money
11	2
12	4
13	8
14	16
15	32
16	64
17	128
18	256

This is an example of an exponential function. Can you see why? The multiplier to go from one pocket money value to the next is 2. When you are 11 the weekly pocket money you will receive is £2 which you can write as 2^1. When you are 12 you will receive double the amount you received when you were 11, so £4, which you can write as 2^2 and so on.

Age	Weekly pocket money	
11	2	2^1
12	4	2^2
13	8	2^3
14	16	2^4
15	32	2^5
16	64	2^6
17	128	2^7
18	256	2^8

The equation of the function is $y = 2^x$, an exponential function.

To draw an exponntial graph, you have two options.

- Make a table of values and plot the curve.
- As you know that this is an exponential graph, you can work out the asymptote and the axes intercepts and then sketch the curve.

Make a table

For the function $y = 2^x$, the table shows the pairs of values.

x	0	1	2	3	4	5	6	7	8
$y = 2^x$	$2^0 = 1$	$2^1 = 2$	$2^2 = 4$	$2^3 = 8$	$2^4 = 16$	$2^5 = 32$	$2^6 = 64$	$2^7 = 128$	$2^8 = 256$

You then use corresponding values of x and y as pairs of coordinates:

(0, 1), (1, 2), (2, 4), (3, 8), (4, 16), (5, 32), (6, 64), (7, 128), (8, 256).

You plot the points and join them with a smooth curve.

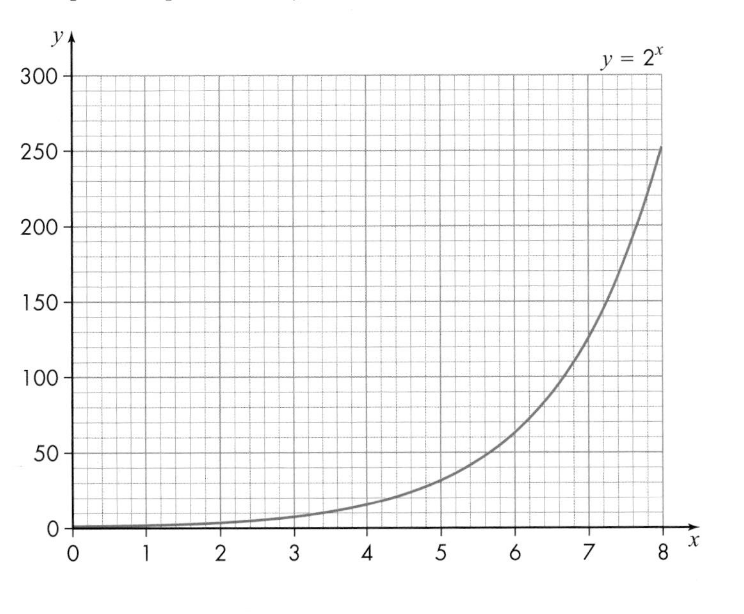

With the pocket money example above, you didn't plot any points for negative values of x. You need to examine what happens when x is less than zero.

x	−4	−3	−2	−1	0
$y = 2^x$	$2^{-4} = \frac{1}{16}$	$2^{-3} = \frac{1}{8}$	$2^{-2} = \frac{1}{4}$	$2^{-1} = \frac{1}{2}$	$2^0 = 1$

Remember that a negative power is the same as the reciprocal:

$$2^{-4} = \frac{1}{2^4} = \frac{1}{16}$$

Is there any value of x where 2^x will equal zero? The answer is no. As x gets smaller, 2^x also gets smaller, but it never reaches zero.

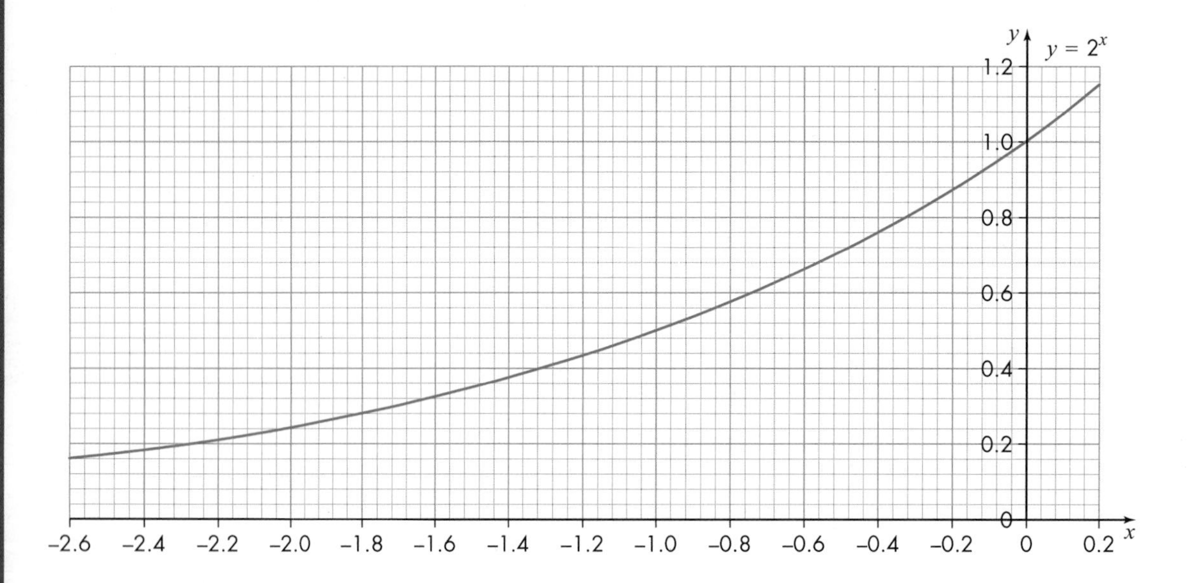

The graph is approaching the x-axis but never actually reaches it. You say that the x-axis is an **asymptote** to the curve.

Asymptote: a line (or curve) which approaches a curve closely but never touches it.

Work out the asymptote and axes intercepts

You have already learned that an axis intercept is where the graph meets or crosses the x-axis or the y-axis.

You have already seen that as x gets smaller, 2^x also gets smaller, but it never reaches zero. The x-axis is an asymptote to the curve.

When the graph crosses the y-axis $x = 0$.

Substituting $x = 0$ into the equation $y = 2^x$ gives:

$$y = 2^0 = 1$$

The graph intercepts the y-axis at (0, 1).

You then need to decide which way the graph will curve.

When x is large and positive what will happen to y?

You have $y = 2^x$. If x is large and positive then y will be large and positive. For example if x is 1 000 000 then y will be very large indeed. Try this on your calculator.

When x is large and negative what will happen to y?

You have $y = 2^x$. If x is large and negative then y will be small and positive. For example if x is –1 000 000 then y will be very small indeed. Try this on your calculator.

The graph looks like this.

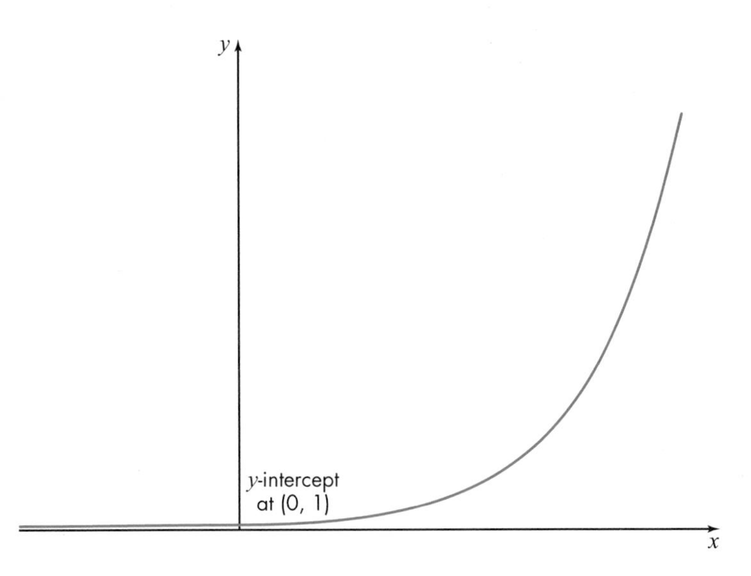

The two different methods have the same outcome except that the second graph does not have the same degree of accuracy as the first and is only a sketch.

Exercise 11C

1 Match the descriptions to the graphs.

a linear **b** quadratic **c** cubic

d exponential **e** none of these

i

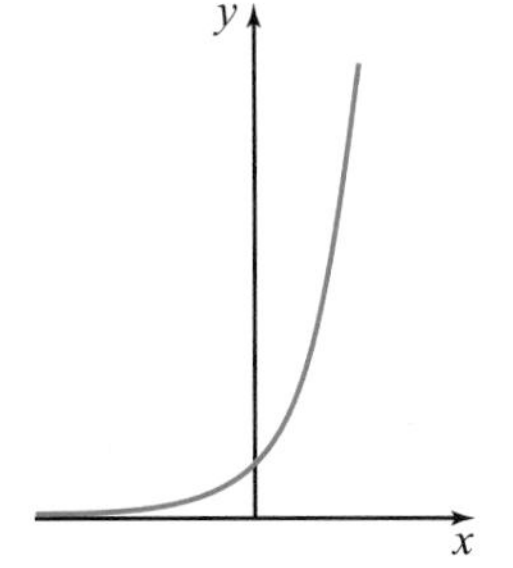

ii

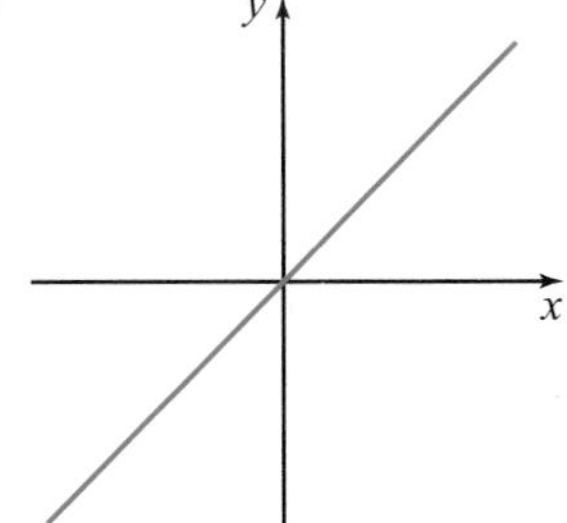

iii

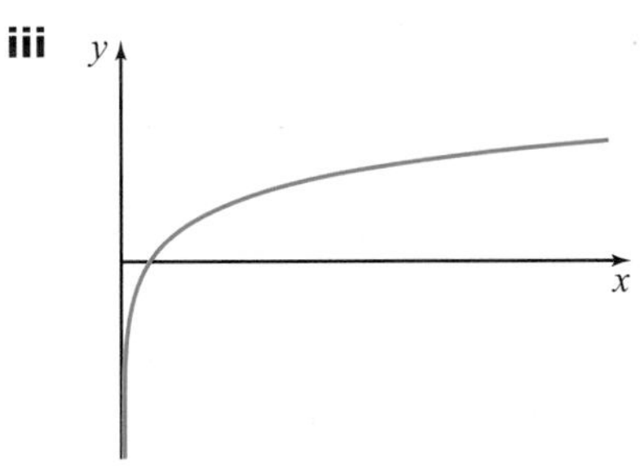

iii

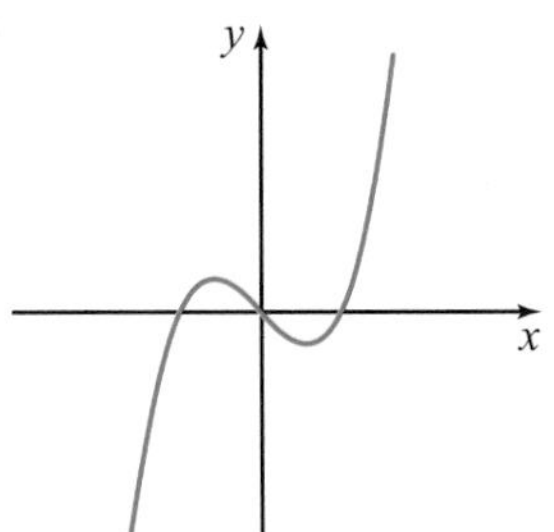

iv

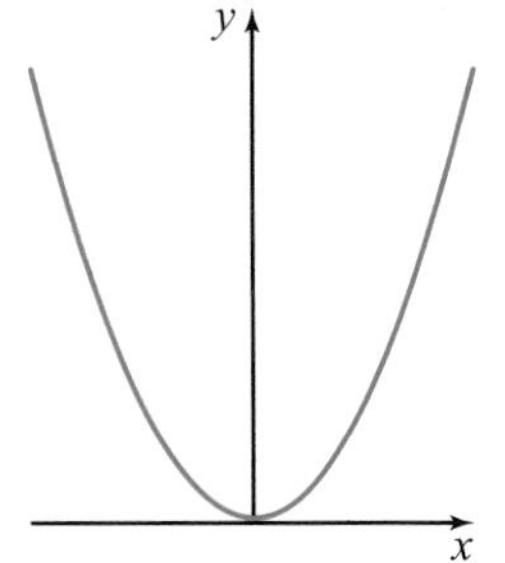

2 At the start of this chapter you saw a graph showing a train trying to travel up a very steep hill.

What type of a graph is this?

3 Sketch the graph of $y = 2^{-x}$.

How does this graph to compare to the graph of $y = 2^x$?

4 Sketch the graph of $y = 2^x + 1$.

(Hint: Think carefully about the position of the asymptote.)

How does this graph to compare to the graph of $y = 2^x$?

5 You have agreed with your parents that, from the ages of 11 to 15, you will receive an element of pocket money that will double each year with the starting amount, when you are 11, being £1.

You will also receive an additional flat rate of £3 per week. This additional amount will stay the same over the five years.

a How much pocket money will you expect to receive when you are 13?

b Taking the year you are 11 as year zero, what is the equation for your pocket money?

c Sketch a graph of this equation, carefully highlighting the position of the asymptote.

6 Exponential *growth* usually occurs when the exponent is positive and exponential *decay* occurs when the exponent is negative. This however is not always the case.

Can you think of an example where this might not be the case.

(Hint: Think about what numbers get smaller as the power that is applied to them gets bigger.)

G2: Intersection points

Learning objectives

You will learn how to:

- Plot and interpret graphs (including exponential graphs) in real contexts, to find approximate solutions to problems including understanding the potential problems of extrapolation.
- Interpret the solutions of equations as the intersection points of graphs and vice versa.

Introduction

A cruise ship is anchored off an island in the Caribbean Sea. Some of the passengers spent the night on the island. One group takes the slow boat back to the cruise ship. A second group leaves a little later and takes a faster boat back to the ship. The two groups get back to the cruise ship at the same time.

The following graph illustrates the journeys of the three different boats.

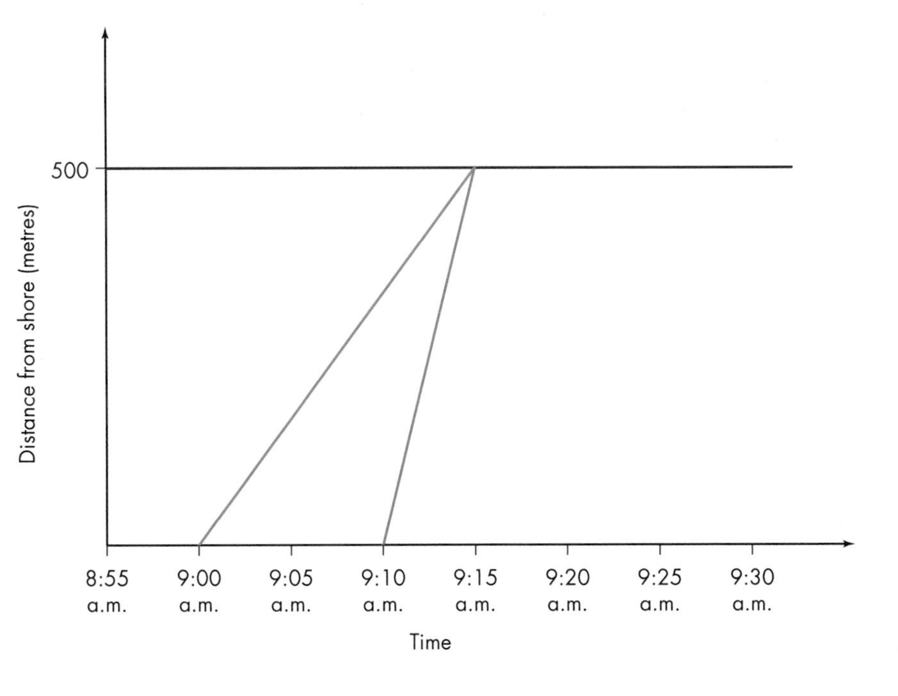

How can you use the graph to answer the following questions.

- Which boat is which?
- How far is the cruise ship from the shore?
- What is the difference in start times between the slow boat and the fast boat?
- What is the difference in speed between the slow boat and the fast boat?

In this section you are going to interpret information from graphs about real-life situations and equations.

Mathematics in the real world

In Japan there is an island that is a haven for a very large population of rabbits. It is called Rabbit Island, Okunoshima, Japan.

So what happens when a population of rabbits starts to breed and there are no predators (cats and dogs have been banned from this island)? Well mathematics has the answer.

Discussion

Choose one of the following and discuss it with your peers.

1. Imagine you plot the journeys to school of some of your friends on a graph.

   Talk about how the distance travelled and time would vary for each person.

   What would the graphs look like?

2. Imagine that a skateboarder goes up and down a curve shaped like a parabola and his path is plotted on a graph.

   Talk about how the graph could be positioned in relation to the axes and what the different intercepts might tell you.

3. Imagine you invest an amount of money at an interest rate of 20% per annum for 10 years.

   At the end of each year the interest is reinvested in the same account.

   Talk about what the graph would look like if you were to plot the amount of money in the account against the investment year.

Plot and interpret graphs in real contexts

Printer ink

Your old printer has just stopper working and sadly, it is cheaper to buy a new one rather than get the old one fixed. You have found a great deal online for the printer that you want: it was originally £100 but it has now been reduced to £50. You have also researched the cost of ink cartridges which retail at a staggering £45. You estimate that you will need four ink cartridges per year. You are going to look at how the costs for running your printer will look over 5 years.

First you need to form an equation.

Let y be the cost and x be the number of years.

So $y = 50 + 180x$

where 50 is the cost of the printer (in pounds) and 180 is the annual cost of ink cartridges (four lots of £45 per year).

Next you need to complete a table of values and then plot the graph.

x	1	2	3	4	5
y	230	410	590	770	950

Here is the graph for the cost of running this printer over a period of 5 years.

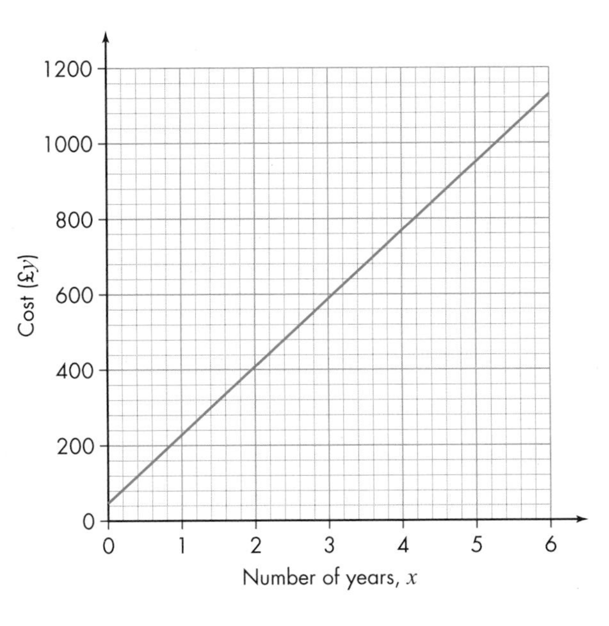

Using the graph, what questions can you answer?

Here are some examples.

- How much will the printer have cost to run, including purchase, after 2 years?
- How much will the printer have cost to run, including purchase, after 5 years?
- When will the cost of running the printer exceed £500?

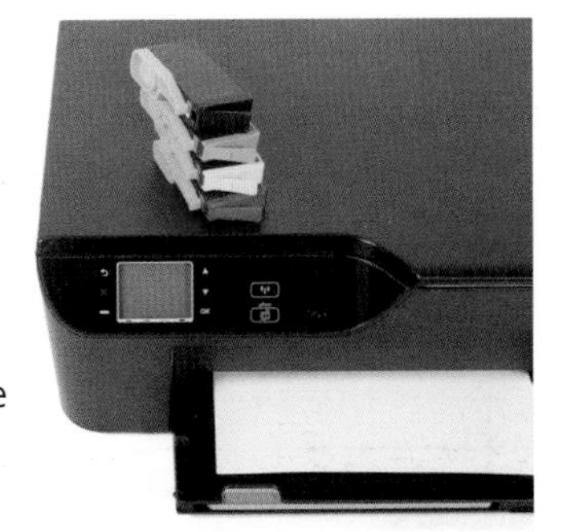

To find the cost after 2 years you need to draw a line from 2 years on the horizontal axis up to the graph then across to the vertical axis. The y value is approximately 400. So the cost of running the printer after 2 years will be approximately £400 (illustrated in red on the graph below).

To find the cost after 5 years you need to draw a line from 5 years on the horizontal axis up to the graph then across to the vertical axis. The y value is approximately 950. So the cost of running the printer after 5 years will be approximately £950 (illustrated in blue on the graph below).

To find when the cost of running the printer will exceed £500 you need to draw a line from 500 across from the vertical axis to the graph then down to the horizontal axis. The x value is approximately 2.6. So the cost of running the printer will exceed £500 after approximately 2.6 years which is around 2 years 7 months (illustrated in green on the graph below).

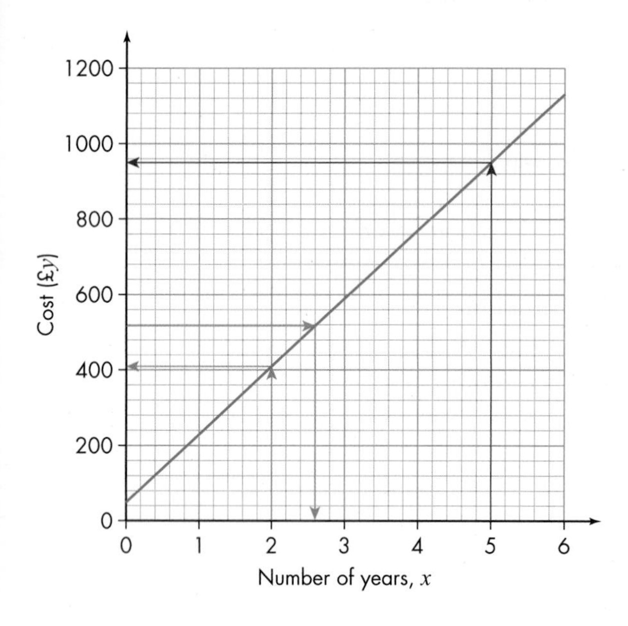

Maybe you need to buy a printer that uses cheaper ink cartridges!

New entrance for school

An architect is designing a new entrance for a school. The head teacher has asked that the design reflects something studied as part of a GCSE course. Fortunately, the architect loves parabolas and designs the arch to be part of a quadratic curve, as studied as part of GCSE Mathematics.

The equation he bases his arch on is $y = -x^2 + 6x - 5$.

The architect completes a table of values and then plots the graph.

x	0	1	2	3	4	5
$-x^2$	0	−1	−4	−9	−16	−25
$+6x$	6 × 0 = 0	6 × 1 = 6	6 × 2 = 12	6 × 3 = 18	6 × 4 = 24	6 × 5 = 30
-5	−5	−5	−5	−5	−5	−5
y	−5	0	3	4	3	0

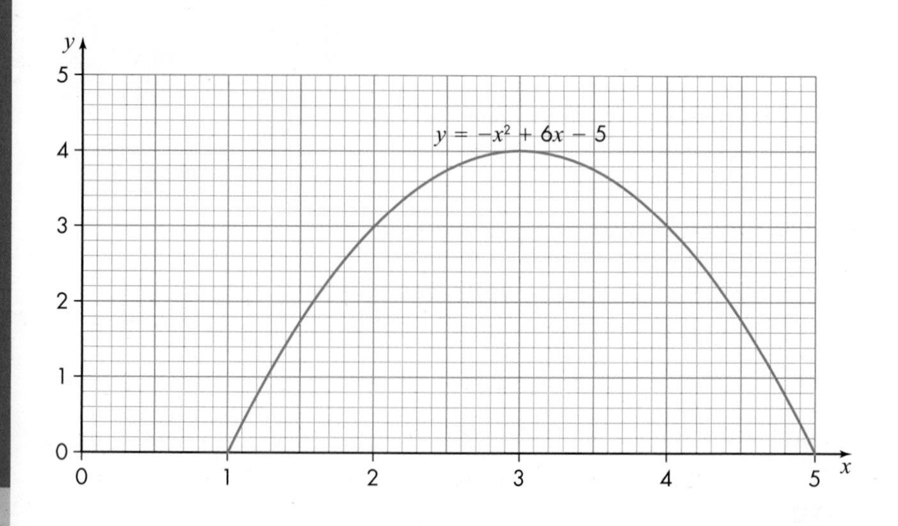

The x-intercepts, $x = 1$ and $x = 5$, are where the graph cuts the x-axis ($y = 0$).

The architect decides that the shape of the school entrance arch will be the same as the graph where y has positive values.

What questions can you answer using the graph?

Here are some examples.

- If the scale on the graph is in metres, how tall will the arch be at its highest point?
- If the scale on the graph is in metres, how wide will it be at its widest point?
- The head teacher decides she wants the whole area of the arch above 2 metres to be glass displaying the school name. How wide will the bottom edge of the glass be?

To find the highest point of the arch you need to look at the highest point on the graph. It is where y is 4 so the arch will be 4 m high (illustrated in red on the graph below).

To find the width of the arch at ground level, you need to look at the x values when $y = 0$. The x values when $y = 0$ are 1 and 5 so the arch runs from 1 m to 5 m and the arch will be 4 m wide (illustrated in blue on the graph below).

To find the width of the bottom edge of the glass you need to draw a line across the graph at $y = 2$ (a height of 2 metres) and then draw lines down to the x-axis where the line $y = 2$ meets the graph. The x values are approximately 1.6 and 4.4. So the width of the bottom edge of the glass will be $4.4 - 1.6 = 2.8$ m (illustrated in green on the graph below).

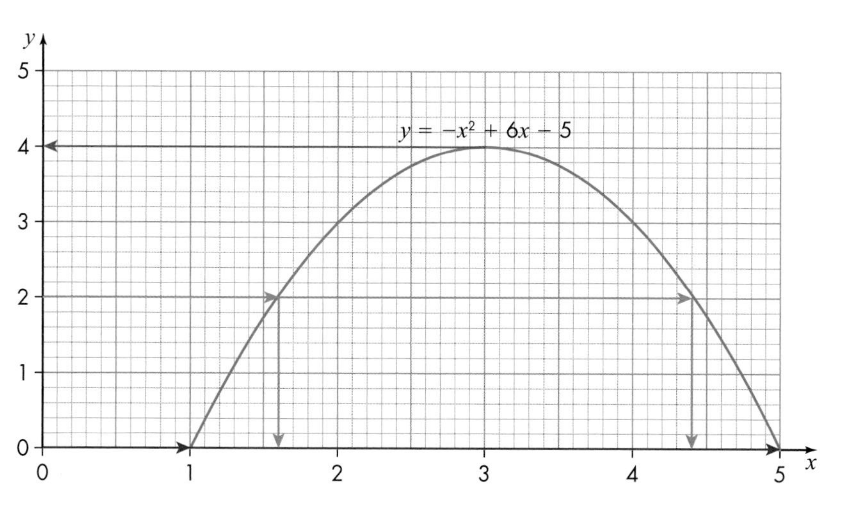

Roller coaster

A new roller coaster is being designed for a theme park. Part of the roller coaster has a section that follows part of the path of the curve $y = x^3 - 2x^2 - 3x$.

The roller coaster designer completes a table of values.

x	−1	−0.5	0	0.5	1	1.5	2	2.5	3
x^3	−1	−0.125	0	0.125	1	3.375	8	15.625	27
$-2x^2$	−2	−0.5	0	−0.5	−2	−4.5	−8	−12.5	−18
$-3x$	3	1.5	0	−1.5	−3	−4.5	−6	−7.5	−9
y	0	0.875	0	−1.875	−4	−5.625	−6	−4.375	0

The designer wants to check the shape of the roller coaster. He decides to do this by reviewing what happens when x is large and positive and when x is large and negative.

When x is large and positive what will happen to y?

You have $y = x^3 - 2x^2 - 3x$. If x is large and positive then y will be large and positive. For example if x is 1 000 000 then y will be very large. Try this on your calculator.

When x is large and negative what will happen to y?

You have $y = x^3 - 2x^2 - 3x$. If x is large and negative then y will be large and negative. For example if x is –1 000 000 then y will be very large. Try this on your calculator.

Using the table of values and the information above, the roller coaster designer plots the graph.

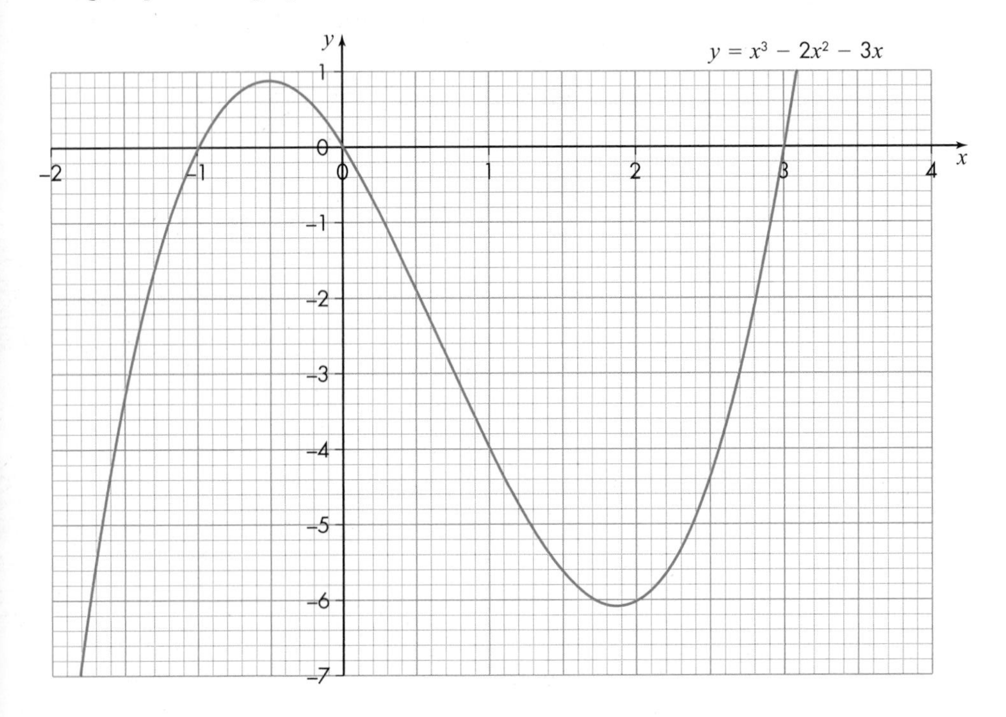

The x-intercepts, $x = -1$, $x = 0$ and $x = 3$, are where the graph cuts the x-axis ($y = 0$).

The roller coaster designer decides that part of the shape of the roller coaster will be the same as the graph from $x = -1$ to $x = 3$.

Using the graph what questions can you answer?

Here are some examples.

- What is the distance between the highest and lowest points of this part of the roller coaster?
- What is the horizontal length of this part of the roller coaster if units on the axes represent metres?
- How many times will the roller coaster be 4 m below its highest point?

To find the highest point of the roller coaster you first need to find out how high the maximum and minimum points are.

The maximum point is at approximately $x = -0.5$ and the minimum point is at approximately $x = 1.9$.

At $x = -0.5$ y is approximately 1 and at $x = 1.9$ y is approximately -6 so the distance between the highest and lowest points will be $1 + 6 = 7$ m (illustrated in red on the graph below).

To find the horizontal length of this part of the roller coaster, you need to look at the x values at the two ends. The x values are -1 at one extreme and 3 at the other extreme so the horizontal length is from -1 to 3 and so is 4 m (illustrated in blue on the graph below).

To find how many times the roller coaster will be 4 m below its highest point you need to draw a line 4 units below the highest point and then see how many times the curve crosses this line. In this case it is two. So the roller coaster will be 4 m below its highest point twice (illustrated in green on the graph below).

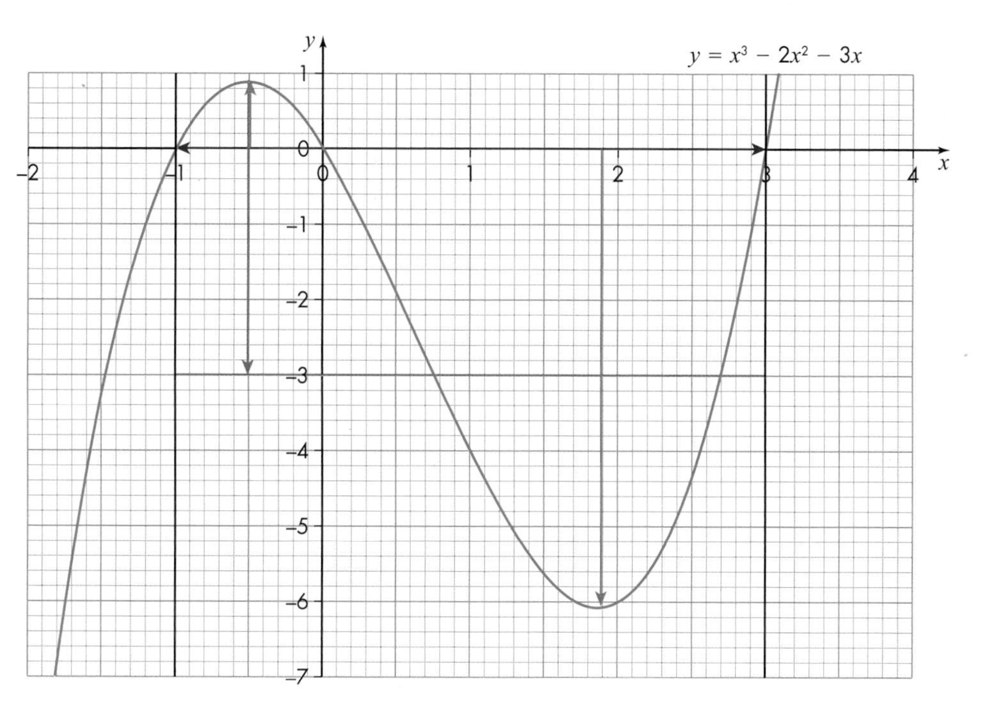

Rabbit population

A friend of yours has a breeding pair of rabbits. The rabbits have a litter of 12 baby rabbits and your friend gives you five of them. You are delighted as you just love rabbits.

However, female rabbits can have between 6 and 7 litters per year and the population for this breed of rabbit increases at a rate of 20% every 2 months.

How many rabbits will you have at the end of two years? Do you think you might need to buy an island?

		Number of rabbits
Let y_0 be the number of rabbits that you are first given:	$y_0 = 5$	5
Let y_1 be the size of your rabbit population after 2 months:	$y_1 = 5 \times 1.2$ (the initial population plus 20%)	6
Let y_2 be the size of your rabbit population after 4 months:	$y_2 = y_1 \times 1.2 = 5 \times 1.2^2 = y_0 \times 1.2^2$	7 (7.2 rounded down to the nearest whole number of rabbits)
Let y_3 be the size of your rabbit population after 6 months:	$y_3 = y_2 \times 1.2 = 5 \times 1.2^3 = y_0 \times 1.2^3$	8
Let y_4 be the size of your rabbit population after 8 months:	$y_4 = y_3 \times 1.2 = 5 \times 1.2^4 = y_0 \times 1.2^4$	10
Let y_5 be the size of your rabbit population after 10 months:	$y_5 = y_4 \times 1.2 = 5 \times 1.2^5 = y_0 \times 1.2^5$	12
Let y_6 be the size of your rabbit population after 12 months:	$y_6 = y_5 \times 1.2 = 5 \times 1.2^6 = y_0 \times 1.2^6$	14
At the end of two years, 24 months:	$y_{12} = 5 \times 1.2^{12} = y_0 \times 1.2^{12}$	44
At the end of five years, 60 months:	$y_{30} = 5 \times 1.2^{30} = y_0 \times 1.2^{30}$	1186
At the end of 10 years, 120 months:	$y_{60} = 5 \times 1.2^{60} = y_0 \times 1.2^{60}$	281 737

In reality some rabbits will breed more rapidly than others and some, sadly, will die. The numbers you have generated do not take these subtleties into consideration.

The general formula for this rabbit population growth is:

$$y_x = 5 \times 1.2^x$$

where x is the number of two-month periods.

The rabbit population growth will be slow to start but it increases rapidly.

What sort of a graph will be produced? In this case the variable is the exponent so this will be an exponential graph.

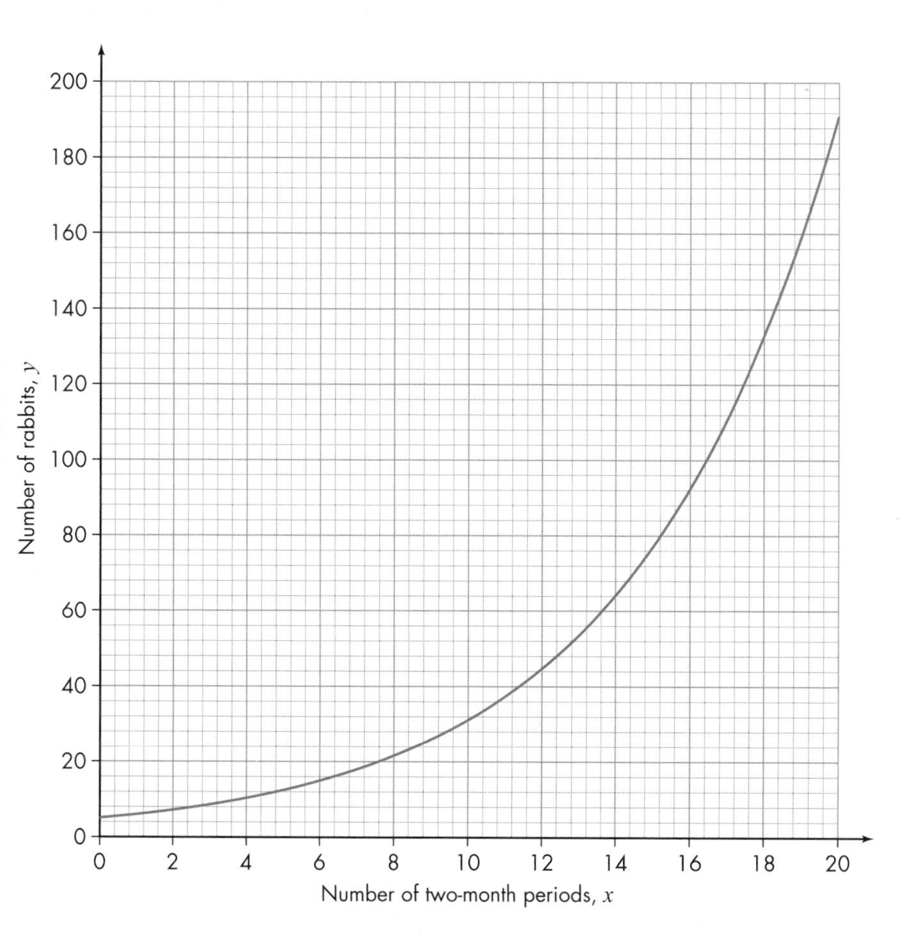

If you had just been given this graph and told that it represents the increase in a population of rabbits over time what questions could you answer?

Here are some examples.

- What was the starting population of rabbits?
- What was the rabbit population after 2 years?
- When did the rabbit population first exceed 100 rabbits?

To find the starting population of rabbits you need to use the graph to find the population when the time was zero. This is the y-intercept, the coordinates of which are (0, 5). So the starting population of the rabbits was 5.

To find the rabbit population after 2 years you need to draw a line from 12 on the x-axis (there are 12 lots of 2 months in 2 years) up to the curve and then draw a line across to the y-axis. The y value is approximately 45. So the rabbit population after 2 years was 45 rabbits (illustrated in red on the graph below).

Finally, to find when the rabbit population first exceeded 100 rabbits you need to draw a line from 100 on the y-axis to the curve and then draw a line down to the x-axis. The x value is approximately 16.5. 16.5 lots of 2 months is 2 years 9 months (illustrated in blue on the graph below).

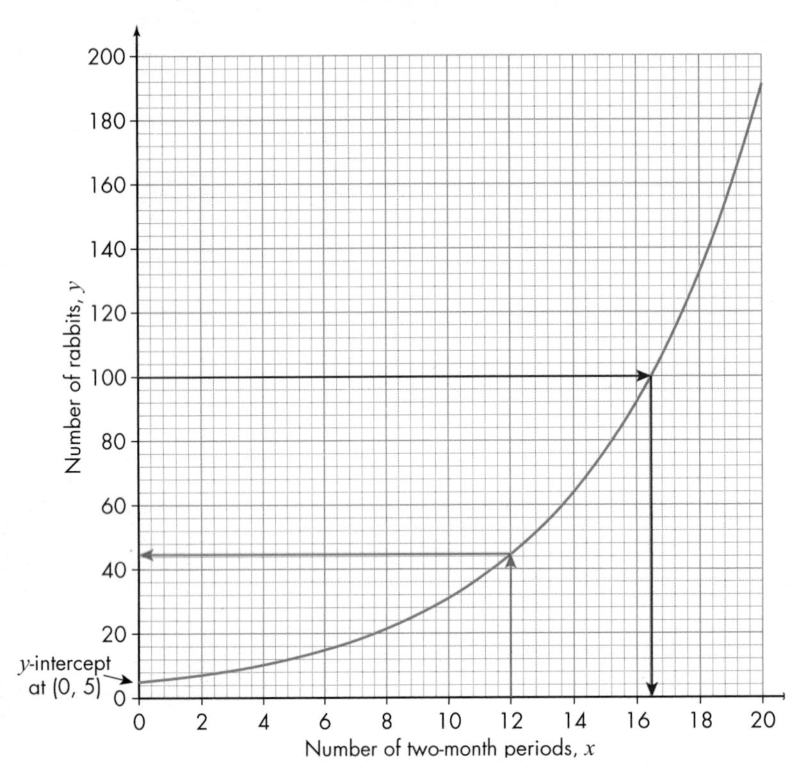

Hooke's law

Robert Hooke was a British physicist who showed that the extension, x (of a spring) is proportional to the force on the spring, F:

$$F = kx$$

where F is the force needed to extend (or compress) a spring, x is the extension (or compression) and k is a constant called the spring constant, related to the stiffness of the spring.

Hooke's law is only true within certain limits. The extension is proportional to the force up to a limit called the limit of proportionality. If the force is increased still further, beyond the spring's elastic limit, the spring will be permanently deformed and it will not return to its original length when the force on the spring is removed.

The graph of Hooke's law $F = kx$ is a linear function, so you should expect part of the graph to be linear until the force exceeds the limit of proportionality.

If you have only been given data for a spring before it had reached its limit of proportionality, the graph would be a straight line.

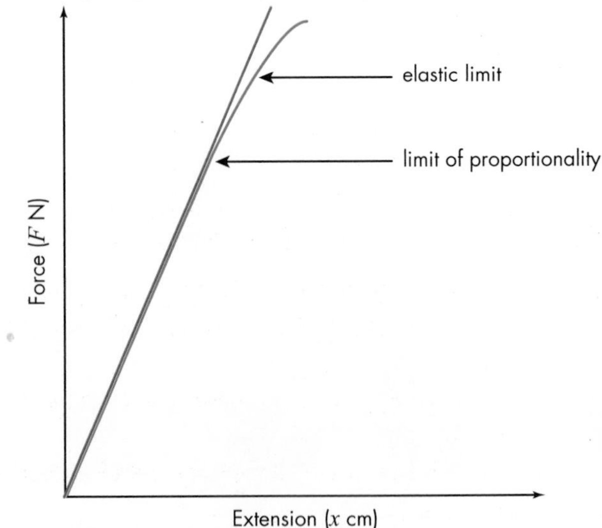

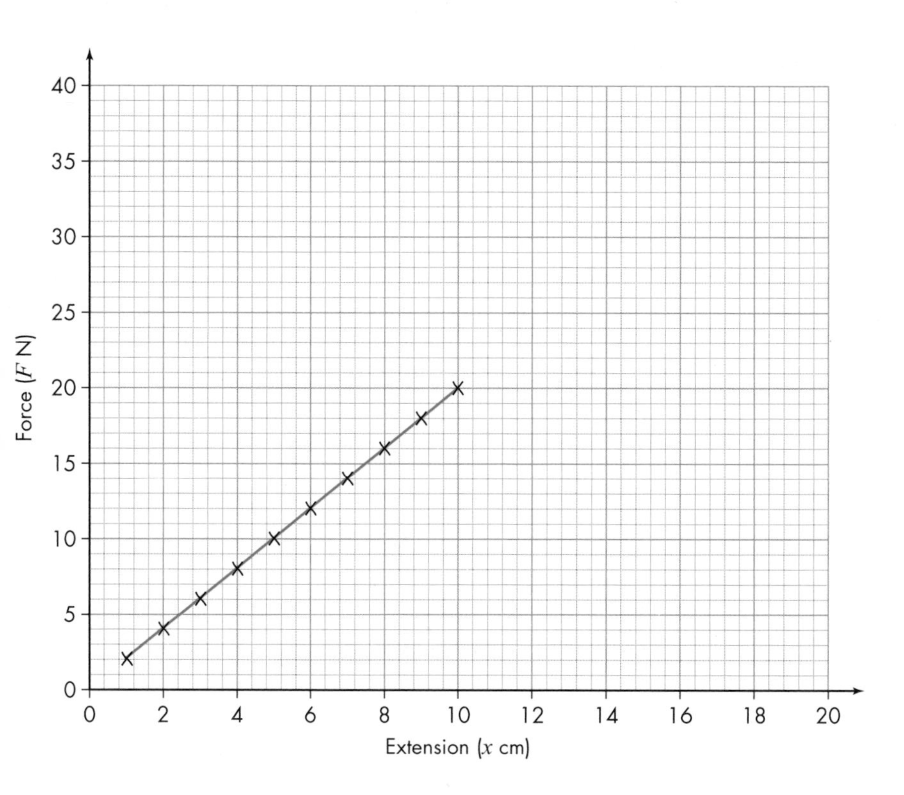

The graph shows a linear relationship for all extensions of the spring up to a distance of 10 cm. There is no data beyond an extension of 10 cm.

To find the force that would give an extension of 7 cm, you need to draw a line up from the horizontal axis to the graph then across to the vertical axis and read off the value. This is called **interpolation** because you are finding an estimate from within the known range of data.

Interpolation: the making of predictions within the range of the known data.

To find the force that would give an extension of 20 cm, you could extend the straight line and read the value of the force from the graph. However, you don't know the value of the limit of proportionality, so an extension of 20 cm could be above or below the limit of proportionality, and the force you read from the graph could be wrong.

Extending the line on a graph and reading a value outside the range of the data is called **extrapolation**. You should always take care with extrapolation because you cannot predict what will happen outside the given range of data as Hooke's law demonstrates. Consequently extrapolation should not be performed. (See Chapter 7 for more on extrapolation.)

Extrapolation: the making of predictions outside the range of the known data.

Exercise 11D

1 You buy a new printer at a cost of £150. You work out that you will use eight ink cartridges annually at a cost of £10 each.

Form an equation to calculate the cost of running the printer over time, complete a table of values and plot the graph.

Use your graph to answer the following questions.

a How much will the printer have cost to run, including purchase, after 2 years?

b How much will the printer have cost to run, including purchase, after 5 years?

c When will the cost of running the printer exceed £500?

2 A well known chain of burger restaurants has asked you to redesign their company logo.

They ask you to draw the graphs of $y = -x^2 + 4x$ between 0 and 4 and $y = -x^2 - 4x$ between −4 and 0 on the same pair of axes.

They need to see a table of values for each equation.

The company would like to know how many units wide and how many units tall their new company logo will be.

3 You have been asked to design the 'S' for a new font.

You have been asked to use the equation $y = -x^3 + 4x$.

Complete a table of values and plot the curve to give the shape of the S for the new font.

At what values of y will the S need to be trimmed to that the start and finish and the S do not extend beyond the maximum and minimum points of the curve?

(Hint: Turn your graph on its side to see the S.)

4 In 2015 it was announced that China's one-child policy was to be relaxed. Its original aim was to reduce China's population to 700 million by 2050.

In 2014 the population of China was approximately 1.367 billion.

If you were to assume that China's population actually declines by 0.5% per year from 2014 onwards will this be sufficient to reach the original goal?

Calculate the population of China for each year from 2014 to 2050 and plot the graph.

What sort of graph do you seem to have even though the function is exponential? Can you explain this?

5 Write an email to your maths teacher explaining interpolation and extrapolation and citing Hooke's law as an example.

Graphical solution of equations

Plumbers

Earlier in this chapter you looked at the fees charged by two plumbers. One plumber charges his clients a £50 call-out fee then £30 per hour. You can represent the total cost by the equation $y = 50 + 30x$. Another plumber doesn't have a call-out fee but charges £50 per hour. You can represent the total cost by the equation $y = 50x$. By examining the graphs of these equations how could you work out at what point the plumbers are charging the same?

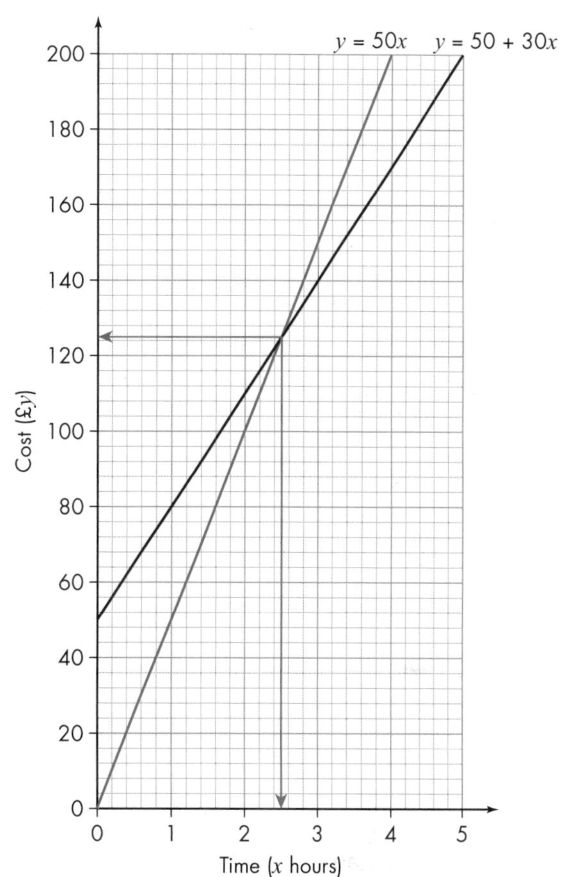

The point at which the plumbers are charging the same is when the graphs cross; the **intersection point**. The graphs intersect at the point (2.5, 125). What does this mean? When the plumbers work for two and a half hours they will both charge the same amount, £125.

You should also notice before the intersection point the plumber who charges £50 an hour is cheaper whereas after the intersection point he is the more expensive plumber.

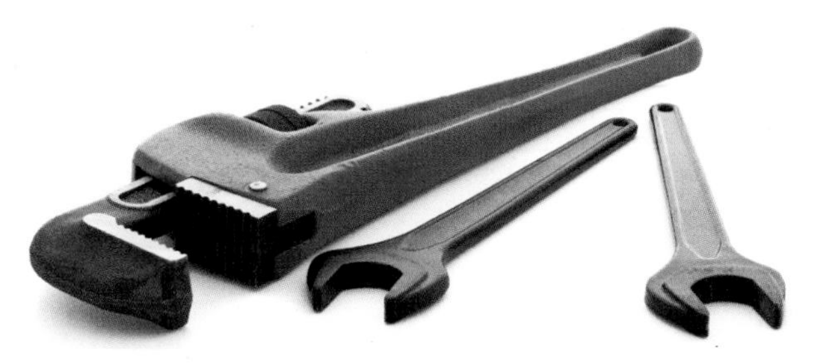

Intersection point: the point at which two or more graphs meet or cross.

Graphical solution of simultaneous equations

$y = x + 3$ and $y = x^2 + 5x + 6$ are two simultaneous equations (equations have the same y value when they share the same x value). If you were to complete a table of values for each equation from –5 to 1 and plot the graphs on the same pair of axes, you would get the following graphs.

How can you now use the graphs find the x values that result in the y value of each equation being the same?

There are two intersection points where the curve and the line cross. These are at (–3, 0) and (–1, 2). So the solutions to these equations when they are solved simultaneously are $x = -3$, $y = 0$ and $x = -1$, $y = 2$. If you aren't convinced, substitute $x = -3$ into each equation. You will get $y = 0$ for both equations. Then substitute $x = -1$ into each equation. You will get $y = 2$ for both equations.

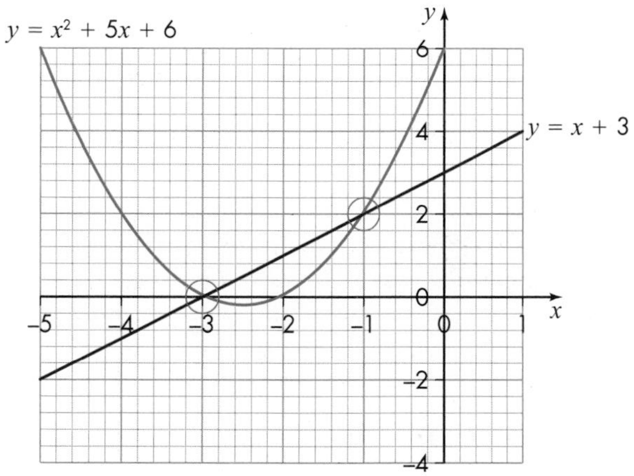

Exercise 11E

1 The graph below shows the very expensive monthly tariffs for two different mobile phone companies.

At what duration of calls do the two companies charge the same amount and what it this amount?

Which one would you choose and why?

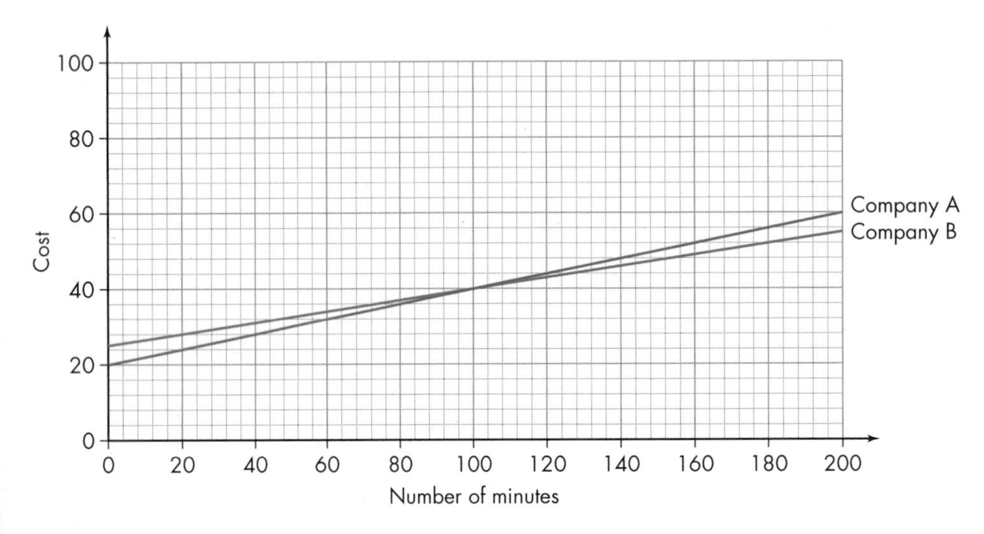

2 $y = 2x + 4$ and $y = -x + 1$ are two simultaneous equations.

Complete a table of values for each equation from –4 to 4 and plot the graphs on the same pair of axes.

Use your graph to find the x value that results in the y value of each equation being the same.

3 Your friends are going to sponsor you to run the London Marathon (26.2 miles). You want to draw a graph showing how much money you will raise if you are paid by the mile.

a One friend has agreed to pay you 50p per whole mile and will give you £10 for just starting the race.

Form an equation for this information, make a table and plot the graph.

(Hint: Convert 50p to £.)

b Another friend has agreed to pay you 75p per whole mile and will give you an extra £5 for just starting the race.

Form an equation for this information, make a table and plot the graph for this equation on the same grid as your graph for part **a**.

(Hint: Convert 75p to £.)

c At what point in the marathon will each friend end up paying you the same amount?

4 $y = x + 1$ and $y = x^2 - 1$ are two simultaneous equations.

Complete a table of values for each equation from –3 to 3 and plot the graphs on the same pair of axes.

Use your graph to find the x values that result in the y value of each equation being the same.

5 The equation of a straight line is $y = x - 2$ and the equation of a curve is $y = x^2 - 1$.

Both the straight line and the curve have been plotted on the same pair of axes.

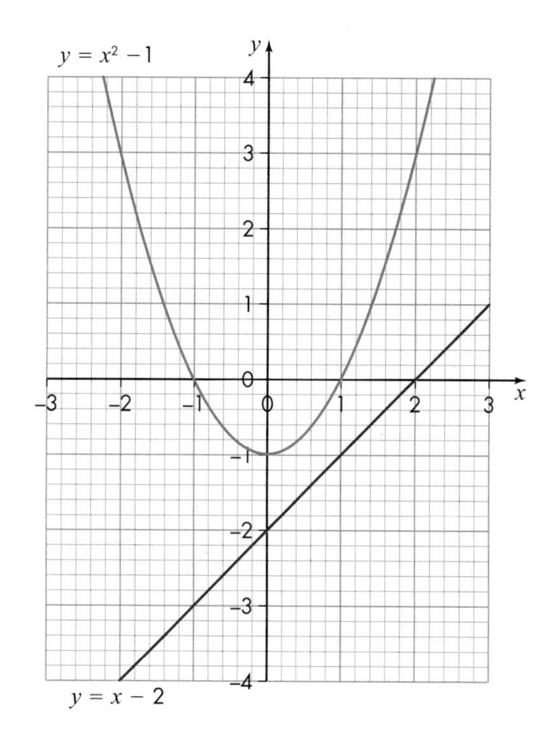

What can you deduce from the graph about the solutions to solving these equations simultaneously?

Case study

Is this the ultimate graph? If you are designing the airspace around an airport or are an air traffic controller then understanding graphs like this could mean the difference between life and death.

This is an image from a tool that airspace designers can use to analyse flight paths into and out of airports. It is particularly complex because this is an image of a three-dimensional space whereas the graphs that you have studied in this chapter have only been two dimensional. If two of the lines on the graph above were to intersect at the same time, what would this mean? Why is getting graphs like this one right so very important?

Project work

You thought this was maths, didn't you? Well, you are going to do a scientific experiment but analyse the results with your knowledge of graphs.

You will need a freshly prepared cup of tea, a watch and a thermometer. (Sadly you are not going to be able to drink the cup of tea until it is cold.)

Once you have made the cup of tea, take its temperature and record the result. Every minute thereafter take the temperature of the cup of tea and record the result until the tea feels quite cold.

Plot the temperature decrease of the cup of tea against time on a graph.

What do you notice?

Check your progress

How confident are you feeling in your level of knowledge? What do you need to practise more?

Spec reference	Learning objective	▶▷▷	▶▶▷	▶▶▶
G1.1	Sketch and plot curves defined by simple equations			
G1.1	Know the shapes of the graphs of linear, quadratic, cubic and exponential functions			
G2.1	Plot and interpret graphs (including exponential graphs) in real contexts, to find approximate solutions to problems including understanding the potential problems of extrapolation			
G2.2	Interpret the solutions of equations as the intersection points of graphs and vice versa			

12 Rates of change

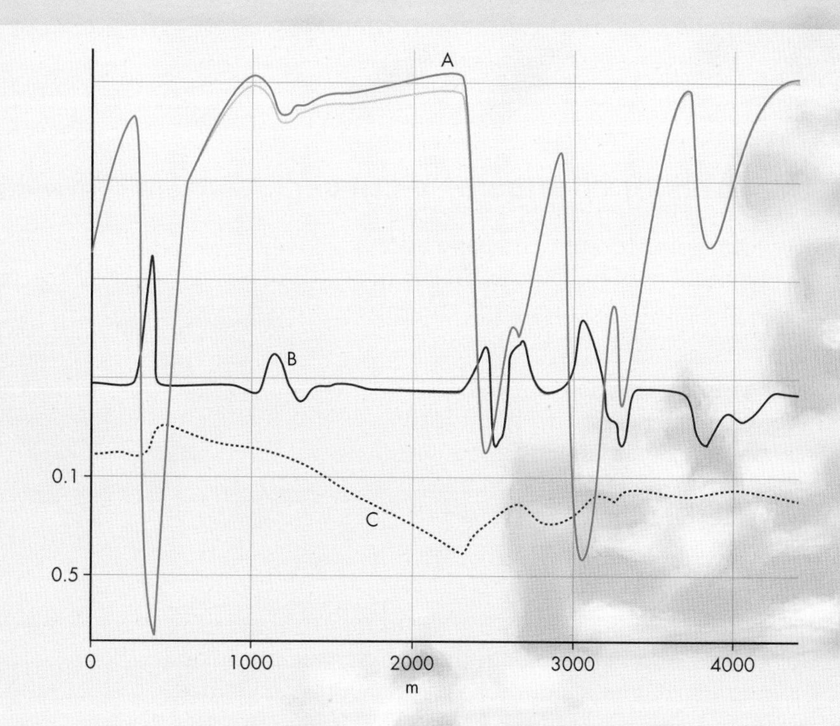

The image above is an example of the type of graph used by Formula 1 teams. The graph shows data sent from two racing cars as they drive around a racing track. Each line represents different data that can be analysed by engineers.

The top line is the speed of the cars. You can see that one car is slightly slower than the other at point A.

The second line, labelled B and overlapping with the top line, is the steering angle with left up and right down. The data from both cars is almost identical as it is difficult to see two separate lines for this graph.

The dotted line, labelled C, shows the time separation between the two cars.

How quickly each of these lines change (their rates of change) is information that the engineers and drivers can use to make the cars go faster.

Discussion

Discuss the following with your peers.

1. In what contexts have you heard the term 'rate of change' used?
2. The word 'derivative' is often used instead of the term 'rate of change'. If you have heard this term before, in what contexts have you heard it used? If not, use a dictionary or search engine to find some examples where it is used and discuss these.

G3: Gradient

Learning objectives

You will learn how to:

- Interpret the gradient of a straight line graph as a rate of change.
- Interpret the gradient at a point on a curve as an instantaneous rate of change.
- Estimate rates of change for functions from their graphs.

Why study the rates of change of a curve?

Most things change with regards to time: the populations of countries change, the number of people unemployed changes, the pass rates of examinations change and the spread of a disease are just some examples.

In many cases what is important is not that change occurs but at what rate the change occurs. What is the rate of change of a population: is it growing at a rate that is not sustainable? How is the unemployment rate changing? How fast is the examination pass rate changing? Is it fair? Is the disease spreading at a faster rate than vaccination can control?

In the previous chapter about graphical methods the graphs that are used to model real-world processes were studied. In this chapter the methods that can be used to find rates of change of those graphs will be explained.

Interpreting the gradient of a straight line graph

The graph shows a taxi firm's charges.

Gradient is a measure of how much a line slopes. It is found by dividing the vertical change by the horizontal change.

The **rate of change** is the same as the gradient.

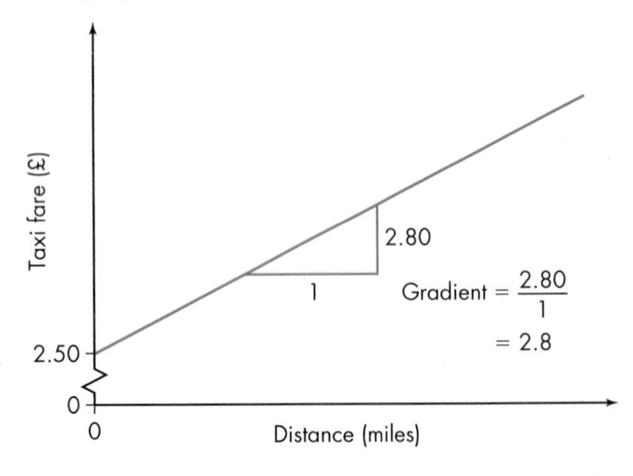

The sketch shows the transaction is modelled by a straight line because for every mile travelled in the taxi the charge increases by £2.80. In this sketch there is a constant rate of change, or gradient, of the line.

The graph of any function that has a constant rate of change will be a straight line.

In this sketch the constant rate of change is £2.80 per mile.

If you write the equation of this straight line in the form $y = mx + c$, then the value of m is 2.8 and the value of c (the intercept on the vertical axis) is 2.5, where you are using y to represent the taxi fare and x to represent the distance travelled.

Constant rates of change are very common in the real-world use of mathematics.

Example 1

A mobile phone company charges a fixed contract rate of £23.00 per month.

When a user travels outside of Europe the company charges £0.40 for every additional minute used.

a Sketch a graph showing how the charge varies depending on the number of minutes used.

b What is the gradient of this line and what does it represent?

c Express the equation of the graph in the form $y = mx + c$.

a

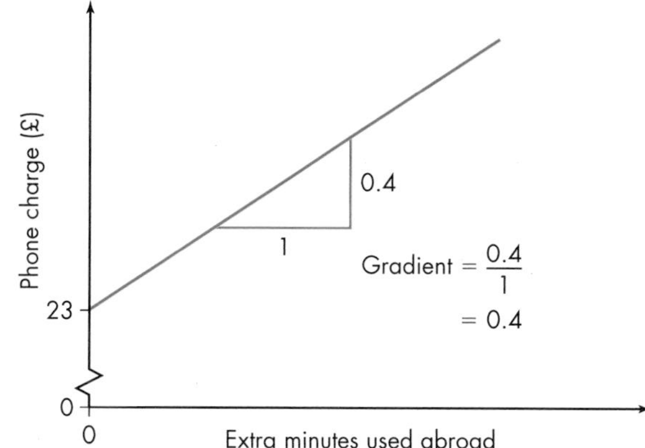

b The gradient is 0.4 and this represents the cost of each minute, in pounds.

c $y = 0.4x + 23$

Exercise 12A

1 The speed of cars on motorways is often restricted by variable speed limits to improve traffic flow.

One day, on a sixteen-mile section of the M25 motorway, the speed was restricted to 60 mph (1 mile every minute).

a Sketch a graph of the journey for a car driving constantly at 60 mph through the section, with time on the horizontal axis and distance travelled on the vertical axis.

(Hint: Distance travelled is zero when at the start of the horizontal axis.)

b Is the gradient constant? What might the gradient represent?

(Hint: speed = $\frac{\text{distance travelled}}{\text{time taken}}$)

2 A mobile phone company offers two different contracts.

Contract A is a standard rental charge of £20 per month and then £5 for every 50 minutes of calls.

Contract B has no rental charge, but charges £7 for every 50 minutes of calls.

a Draw two straight line graphs to model the contracts.

b What does the gradient of the graph for these straight lines represent?

c Which contract would suit you best? Why?

3 Anton and Becky are planning their summer holidays.

Anton decides to go to Germany and takes €600, expecting to spend €60 daily on meals and accommodation.

Becky will go to Hungary and takes €500, expecting to spend only €40 each day.

They both start their holidays on 1st August.

a Draw two straight line graphs on the same set of axes to model their spending.

b What does the gradient of the graph for these straight lines represent?

c How does the gradient of these lines differ from the gradients in questions **1** and **2**?

d On what date will Anton and Becky both have the same amount of money left?

4 Retailers study cumulative footfall, the number of people entering their shops over several weeks. These footfalls are sometimes modelled by straight line graphs.

Draw on the same set of axes the straight lines that represent three branches of a shop where F (on the vertical axis) represents the number people who cumulatively visit the shop and W (on the horizontal axis) represents the week number.

Week (*W*)	**Footfall (*F*) (hundreds of visitors)**		
	Bristol	**Manchester**	**Newcastle**
1	35	10	5
2	55	40	15
3	75	70	25
4	95	100	35
5	115	130	45

a What does the gradient of each of these lines represent?

b How would you decide which shop is increasing its footfall at the greatest rate?

5 Hooke's law relates the extension (or compression) of a spring to the force on it (see Chapter 11).

The diagram shows the apparatus used in an experiment carried out by four groups of students using different springs to investigate Hooke's law. The table shows their results.

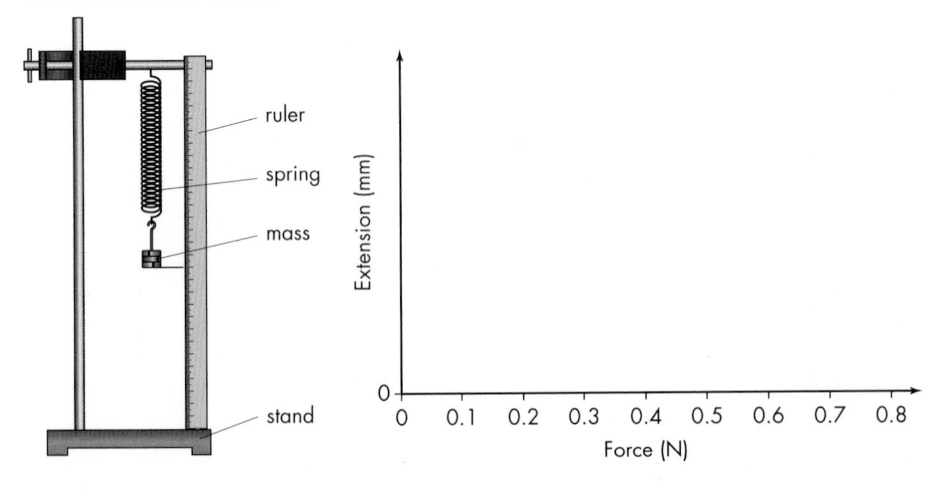

		Force (newtons)							
		0.1	**0.2**	**0.3**	**0.4**	**0.5**	**0.6**	**0.7**	**0.8**
Group A	**Extension (mm)**	10	20	30	40	50	60	70	80
Group B		14	29	41	56	71	83	98	113
Group C		21	20	22	21	23	22	23	21
Group D		8	15	22	30	38	45	52	60

a Draw graphs of the force and extension of the four different springs.

b Work out the gradient of each line. This will be the rate of change of the extension with respect to the force.

(Remember that experimental data does not always lie exactly in a straight line. You will have to draw a line of best fit in some cases.)

c Which is the stiffest spring? The slackest spring? Explain how you deduced your answers.

6 For the subjects you are studying, find examples of straight line graphs, and find out what the gradients (the rates of change) represent.

Mathematics in the real world

Social media companies such as Facebook and Twitter make their money from charging advertisers to advertise to the users of the social media. The amounts they can charge are based on the numbers of users. The more people that use the social media, the more the social media companies can charge the advertisers.

The graph shows the type of chart advertising companies use to make decisions about where to place advertisements. The advertising companies will have to balance the cost the social media companies charge against the number of people they could reach with their advertisement, so they will be looking for the type of social media that promises great growth.

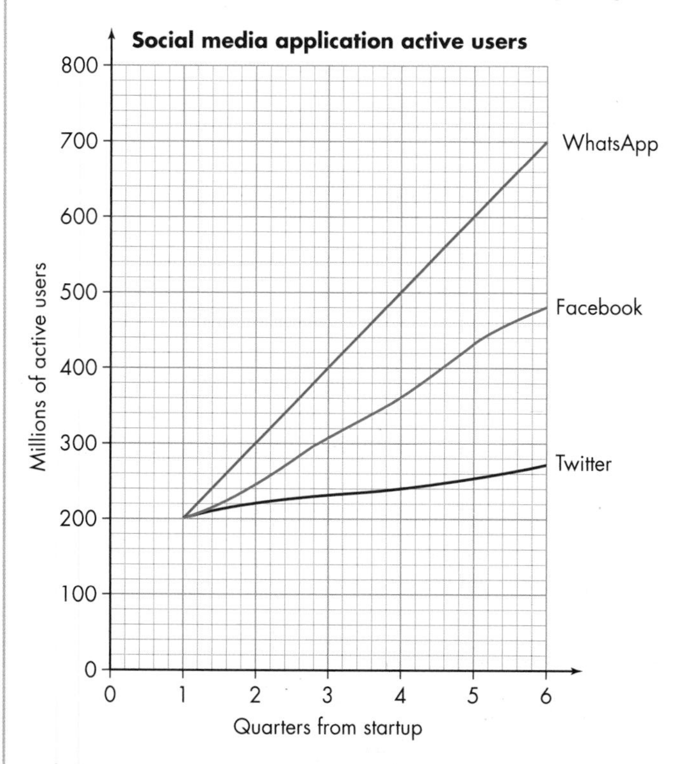

The increase in users for all three social media applications can be clearly seen in this graphical visualisation of the data.

The number of users can be read directly from this graph and the rate of change of users can be found from the gradient of the lines on the graph.

In the second to sixth quarters after startup, the average rate of change for WhatsApp was (700 – 200) ÷ (6 – 1) = 100 million new users every quarter. This means that on average, 100 million new users joined WhatsApp every three months.

In the second to sixth quarters after startup, the average rate of change for Twitter was (267 – 200) ÷ (6 – 1) =13.4 million new users every quarter.

The advertising companies can therefore plan on where to spend their advertising budget, balancing the rate of change with the cost (unknown) of advertising. More rapid growth means the potential of more people responding to their adverts!

Interpreting the gradient at a point on a curve

If the graph of a function is not a straight line, then the gradient at points on the curve can be found by drawing **tangents** to a plot of the curve at any point. The gradient of a tangent to a curve is the **instantaneous rate of change** at the point where the tangent touches the curve.

Tangent: a straight line that just touches a curve at a given point

The **instantaneous rate of change** at a point is found by calculating the gradient of the tangent at that point.

The graph shows the increase in the number of Ebola cases in Sierra Leone from September 2014 to November 2015.

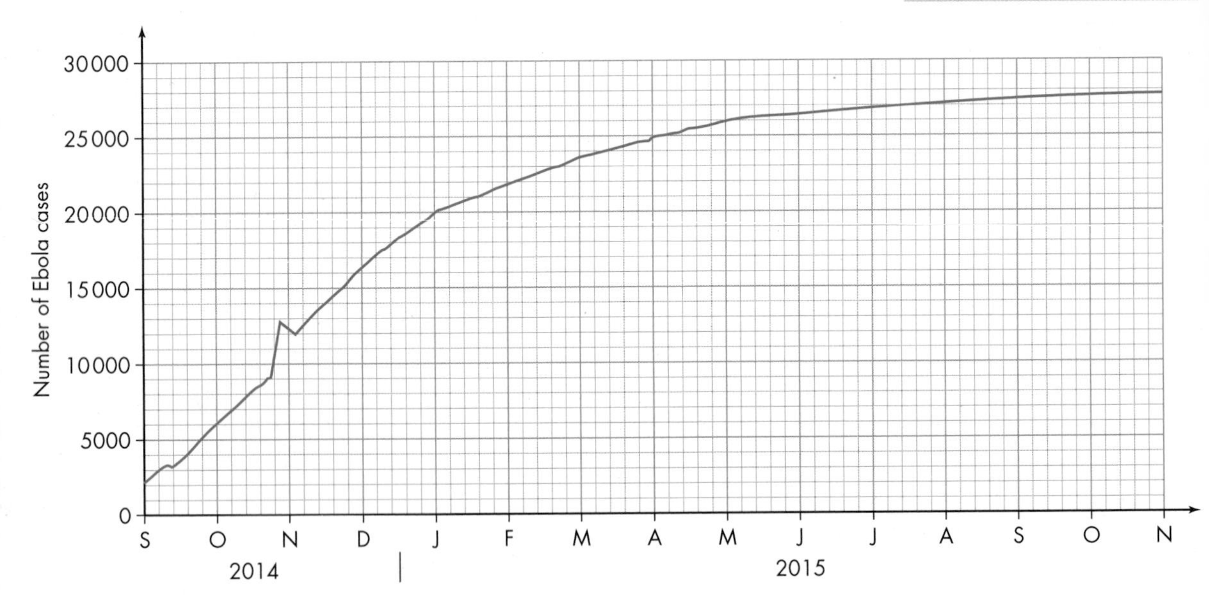

Discussion

Discuss the following with your peers.

What does the graph show? Is the number of cases increasing by the same amount each time? How do you know? Tell someone in your own words what is happening? Do they agree? Does the graph worry you or not? Why?

To find the rate of change on 1st January, draw the tangent to the curve at the point on the curve corresponding to 1st January on the horizontal axis. The tangent should just touch the curve at that point. To draw such a line, rotate your ruler around the 1st January point on the graph until you can see the graph close to that point below the ruler.

On the graph shown below, the red line is not a tangent because it cuts the graph at the January point; the blue line is not a tangent because it does not touch the graph at all. The green line is a tangent because it just touches the graph at the January point.

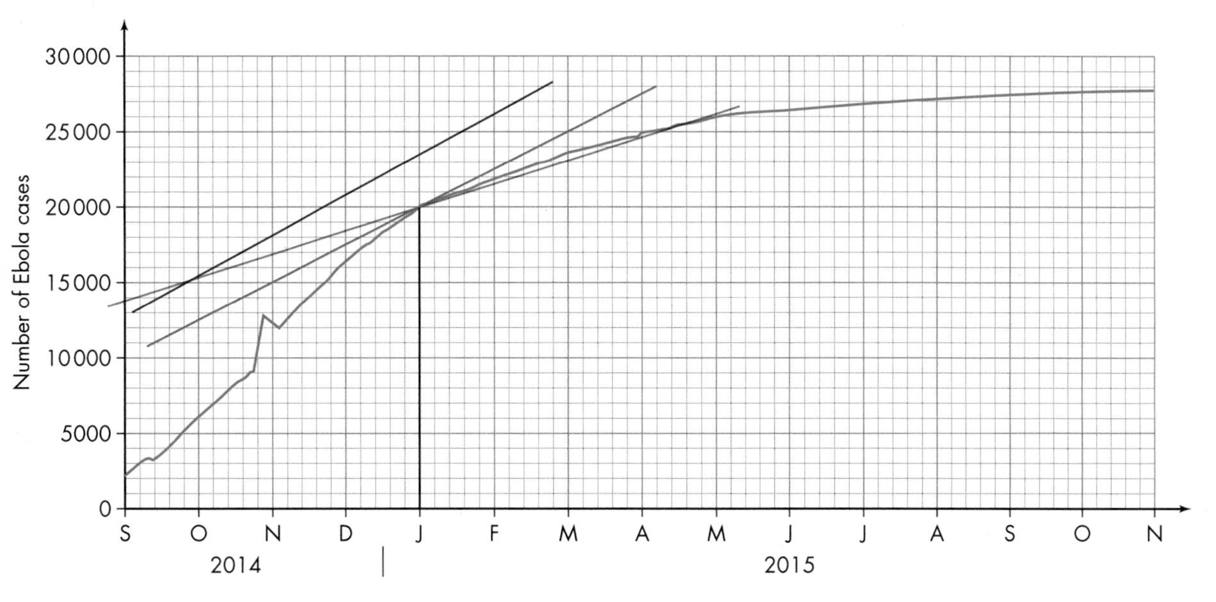

The instantaneous rate of change on 1st January is the gradient of the tangent.

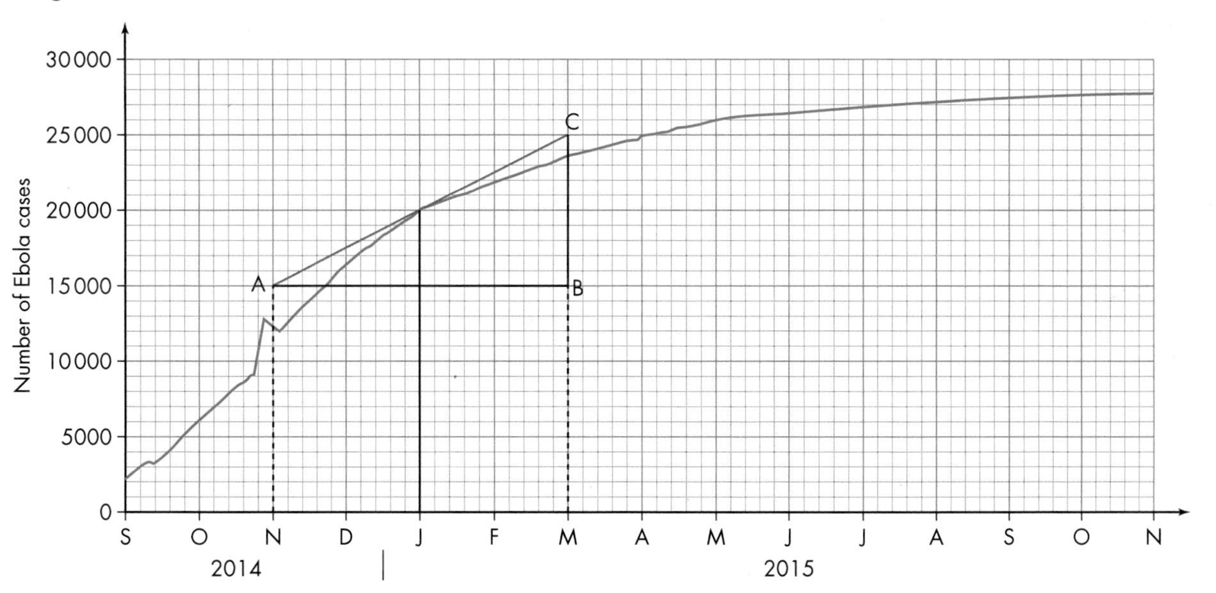

The line AC is the tangent to the graph on 1st January.

The gradient here is calculated by

$$\text{gradient} = \frac{\text{change in number of cases}}{\text{number of months for the change in number of cases}} = \frac{BC}{AB}$$

$$\text{gradient} = \frac{25\,000 - 15\,000}{4} = \frac{10\,000}{4} = 2500 \text{ cases per month}$$

The gradient is 2500, so you can say the instantaneous rate of change is 2500 cases per month on 1st January.

Since the gradient is positive, this means that the number of Ebola cases is increasing by 2500 cases per month at that instant.

Discussion and group exercise

Ask for copies of the Ebola graph and in a small group share out the work.

Draw the tangents to the graph on the first of December 2014 (ignore the 'blip' in November), and February, March, April and May 2015 and calculate the instantaneous rates of change on those dates.

Discuss with your group what is happening to those rates of change as time goes on and then present your findings to one of the other groups or the class, either as a poster or a PowerPoint presentation.

Maximum and minimum points

The rates of change of a function at local **maximum** points and at local **minimum** points are always zero; the gradients at these points are zero. At these points on curves the tangents are horizontal.

A **maximum** is the largest value in an interval where the gradient is zero and the gradient changes from positive to negative.

A **minimum** is the smallest value in an interval where the gradient is zero and the gradient changes from negative to positive.

A horizontal tangent has a gradient of zero.

Estimating the rate of change of a function at a point

To estimate the rate of change of a function at a point:

- plot a graph of the function
- draw a tangent at the point where the rate of change is to be found
- calculate the gradient of this tangent to find an estimate of the instantaneous rate of change.

The graph shows a quadratic graph with a negative term in x^2.

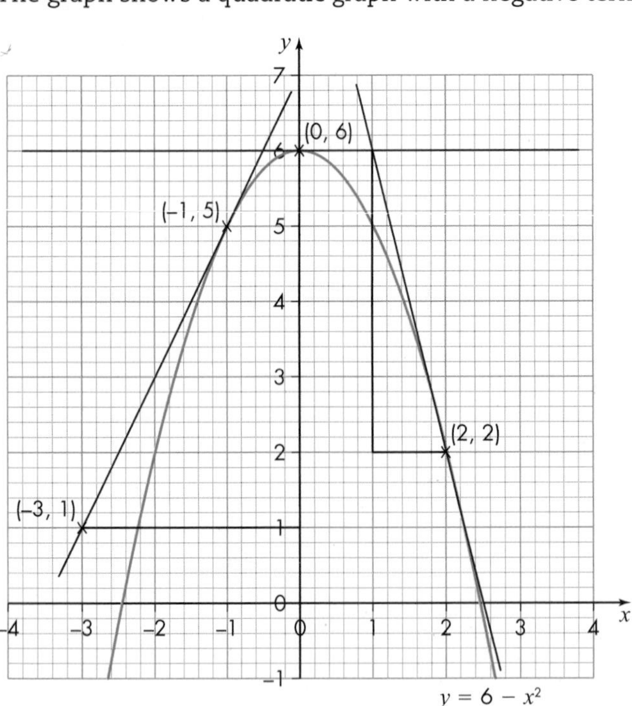

Three tangents have been drawn on the graph, at the points (–1, 5), (0, 6) and (2, 2).

- At the point (–1, 5), the gradient is $\frac{6}{3} = 2$.
- At the point (0, 6) the gradient is 0 because the tangent is horizontal.
- At the point (2, 2), the gradient is $-\frac{4}{1} = -4$.

The gradient (or rate of change) has gone from being positive (+2) to negative (–4), so at some point the gradient must have been 0. This is always the case for continuous graphs such as the graph shown.

Discussion

The graph shows a quadratic graph with a positive term in x^2.

Discuss with your peers how to calculate the gradient at the points (1, 3), (3, 1) and (6, 5.5) and then work out the gradients at those points.

What can you say about the rate of change of y with respect to x at these points?

What are the distinguishing features of maximum and minimum points?

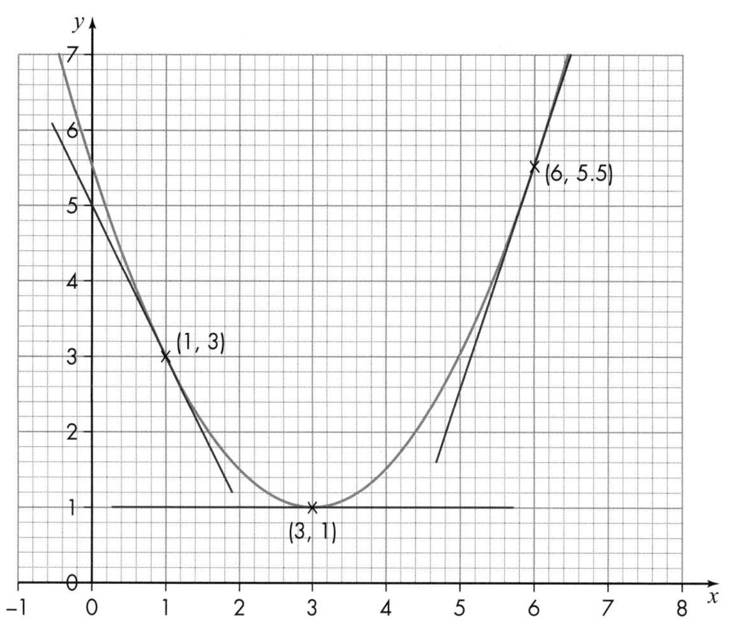

Example 2

The diagram shows the graph of the quadratic function

$y = -x^2 + 100x - 2000.$

Use the graph to estimate the instantaneous rate of change when $x = 40$ and when $x = 55$.

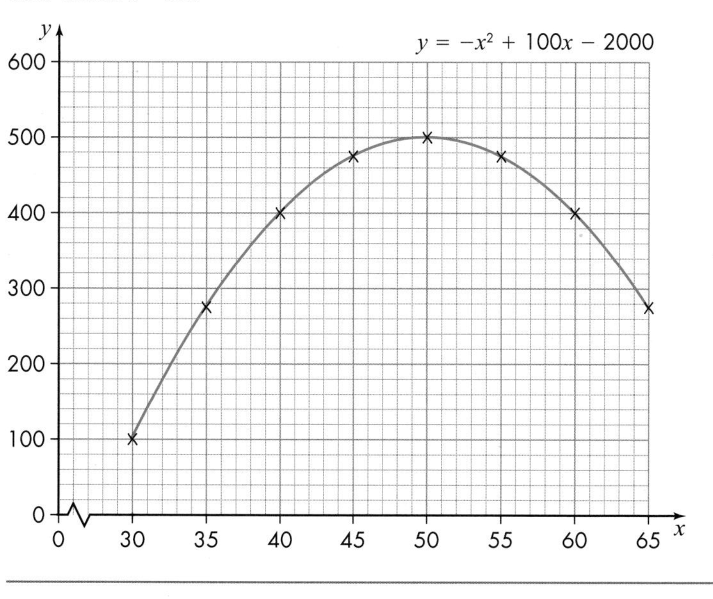

To find the rate of change of y when x is 40, use a ruler to draw a tangent (green on the graph below).

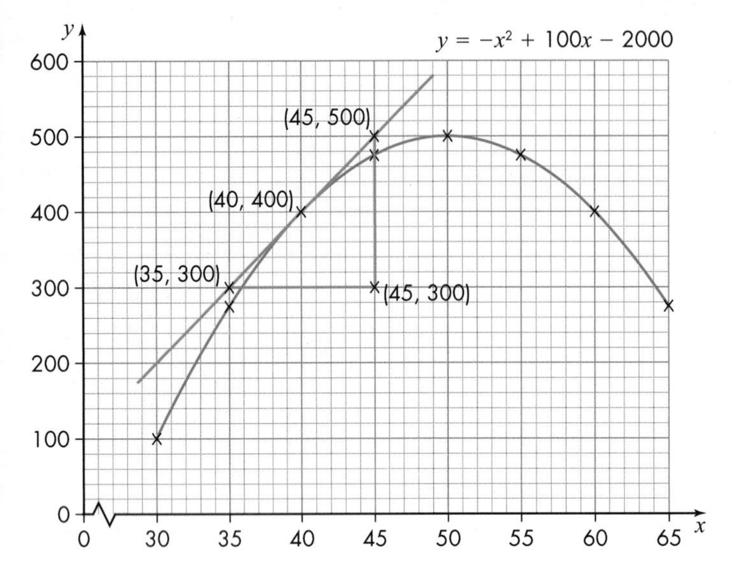

The instantaneous rate of change of y, when x is 40, is given by the gradient.

$$\text{gradient} = \frac{500 - 300}{45 - 35} = \frac{200}{10} = 20$$

The gradient is positive (increasing).

If the tangent goes from high on the left of a graph to low on the right of a graph, the instantaneous rate of change, or gradient, is negative (decreasing).

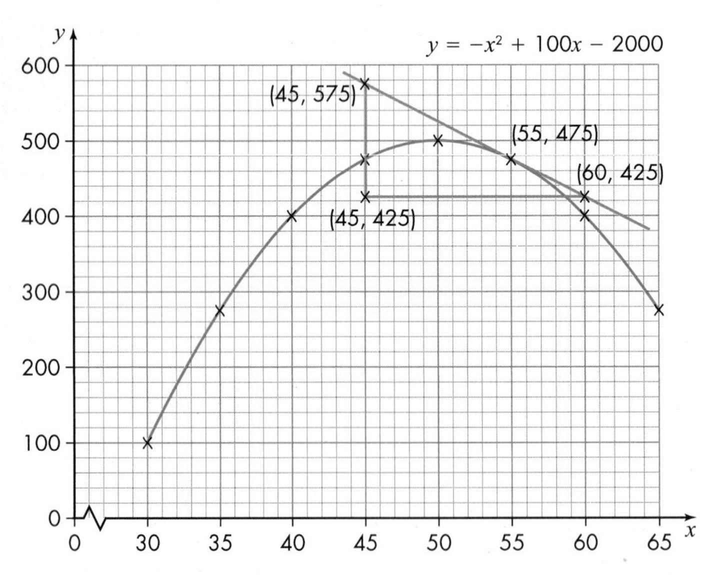

The instantaneous rate of change of y, where x is 55, is given by the gradient.

$$\text{gradient} = \frac{425 - 575}{60 - 45} = \frac{-150}{15} = -10$$

The gradient is negative (decreasing).

Exercise 12B

1 The graph shows the height above the ground of a projectile fired from a cannon.

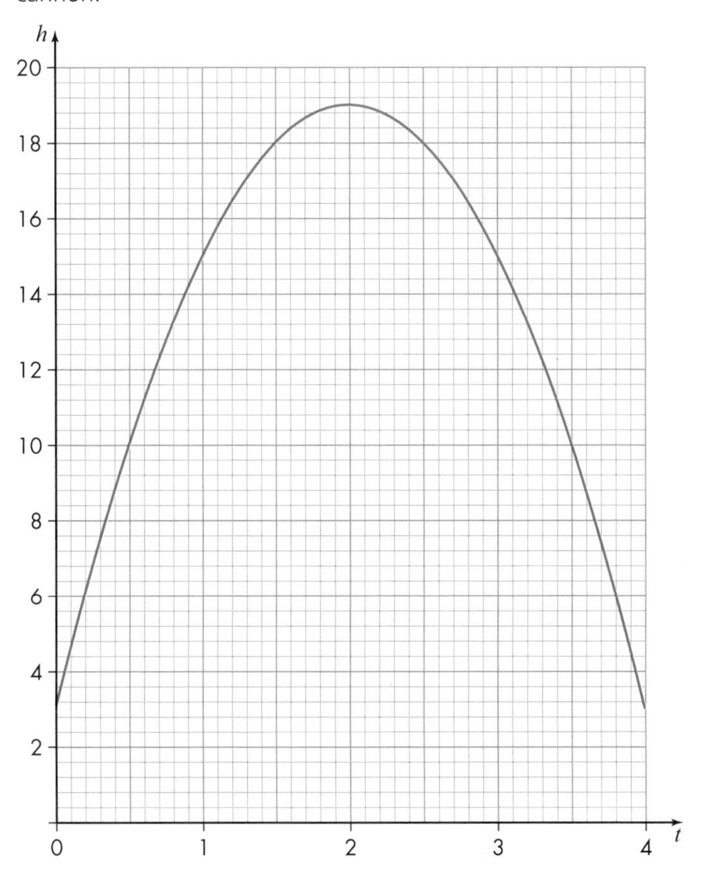

a On a copy of the graph, draw a tangent at the point (0.5, 10).
Use this to work out the rate of change of height, h, in metres, with respect to time, t, in seconds.

b Repeat for the point (3, 15).

c What is the rate of change when $t = 2$?

d Between what two times is the projectile above 18 m high?

2 The profit P in $ millions earned by an internet company after t years is given by the equation $P = t^3 - 6t - 6$ with t from 0 to 6 years.

a What is the instantaneous rate of change of the company's profit after one year? What is the rate after three years?

b Are these positive or negative rates of change?

c Describe what happens to the internet company's profit over the six years.
(Hint: Copy and complete the table below before plotting the graph and drawing a tangent at $t = 3$, or use software to draw the graph and print it out.)

t	0	1	2	3	4	5	6
t^3	0		8			125	
$-6t$	0		−12			−30	
-6	−6		−6			−6	
P	−6		−10			89	

3 The flow of water in rivers is monitored by the Environment Agency.

The flow of water after a storm is modelled by a cubic equation, where F is the flow in $m^3 s^{-1}$ at time t hours after a storm.

The equation that models the flow for the first 4 hours after the storm is

$F = t^3 - 3t^2 + 6t + 10.$

a Draw a graph of F (on the vertical axis) against t (on the horizontal axis) for the first 4 hours after the storm.

(Hint: Use a table of values and plot points at intervals of 0.5 hours.)

b Draw a tangent at $t = 1$ to find the rate of change of the flow at that point.

c Repeat part **b** for $t = 2$ and $t = 3$.

d Use the graph to find the minimum flow and the time after the storm when this occurs.

e Why might the cubic equation not be a valid model for the river for times more than 4 hours after the storm?

4 After braking, the distance taken to stop, d, in metres, by a car travelling at v km h^{-1} is modelled by the equation

$d = 0.006v^2 + 0.2v.$

a Plot a graph for stopping distance, with values of v between 0 and 100 km h^{-1}.

b What is the instantaneous rate of change of distance with speed when braking from 50 km h^{-1}?

c What does this model imply about the rate of change of braking distance with speed as the speed increases?

5 An exciting candidate decided to stand in an election and lots of people registered to vote.

The number of people registered to vote, r, in the days, t, after she decided to stand was modelled by the quadratic equation

$r = 10t^2 + 20t + 15.$

a Plot a graph for this model of elector registration for the first 6 days.

b What is the instantaneous rate of change of those registered after 2 days? What is the rate after 4 days? What is the rate after 5 days?

c If this quadratic model is used for a large number of days, what would it imply about the rate of registration?

6 A golfer is a keen mathematician. She knows that the path of a golf ball can be modelled by a quadratic equation with h the height in metres above ground and d the horizontal distance in metres. The equation is

$$h = 0.3d - 0.001d^2$$

The golfer knows the best shot to the 18th green is a drive over the club house. The club house is 100 m from the tee.

a Draw a graph for the path of the golf ball with d between 0 and 300 metres.

b Use the graph to find the maximum height the ball will reach.

c Will the ball be able to travel over the club house, which is 20 metres high?

7 A nurse studying a patient's temperature graph sees it has a zero rate of change.

What might be happening to the patient? (There are a number of possibilities.)

8 A financial advisor sees that the share value of her client's stocks had a zero rate of change over the past two days.

How might her advice differ if the rate of change was at a minimum or at a maximum?

9 For the subjects you are studying, find examples of non-straight line graphs.

Sketch them and indicate on them the maximum and minimum points that occur.

How is knowing the maximum and minimum points useful in the examples you find?

G4: Average speed

Learning objectives

You will learn how to:

- Calculate average speeds.

Calculating average speed from a formula

The **average speed** is calculated by dividing the distance travelled by the time taken.

Average speed is calculated using the equation

$$\text{Average speed} = \frac{\text{distance travelled}}{\text{time taken}}$$

The units for speed are usually:

- miles per hour (mph)
- kilometres per hour (km/h, or km h^{-1}).

In physics, the units metres per second (m/s, or m s^{-1}) are also often used.

Example 3

A car travels 16 miles in 25 minutes. What is the average speed?

Before using the formula the time must be converted to hours in order to give an answer in mph.

$$25 \text{ minutes} = \frac{25}{60} \text{ of an hour}$$

so

$$\text{Average speed} = \frac{16}{25 \div 60} = 38.4 \text{ mph}$$

Exercise 12C

1 There are various expressions you say when things happen quickly.

Here are some examples. Work out the speeds and put them in order from the slowest to the fastest.

To compare the speeds you will have to express them in the same units, so remember that 1 mile is about 1.6 km.

a Bat out of hell: a noctule bat can fly about 70 metres in 5 seconds

b Blink of an eye: the centre of an eyelid travels about 1.2 centimetres in 200 milliseconds

c Like a speeding bullet: a bullet can travel 600 metres in half a second

d Lightning fast: a lightning flash takes about 0.07 seconds to travel 4 miles to the ground

e Flash in the pan: a flame takes about 0.1 seconds to travel across a 30 centimetre diameter pan

f Drop of a hat: if you drop a hat, it travels about 1.7 metres in about 0.6 seconds

g Like wildfire: without being wind assisted, a forest fire moves about 3 miles in 15 minutes

h Make it snappy: the end of your finger moves about 8 centimetres in 12 milliseconds

i Snail's pace: a common garden snail can cover 1 metre in 3 minutes

2 A lorry travels 120 miles in 3 hours and 20 minutes.

What is the lorry's average speed in mph?

3 A courier van makes four deliveries with the distances travelled and times to drive the distances given in the table below.

Drop off	Distance to drop off (miles)	Time travelling to drop off (hours and/or minutes)
1	30	1 hr
2	15	30 mins
3	40	1 hr 20 mins
4	35	1 hr 10 mins

a What are the four average speeds of the van while delivering to the four locations?

b What is the average speed over all four journeys in total?

4 The 60-mile journey from Salisbury to London is completed in a car at an average speed of 45 mph.

The return journey from London to Salisbury is completed at an average speed of 60 mph.

What is the average speed for the whole journey?

5 Jane and Ben live 15 km from college. They leave home at the same time to ride to their first lecture.

Jane arrives after 40 minutes and Ben arrives 20 minutes later.

How much faster is Jane riding than Ben?

6 Here are the distances, in miles, between four places in the UK.

These distances are based on the fastest journey times between these places.

Caernarfon	Oxford	Southampton	
159	78	143	**Birmingham**
	234	299	**Caernarfon**
		67	**Oxford**

Claire drives from Southampton to Birmingham to do a $2\frac{1}{2}$ hour mathematics masterclass and then goes on to Caernarfon to see a friend.

She drives on motorways or dual carriageways for 80% of the distance and on urban roads for the rest.

She maintains an average speed of 70 miles per hour on the motorways or dual carriageways and 30 miles per hour on urban roads.

As well as the masterclass, she plans to have breaks totalling 2 hours which can be taken whenever she likes.

a If she leaves Southampton at 06:00, what time does she expect to arrive at Birmingham if she travels there without a break?

b Would she arrive in time for afternoon tea at 16:00 in Caernarfon?

Show all your working to support your answer.

7 Claire decides to take the shortest distance back from Caernarfon to Southampton.

That distance is 252 miles, but she can only average 45 miles per hour while driving for the whole journey.

She stops every 2 hours for a half hour break.

a If she leaves Caernarfon at 09:00, what is the earliest time she could arrive at Southampton?

b Would she be any quicker using the distances and speeds stated in question **6**, assuming she still stops every 2 hours for a half hour break?

Show all your working to support your answer.

Mathematics in the real world

Modern science lives with uncertainty. The equation for average speed is referenced in the Heisenberg Uncertainty Principle that is at the foundation of quantum mechanics.

The equation for average speed of an object includes both the distance travelled and the time taken to travel that distance.

The Heisenberg Uncertainty Principle states that if the exact position of an object is known then it is impossible to know its speed with great certainty. Alternatively, if the exact speed of an object is known it is impossible to know its position accurately. In other words, you can't know exactly both an object's speed and its position. However, the uncertainties are only noticeable for objects at the scale of sub-atomic particles so you can't use the Uncertainty Principle to argue that measurement of a car's speed by an average speed camera is not accurate.

However, electrons would be able to overturn a speeding fine as it would not be possible to say with sufficient accuracy what their average speed is (even if you had a camera capable of measuring very high speeds). The world of quantum mechanics is full of such fuzzy edges and beautiful mathematics.

Discussion

Discuss the following with your peers.

Two drivers entered a 12-mile section of motorway monitored by speed cameras to maintain the average speed of cars at or below 40 mph.

Driver A completed the 12-mile section by driving at a constant speed for $\frac{1}{4}$ of an hour.

Driver B completed the 12-mile section by first driving at 80 mph for $\frac{1}{10}$ of an hour and then at 20 mph for the rest of the journey.

Which driver would get fined? Why?

Who do you think is the more dangerous driver? Why?

G5: Speed and acceleration

Learning objectives

You will learn how to:

- Understand that the gradient of a distance–time graph represents speed.
- Understand that the gradient of a velocity–time graph represents acceleration.

Distance–time graphs

Average speed is calculated using the equation

$$\text{Average speed} = \frac{\text{distance travelled}}{\text{time taken}}$$

When speed is displayed graphically using distance–time graphs, speeds can be calculated from the gradients of the lines. For linear graphs, the gradients are calculated simply using two points on the graph. If the graph is not linear then you need to draw a tangent at the point you are interested in and calculate the gradient of the tangent.

Example 4

This distance–time graph shows the progress of a cyclist who cycles at a steady speed to see a friend, then cycles home.

Find the gradient of each of the line segments and hence the speed of the cyclist for each segment.

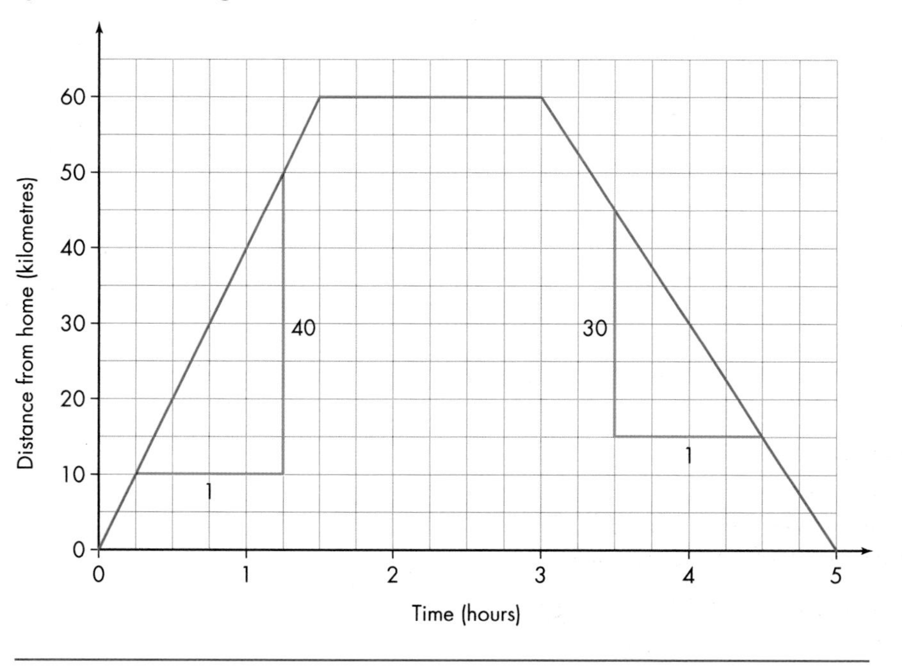

For the first line segment you can find its gradient by drawing the green triangle anywhere, provided its hypotenuse lies along the segment. Choosing the horizontal distance to be 1 hour makes the calculation easier!

$$\text{Gradient} = \frac{\text{distance travelled}}{\text{time taken}} = \frac{40}{1} = 40$$

So the speed is 40 km h^{-1}.

The second line segment is horizontal, so its gradient is 0 and the speed is 0 km h^{-1}.

The third line segment slopes downwards so its gradient is negative.

$$\text{Gradient} = \frac{\text{distance travelled}}{\text{time taken}} = \frac{-30}{1} = -30$$

So the speed is 30 km h^{-1}.

The negative sign in the gradient occurs because the cyclist is coming back home, not going away from it. The negative sign is important in describing the cyclist's velocity, as you shall see later, but does not matter when describing speed as the direction of travel is not important. Speed only has positive values.

Example 5

The distance–time graph shows the movement of a student running 1000 metres.

Calculate the **instantaneous speeds** at, and the average speed after

a 0.5 minutes

b 2 minutes.

Instantaneous speed is calculated by working out the gradient of a tangent to the distance–time graph at that instant.

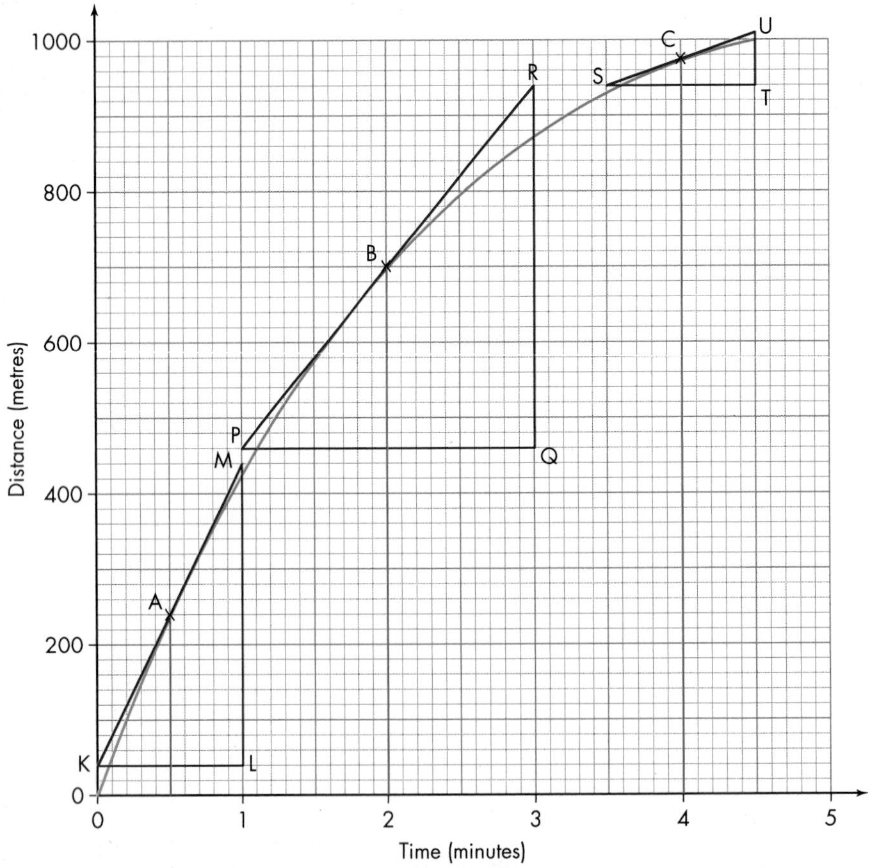

a To find the instantaneous speed, you need to find the gradient when the time is 0.5 minutes.

Draw the tangent at $t = 0.5$.

This is shown on the graph above as line KM at point A.

Draw the right-angled triangle KLM and identify the lengths KL and ML. Make KL equal to 1 minute to make the calculations easy.

$ML = 440 - 40 = 400$

$$\text{Gradient} = \frac{\text{distance travelled}}{\text{time taken}} = \frac{400}{1} = 400 \text{ metres/minute}$$

You know that distance divided by time gives you speed, so you can say that the student was running at an instantaneous speed of 400 metres per minute 0.5 minutes after the start.

To find the average speed for the first 0.5 minutes, use the graph to find the distance run after 0.5 minutes.

After 0.5 minutes the student is 240 metres from the start, so:

$$\text{Average speed} = \frac{240}{0.5} = 480 \text{ metres per minute}$$

b Draw the tangent at $t = 2$ (the line PR at point B on the graph above).

Draw the right-angled triangle PQR and identify the lengths QR and PR.

Make PQ equal to 2 minutes.

$QR = 940 - 460 = 480$

$$\text{Gradient} = \frac{\text{distance travelled}}{\text{time taken}} = \frac{480}{2} = 240 \text{ metres/minute}$$

Instantaneous speed = 240 metres/minute

After 2 minutes the student is 700 metres from the start, so:

$$\text{Average speed} = \frac{700}{2} = 350 \text{ metres per minute}$$

Discussion

Discuss the following with your peers.

How would you find the instantaneous speed at 4 minutes?

How do you think this compares with the average speed after 4 minutes?

What is happening to the instantaneous speed of the student as time passes?

Exercise 12D

1 Here is a distance–time graph of Jessica's walk home from college.

On her way home she looked for her phone and realised she had left it at college, so went back, found it and eventually returned home.

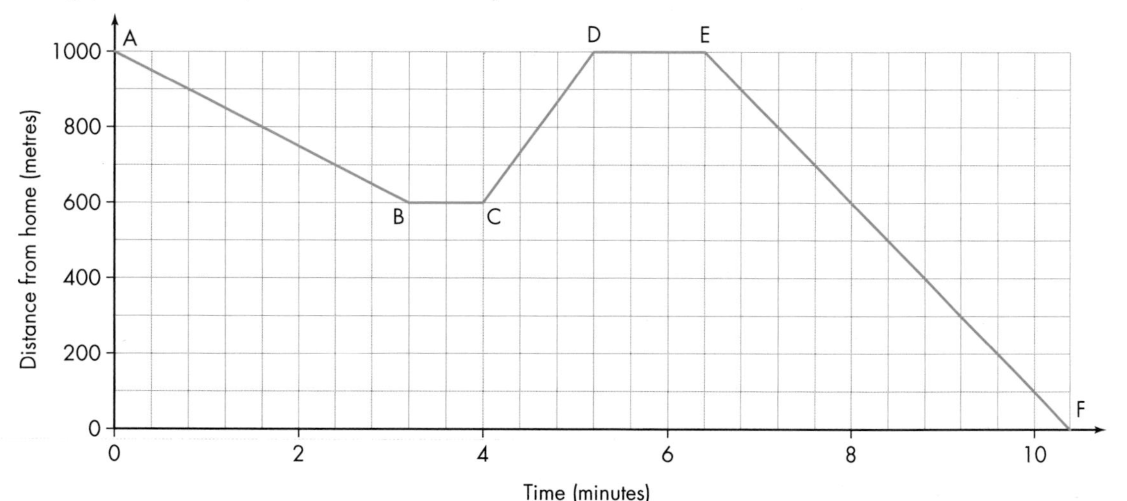

a Calculate the gradient of line segment AB. What speed does this represent?

b Calculate the gradient of line segment CD. What speed does this represent?

c Calculate the gradient of line segment EF. What speed does this represent?

2 You will need a copy of this graph to draw on.

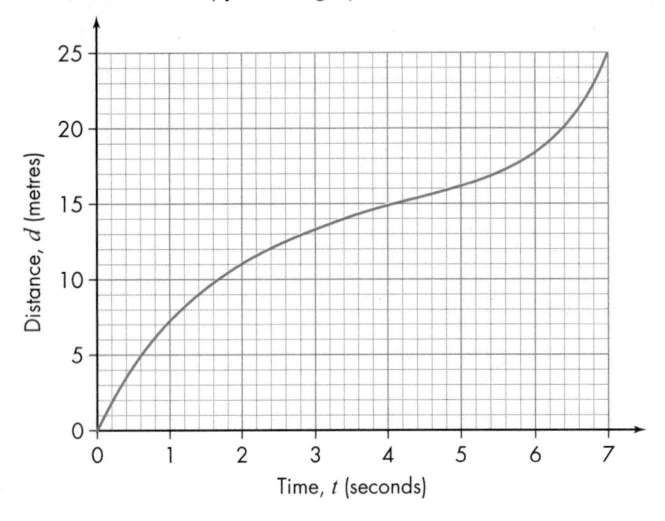

a Estimate the instantaneous speed when $t = 2$ and $t = 6$.

b Estimate the average speed after 2 seconds and after 6 seconds.

c Comment on your answers to parts **a** and **b**.

3 You will need a copy of this graph to draw on.

It shows the distance from home of a long-distance cyclist.

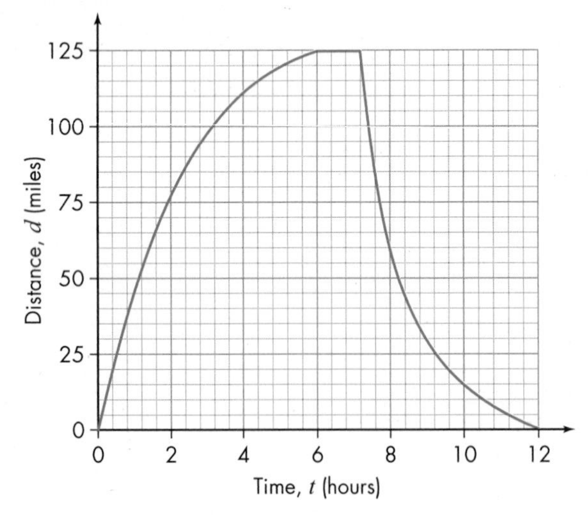

a Estimate the instantaneous speed when $t = 2$, $t = 6.5$ and $t = 8$.

b Estimate the average speed after 2 hours, after 6.5 hours and after 8 hours.

c Comment on your answers to parts **a** and **b**.

d Describe how the cyclist's speed varied throughout his 12-hour ride.

4 Adele timed her progress in a lift as it went from the ground floor to the tenth floor.

She knew that there was a distance of 2.5 metres between each floor so she knew how far she had travelled in that time.

Here are her results.

Time taken, *t* (seconds)	8	15	23	28	33	39	47	55	65
Distance travelled, *d* (metres)	2.5	5	7.5	10	12.5	15	17.5	20	22.5

a Draw a distance–time graph showing Adele's progress in the lift.

b Estimate the instantaneous speed at $t = 20$.

c Estimate the instantaneous speed at $t = 40$.

d When was the lift travelling fastest? Estimate its speed at that point.

5 Rapunzel dropped her long hair down from the top of a tower to her prince at the bottom.

Fortunately the tower had a very long ruler attached to it and so the prince was able to see how far the end of Rapunzel's hair had travelled after each second.

Here are his results.

Time taken, *t* (seconds)	1	2	3	4	5	6	7
Distance travelled, *d* (metres)	5	20	45	80	125	180	245

a Draw a distance–time graph showing how far the end of Rapunzel's hair had travelled.

b Estimate the instantaneous speed at $t = 2$.

c Estimate the instantaneous speed at $t = 5.5$.

The prince now climbed up Rapunzel's hair, calling out to her every 20 metres.

Rapunzel used the stopwatch on her mobile to keep a record of those times.

Here are her results.

Time taken, *t* (minutes)	5	12	21	32	45	60	77	96
Distance travelled, *d* (metres)	20	40	60	80	100	120	140	160

d Draw a distance–time graph showing the prince's progress.

e Estimate the instantaneous speed after 25 minutes.

f Estimate the instantaneous speed after 65 minutes.

Mathematics in the real world

Sports scientists analyse data relating to their athletes to help improve performances. Usain Bolt holds the world record and the Olympic gold medal for the 100-m sprint, and his races are analysed by sports scientists and his coach.

The distance–time graph shows his motion when he won the 100-m Olympic gold medal in Beijing in 2008.

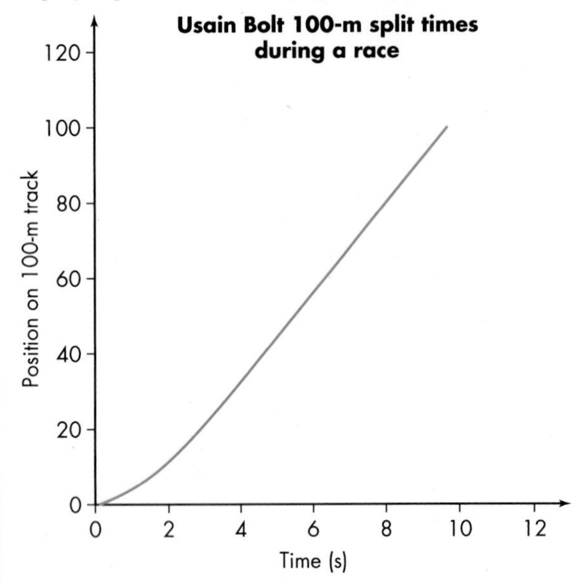

The graph shows that Bolt's speed changed through the race. Over the first 2.5 seconds, the tangents drawn to the curve have increasing gradients. After about 2.5 seconds, the graph is a straight line, so Bolt ran at a constant speed until the end of the race.

Analysis of the graph suggests that Usain Bolt could possibly run the 100 m faster. His reaction time at the start is quite slow. This is shown by the graph intersecting with the time axis sometime after the start of the race.

Velocity–time graphs

The words speed and **velocity** are often used synonymously. This is unfortunate, as in the sciences they have particular and different meanings. Two objects can have the same speed, but in the opposite direction. For objects moving in a straight line, if the velocity becomes negative, this means the direction of motion has reversed.

The **velocity** of an object is its speed in a stated direction.

Average acceleration is calculated by dividing the change in velocity by the time taken.

You will use the terms velocity and velocity–time graph when discussing acceleration, because the direction of movement as well as the size of the speed is important when calculating **acceleration**.

When you calculate the gradient of a velocity–time graph, you are dividing the velocity (in metres per second, or m s^{-1}) by time (in seconds, or s). In other words,

$$\text{Gradient} = \frac{\text{change in velocity}}{\text{time taken}} = \text{acceleration.}$$

This means the units of the gradient are metres per second per second, or m s^{-2}. These are the units of acceleration and that is what the gradient of a velocity–time graph represents.

If the gradient is positive then the object being considered is increasing its velocity: i.e. it is accelerating. If the gradient is negative then the object being considered is decreasing its velocity: i.e. it is decelerating. Notice that on velocity–time graphs the velocity can be negative. When a negative velocity becomes more negative (a negative gradient) the speed is actually increasing even though the velocity is decreasing!

Example 6

Draw a velocity–time graph which describes the following journey.

- A cyclist joins a straight road travelling at 4 m s^{-1},
- accelerates at a constant rate to a velocity of 10 m s^{-1},
- continues at a constant velocity for 20 s,
- then decelerates at a constant rate until at rest.

Now calculate the gradient of each part of the journey and interpret what the gradient represents.

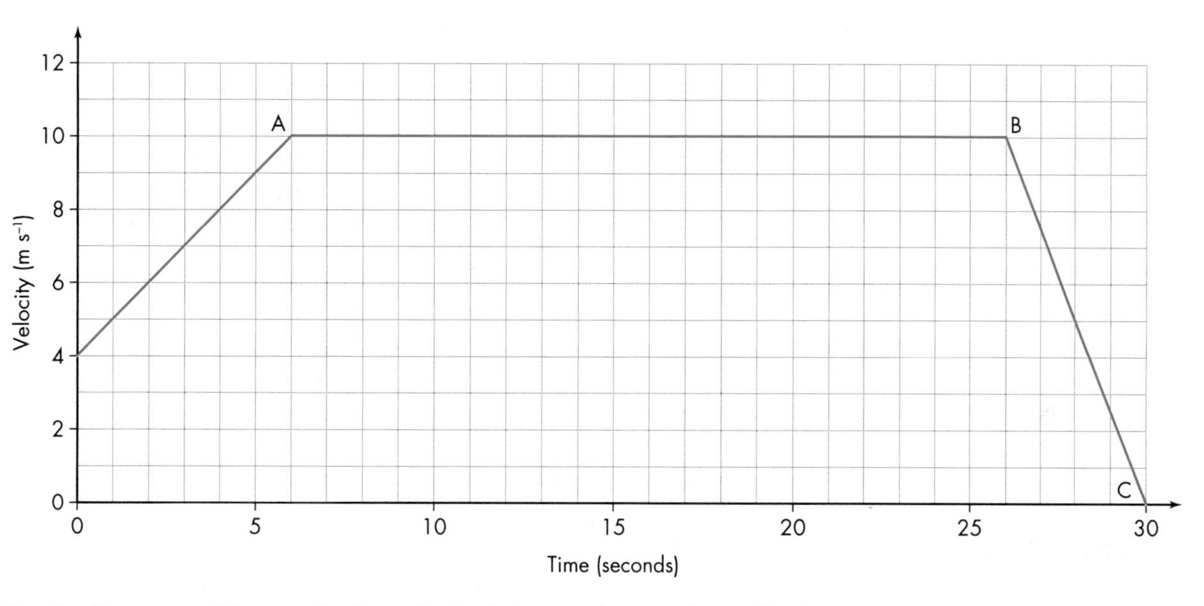

For the first part of the graph, the velocity is increasing, so the cyclist is accelerating.

$$\text{average acceleration} = \frac{(10-4)}{6} = \frac{6}{6} = 1 \text{ m s}^{-2}$$

Velocity here is measured in metres per second (m s^{-1}) and time is measured in seconds (s), so acceleration is measured in metres per second squared (m s^{-2}).

While the cyclist is accelerating, the gradient is positive.

For the second part of the graph, the velocity is constant at 10 m s^{-1}.

$$\text{average acceleration} = \frac{0}{26-6} = 0 \text{ m s}^{-2}$$

The cyclist is moving at a constant velocity and the graph has zero gradient.

For the third part of the graph, the velocity is decreasing, so the cyclist is slowing down.

$$\text{average acceleration} = -\frac{(10-0)}{(30-26)} = -\frac{10}{4} = -2.5 \text{ m s}^{-2}$$

The negative gradient represents a negative acceleration, or deceleration.

The cyclist's motion between points B and C can be described as either an acceleration of −2.5 m s^{-2} or as a deceleration of 2.5 m s^{-2}. Be careful with the signs of accelerations.

Exercise 12E

1 Here is a velocity–time graph.

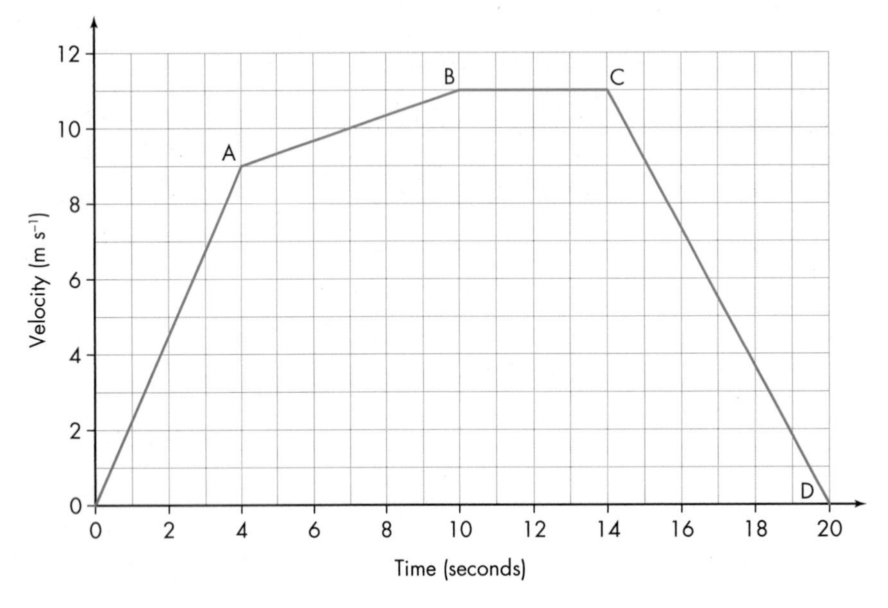

a Calculate the average acceleration from A to B.

b Calculate the average acceleration from B to C.

c Calculate the average acceleration from C to D.

2 Here is a velocity–time graph.

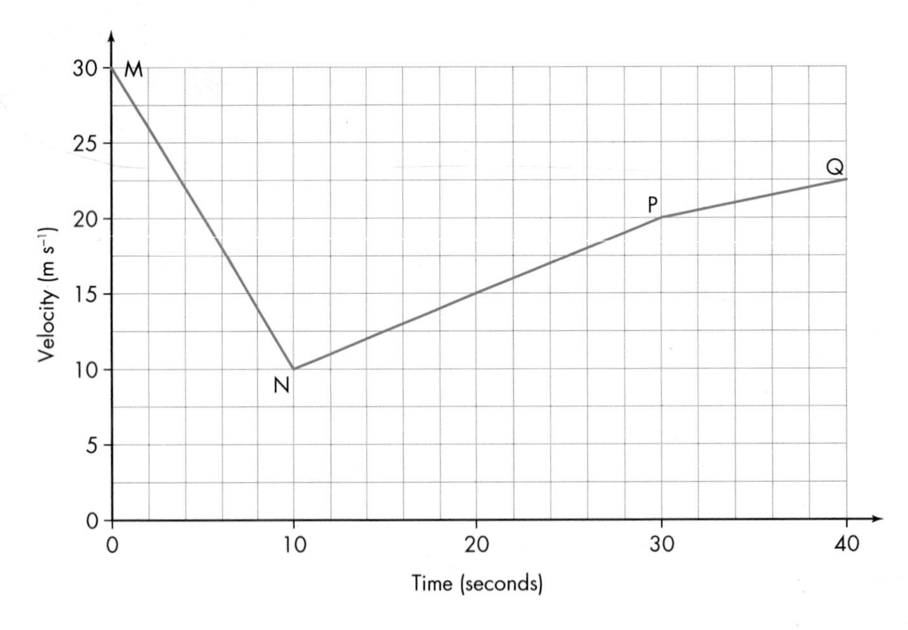

a Calculate the average acceleration from M to N.

b Calculate the average acceleration from N to P.

c Calculate the average acceleration from P to Q.

Instantaneous acceleration

In Example 5 you looked at instantaneous speeds and average speeds on a distance–time graph.

You can use similar techniques to estimate **instantaneous acceleration** and average acceleration on a velocity–time graph. To find the instantaneous acceleration you draw a tangent at the relevant point on the graph and calculate its gradient. To find the average acceleration you divide the change in velocity by the time interval in which that change has taken place.

The **instantaneous acceleration** is calculated by working out the gradient of a tangent to the velocity–time graph at that instant.

Example 7

The velocity–time graph shows the velocity of a racing car as it goes round a bend on the track.

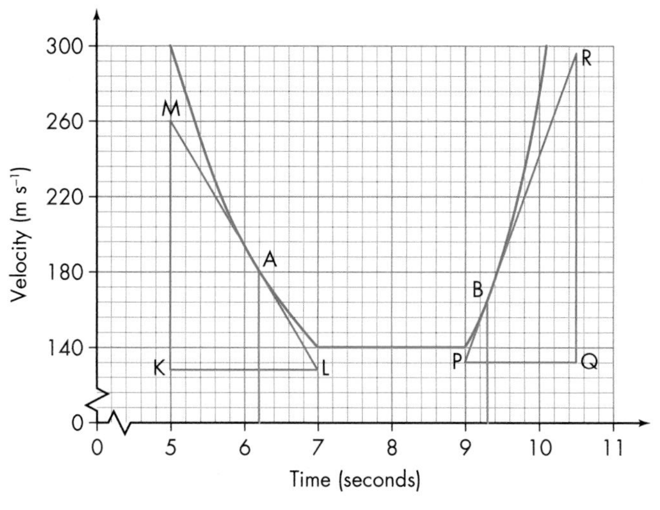

a Calculate the instantaneous accelerations at

 i 6.2 seconds

 ii 9.3 seconds.

b Calculate the average acceleration between 5 and 9 seconds.

---

a i To find the instantaneous acceleration, you need to find the gradient when the time is 6.2 seconds.

Draw the tangent after 6.2 seconds.

This is shown on the graph as line ML at point A.

Draw the right-angled triangle KLM and identify the lengths KL and MK.

Here KL is made equal to 2 seconds to make the calculations easy.

$$\text{MK} = 260 - 128 = 132$$

$$\text{Gradient} = \frac{\text{change in velocity}}{\text{time taken}} = \frac{-132}{2} = -66 \text{ m s}^{-2}$$

Since this is negative it represents a deceleration of 66 m s^{-2}.

ii Now you need to find the gradient when the time is 9.3 seconds.

Draw the tangent after 9.3 seconds.

This is shown on the graph as line PR at point B.

Draw the right-angled triangle PQR and identify the lengths PQ and QR.

Here PQ is made equal to 1.5 seconds. (You could make the calculation simpler by making the triangle smaller and drawing PR from just 9 to 10 seconds, but when the tangent is drawn by eye, a larger triangle can reduce the inaccuracies resulting from a small triangle).

$QR = 296 - 132 = 164$

$$\text{Gradient} = \frac{\text{change in velocity}}{\text{time taken}} = \frac{164}{1.5} = 109 \text{ m s}^{-2} \text{ (to 3 s.f.)}$$

Since this is positive it represents an acceleration of 109 m s^{-2}.

b In the interval from 5 to 9 seconds the velocity has changed from 300 m s^{-2} to 140 m s^{-2}.

So

$$\text{Average acceleration} = \frac{-160}{4} = -40 \text{ m s}^{-2}$$

This is interpreted as an average deceleration of 40 m s^{-2}.

Example 8

A velocity–time graph is shown.

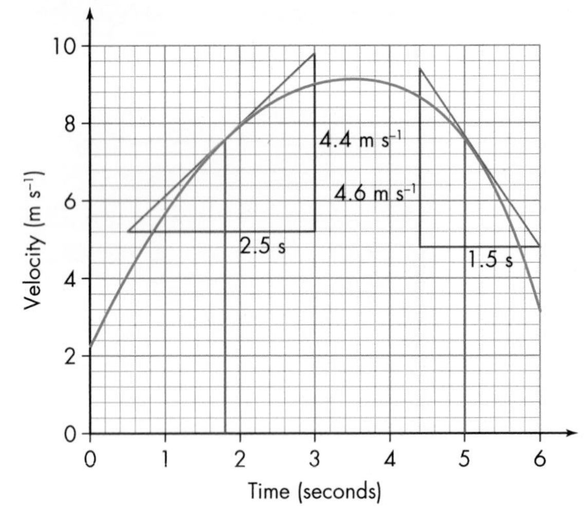

Estimate the instantaneous acceleration at

a 1.8 seconds

b 5 seconds.

a To find the instantaneous acceleration, you need to find the gradient when the time is 1.8 seconds.

Draw the tangent at $t = 1.8$.

$$\text{Gradient} = \frac{\text{change in velocity}}{\text{time taken}} = \frac{4.4}{2.5} = 1.76 \text{ m/s}^2$$

Change in velocity divided by time gives acceleration, so the instantaneous acceleration at $t = 1.8$ s is 1.76 m/s^2.

b Draw the tangent at $t = 5$.

$$\text{Gradient} = \frac{\text{change in velocity}}{\text{time taken}} = \frac{-4.6}{1.5} = -3.07 \text{ m/s}^2 \text{ (to 3 s.f.)}$$

The instantaneous acceleration is −3.07 m/s^2. The negative sign indicates this is a deceleration.

Discussion

Discuss the following with your peers.

How would you find the instantaneous acceleration at 3.3 seconds?

What is happening to the acceleration as time passes?

Compare the two examples and discuss the way the units of acceleration have been written.

Exercise 12F

1 You will need a copy of this graph to draw on.

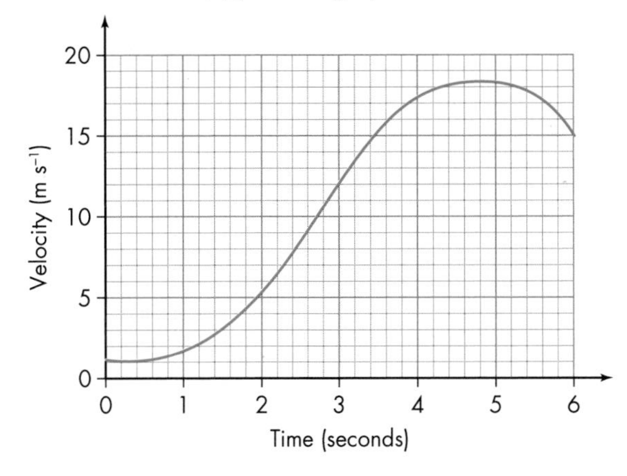

a Estimate the instantaneous acceleration when $t = 2$ and $t = 4$.

b Estimate the velocity when the acceleration is zero.

c Give the time interval when the acceleration is negative.

2 You will need a copy of this graph to draw on.

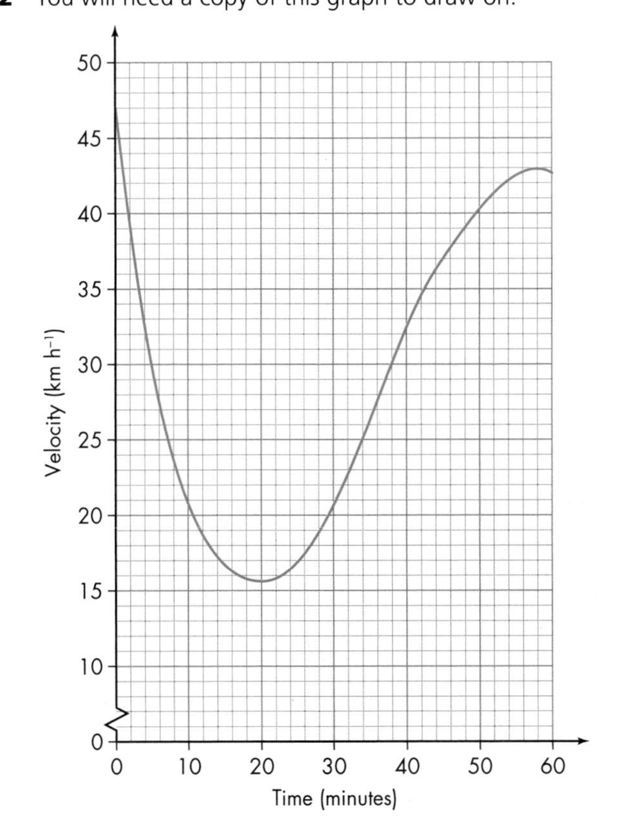

a Estimate the acceleration at 50 minutes.

b Estimate the values of the acceleration when the velocity is 30 m s^{-1}.

c Estimate the velocities when the acceleration is zero.

d Describe how the acceleration changes over the hour.

3 The velocity of a baseball, v m s^{-1}, t seconds after being hit is modelled by the quadratic equation

$$v = 25 - 5t^2$$

a Plot a velocity–time graph for this, for values of t every 0.5 seconds between 0 and 5 seconds.

b Estimate the acceleration of the baseball after 3 seconds.

4 The velocity of a shuttlecock, v m s^{-1}, t seconds after being hit is modelled by the quadratic equation

$$v = 15 + 1.5t - 0.2t^2$$

a Plot a velocity–time graph for this quadratic model, for $0 \leqslant t \leqslant 12$.

b For what values of t is the shuttlecock accelerating?

c At what time is the acceleration zero?

d Estimate the instantaneous acceleration after 9 seconds.

5 An astronaut on the moon is calculating the value of acceleration due to gravity.

He drops a ball down a crater so that it falls vertically without touching anything on its descent.

The data from this experiment is recorded in the table.

Time after the ball is dropped, *t* (seconds)	0	0.5	1	1.5	2	2.5	3	3.5	4
Vertical distance travelled, *d* (metres)	0	0.4	1.6	3.6	6.4	10	14.4	19.8	25.6

a Draw a distance–time graph showing this data.

b Draw tangents at the non-zero values of t and calculate their gradients.

c Draw a velocity–time graph using your results from part **b**.

d What do you notice about your velocity–time graph?

e Write a report about the acceleration of the ball on the moon.

Case study

Linear accelerators used in medicine

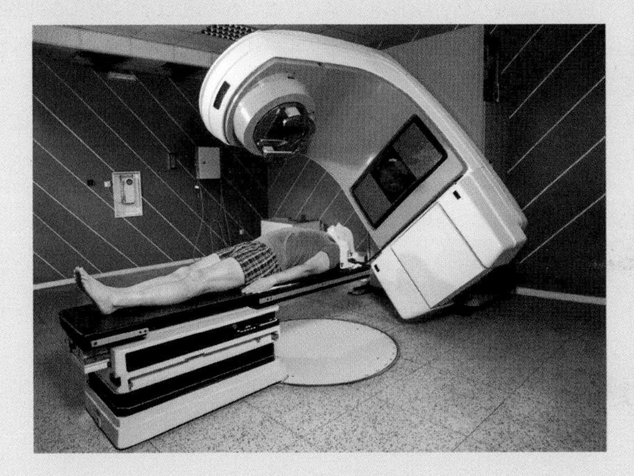

Some radiation therapy equipment for the treatment of cancer makes use of the equations for acceleration studied in this chapter.

The patient lies under a linear accelerator that accelerates electrons to very high velocities. By adjusting the acceleration of the electrons, oncologists are able to change the energy levels of the X-ray radiation used to treat different cancerous tumour cells.

The diagram shows the main parts of a linear accelerator used in radiation therapy.

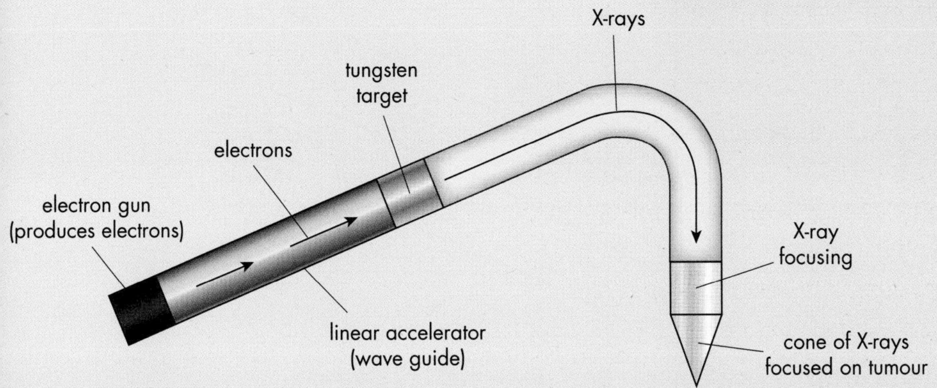

The energy of the X-rays produced depends on the speed at which electrons leaving the linear accelerator hit the tungsten target.

Electrons enter the linear accelerator at relatively low speeds but leave it at nearly the speed of light. The speed of light is 300 000 000 m s^{-1}.

Discussion

Discuss the following with your peers.

1 What would the average acceleration be in the accelerator if the electrons enter at 40% of the speed of light, stay in the wave guide for 0.000 005 seconds, and leave at a speed of

a 50% of the speed of light?

b 75% of the speed of light?

c 99.9% of the speed of light?

Electrons actually leave the linear accelerator at 99.9% of the speed of light. To achieve this acceleration very large amounts of energy are needed.

2 To obtain high speeds long linear accelerators are needed. Instead of using a linear accelerator can you think of another way of achieving the required acceleration that would use less space for the radiation therapy equipment?

(Hint: Find out how particles are accelerated at CERN, The European Organisation for Nuclear Research in Geneva.)

Project work

The changes in an infant's length during their first 36 months are monitored by midwives, health visitors and doctors. The graph shows the average lengths for boys at different ages for different segments of the population.

The line labeled A shows the average lengths for the 25th percentile. This means that 25% of the population at that age will have the same length or less, so 75% of the population will be longer.

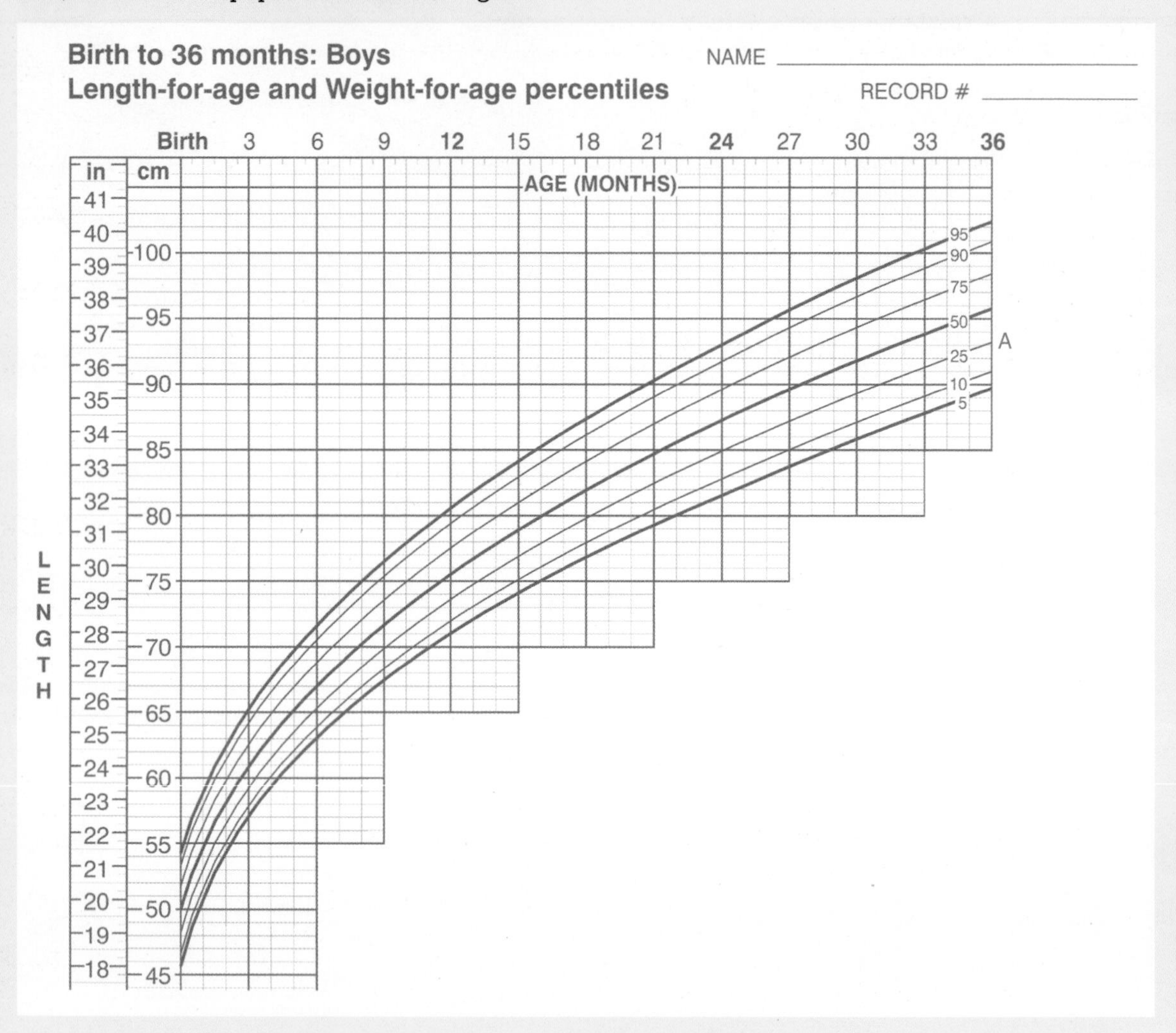

The rate of change of these graphs shows the increase in length per month.

Look at the 5% and 95% lines. How does the rate of change (gradient) differ between boys on the 5% line and boys on the 95% line at 3 months old and 30 months old?

Calculate the rates of change of length of a boy (i.e. the rate at which he is growing) at 3 months and 30 months.

Discussion

Consider two sets of parents with new born boys.

John is on the 5% line. Hugh is on the 95% line.

Discuss with your peers what differences they would notice as the boys grew up.

Check your progress

How confident are you feeling in your level of knowledge? What do you need to practise more?

Spec reference	Learning objective			
G3.1	Interpret the gradient of a straight line graph as a rate of change			
G3.2	Interpret the gradient at a point on a curve as an instantaneous rate of change			
G3.3	Estimate rates of change for functions from their graphs			
G4.1	Know that the average speed of an object during a particular period of time is given by $\frac{\text{distance travelled}}{\text{time taken}}$			
G5.1	Know that the gradient of a distance–time graph represents speed and that the gradient of a velocity–time graph represents acceleration			

13 Exponential functions

Exponential function equations look very similar to some non-linear functions you have seen before, such as $y = x^2$. In the equation $y = x^2$, x is the base and the power is 2. The difference is that in exponential functions it is the power that is the variable, not the base, so $y = 2^x$ is an exponential function where 2 is the base and the power is the variable x.

An **exponential function** has the variable as the power.

The **base** of an exponential function is the number which is raised to a given power.

Exponential functions are very common when mathematics is used to model processes in the real world. You met some exponential functions when you dealt with compound interest in Chapter 2 on personal finance and looked at their graphs in Chapter 11.

In Chapter 2 you found that the formula for calculating the amount accumulated using compound interest is

$$A = P(1 + r)^t,$$

where:

A is the amount accumulated

P is the principal

r is the (decimal equivalent of the) interest rate

t is the time, usually measured in years.

This might not look like a function such as $y = 2^x$ but this example shows how it compares.

William invested £1 in the Hastings Building Society in 1966 when it agreed to pay 10% per annum.

The amount in the building society after t years will be

$$A = 1(1 + 0.1)^t \text{ or } A = 1.1^t.$$

If you plot the graph of $A = 1.1^t$ you will see how William's investment grows.

Notice that it grows very slowly at first – it takes just over 7 years to double (the Rule of 72!) and then the graph gets steeper and steeper. After 20 years William's investment is worth £6.73, after 30 years £17.45, after 40 years £45.26 and after 50 years it is worth 1.1^{50} pounds, that is £117.39.

After 100 years it would be worth 1.1^{100} pounds, that is £13 780.61.

Of course interest rates vary year on year so this example is not a real-life scenario.

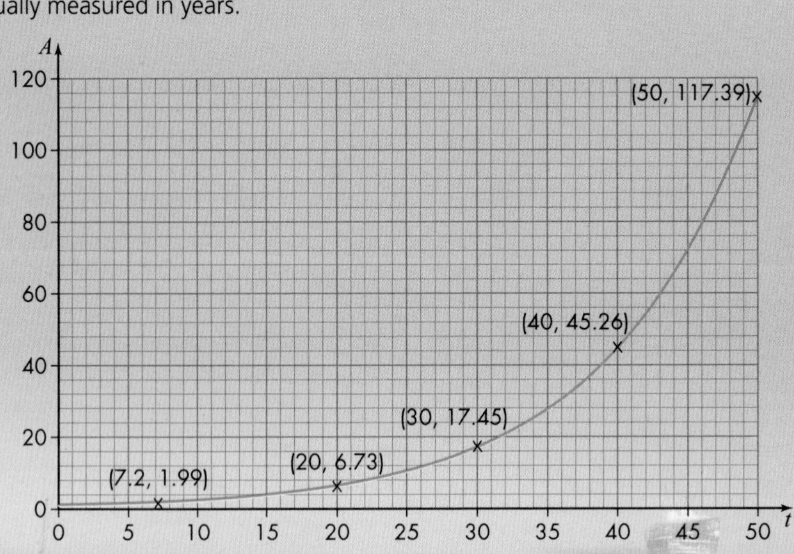

Mathematics in the real world

In 1975 Gordon E. Moore suggested that the number of components in an integrated circuit would double every two years. This prediction proved accurate for several decades and was the basis the semiconductor industry used for long-term planning and to set targets for research and development. Other digital advancements such as memory capacity and the number of pixels in digital cameras are linked to what is now known as Moore's Law. Strictly speaking, this is not a law, but an observation.

In the mid 19th century there was a rapid growth in the rabbit population in Australia. Rabbits had been introduced as a food source, but some were released into the wild in 1859 for hunting purposes and with mild winters and few, if any, natural predators, they bred freely all year round. This meant that their numbers increased like the numbers in an exponential function: slowly at first and then faster and faster.

Both these real-world examples (and those mentioned later in this chapter) follow exponential growth for a while. However, there often becomes a point when modelling natural growth as an exponential function breaks down. In the case of Moore's Law, the technology company Intel confirmed in 2015 that the rate of advancement had slowed. In Australia, various control measures have been put in place to try and keep the number of rabbits down, so the population is not continuing its exponential growth.

Discussion

Choose one or more of the following and discuss it with your peers.

1. What other examples can you think of where you have seen similar growth?
2. Why do you think some examples of exponential growth in the natural world break down after a time?

G6: The function a^x

Learning objectives

You will learn how to:

- Use a calculator to find values of exponential functions.
- Use a calculator's log and ln functions to solve equations that include exponential functions.

Using a calculator to calculate values of exponential functions

In the opening section you saw the graph of $y = 1.1^x$ for $x \geqslant 0$. You also met the graphs of exponential functions in Chapter 11.

You will now look at how to find values for y and how to draw and sketch the graphs of exponential functions.

You are expected to use a graphics calculator when drawing the graphs of exponential functions. You will be able to use one in the examinations (though it is not a requirement) and it is important to know how to use the functions quickly and correctly. Developing your techniques now will make sure you can sketch graphs swiftly and competently.

Example 1

Use a calculator to work out the following.

a 4^5 b 3^0

c 2^{-2} d 1^{-5}

To work these out, enter the first number (the base), then press the key for raising a number to a power (or exponent) followed by the second number (the exponent).

On some calculators the key for raising a number to a power is marked with ∧ like the symbol used in spreadsheets but calculators vary. Make sure you know which key you need to use on your calculator.

For the negative exponents you might have to use the sign change key, often marked (–), rather than the minus key. Some calculators will give the answer as a fraction, so if you want the answer as a decimal, convert your answer using the appropriate keys or menu.

a $4^5 = 1024$ b $3^0 = 1$

c $2^{-2} = \frac{1}{4}$ or 0.25 d $1^{-5} = 1$

Example 2

Use a calculator to work out the following.

a 4.1^3 b 7.2^0

c $81^{0.5}$ d $0.25^{-0.5}$

a $4.1^3 = 68.921$ b $7.2^0 = 1$

c $81^{0.5} = 9$ d $0.25^{-0.5} = 2$

Exercise 13A

Give your answers to three significant figures where appropriate.

1 Use a calculator to work out the following.

a 5^4 **b** 6^0

c 7^2 **d** 8^5

2 Use a calculator to work out the following.

a 4.1^0 **b** 234^0

c $(-8)^0$ **d** 0.09^0

What do your results suggest about any number, except 0, to the power 0?

3 Use a calculator to work out the following.

a 4^{-3} **b** 23^{-5}

c $(-8)^{-6}$ **d** $2.1^{-3.1}$

What do you notice about your results?

4 This question can be done using a graphics calculator or a spreadsheet. If you use either of these, then you should make a sketch drawing of the graphs.

a Copy and complete the table below for the exponential function $y = 2^x$.

x	−3	−2	−1	0	1	2	3
y	0.125		0.5			4	

b Copy and complete the table below for the exponential function $y = 1.2^x$.

x	−3	−2	−1	0	1	2	3
y		0.694…			1.2		

c Copy and complete the table below for the exponential function $y = 1.5^x$.

x	−3	−2	−1	0	1	2	3
y				1			

d Draw the three graphs represented by the functions and tables of values on the same set of axes for $-3 \leqslant x \leqslant 3$.

You can do this on a sheet of graph paper, or show them on a graphics calculator. If you use a graphics calculator, sketch a screenshot of all three graphs on the same diagram.

e What are the coordinates of the point that is common to all three graphs?

f As x becomes large what happens to the value of y in each of the exponential functions of parts **a**, **b** and **c**?

5 This question can be done using a graphics calculator or a spreadsheet. If you use either of these, then you should make a sketch drawing of the graphs.

a Copy and complete the table below for the exponential function $y = 0.5^x$.

x	−3	−2	−1	0	1	2	3
y	8		2			0.25	

b Copy and complete the table below for the exponential function $y = 0.8^x$.

x	−3	−2	−1	0	1	2	3
y		1.5625			0.8		

c Copy and complete the table below for the exponential function $y = 0.4^x$.

x	−3	−2	−1	0	1	2	3
y				1			

d Draw the three graphs represented by the functions and tables of values on the same set of axes for $-3 \leqslant x \leqslant 3$.

You can do this on a sheet of graph paper, or show them on a graphics calculator. If you use a graphics calculator, sketch the screen shot of all three graphs on the same diagram.

e What are the coordinates of the point that is common to all three graphs?

f As x becomes large what happens to the value of y in each of the exponential functions of parts **a**, **b** and **c**?

Using the calculator log function to solve problems with exponential equations

The **logarithm** or **log** of a number to a given base is the power of that base that gives the number.

For example, $10^2 = 100$.

Therefore, the logarithm of 100 to the base 10 is 2.

Ideally you write $\log_{10} 100 = 2$, but since most of the time you are using logarithms to the base 10, you can omit the subscript 10.

If a process is modelled by an exponential equation of the form $a^x = b$, then x can be found using the log function on your calculator as follows.

$$x = \frac{\log b}{\log a}$$

The **logarithm** or **log** of a number to a given base is the power of that base that gives the number.

The **log function** on a calculator gives the logarithm of a number to the base 10.

Example 3

Solve the equation

$$8^x = 25$$

Here $a = 8$ and $b = 25$, so substituting these values into $x = \frac{\log b}{\log a}$, you have

$$x = \frac{\log 25}{\log 8} = \frac{1.3979...}{0.9031...} = 1.5479... = 1.55 \text{ to 3 s.f.}$$

Alternatively, using the laws of logarithms:

If $8^x = 25$ then you can take logarithms of both sides.

$$\log(8^x) = \log(25)$$

$$\log(8^x) = x \times \log 8, \text{ so } x \times \log 8 = \log 25$$

Divide each side by log 8 to calculate x as shown above.

Different models of calculators have slightly different keys to access these functions, so make sure you know the correct keys to use.

When you do calculations like this, you can check by substituting for x in the original equation.

$8^{1.55} = 25.1066...$ which rounds to 25

Using the calculator ln function to solve problems with exponential equations

The **ln function** on a calculator gives the logarithm of a number to the base e. e is the number 2.718 281....

For example, $e^{3.5} = 33.12...$

Therefore, the logarithm of 33.12 to the base e is 3.5.

Since e is such a special number in mathematics, instead of $\log_e 33.12 = 3.5$, a special notation, ln, is used. In this case you write ln 33.12 = 3.5.

If a process is modelled by an exponential equation of the form

$$e^{kx} = b$$

where k is a constant, then x can be found using the ln function on your calculator as follows.

$$x = \frac{\ln b}{k}$$

The **ln function** on a calculator gives the logarithm of a number to the base e.

e is the number 2.718 281....

Example 4

Solve the equation

$$e^{2x} = 4$$

Here $k = 2$ and $b = 4$ so substituting these values into $x = \frac{\ln b}{k}$, you have

$$x = \frac{\ln 4}{2} = \frac{1.3862...}{2} = 0.6931... = 0.693 \text{ to 3 s.f.}$$

More on solving exponential equations

It is important to take care with the log and ln functions on your calculator. The log function uses logarithms to the base 10 and the ln function uses logarithms to the base e.

The flowchart shows a summary of the different types of functions that you have met so far and the formulas used to solve them. The flowchart will be extended later in the chapter.

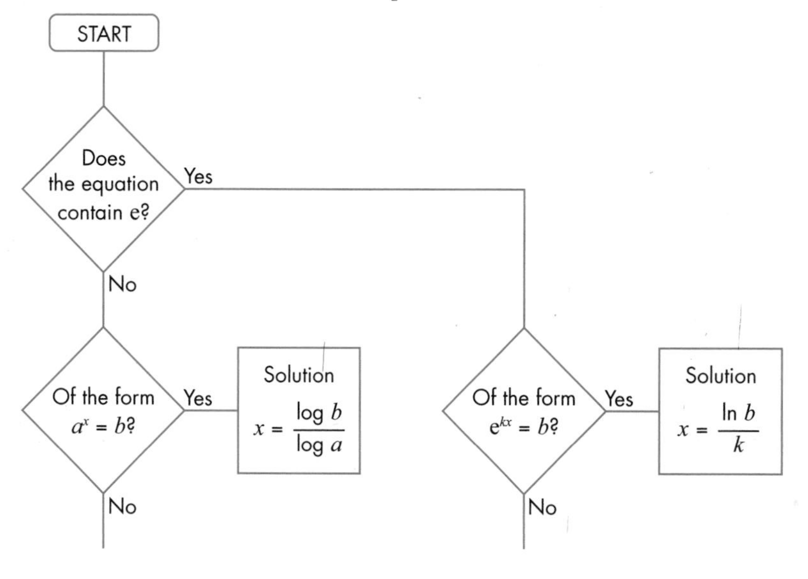

If you have negative values in these equations, do not worry: the formulae still work. Logarithms of negative numbers do not exist, so a and b will always be positive numbers. However k can be negative and x can turn out to be negative. Your calculator will take care of the signs for you. Remember always to work with all the digits in your calculator display, only rounding the final answer. Round to three significant figures if no other accuracy is stated. Here are three examples that involve negative numbers.

Example 5

Solve the equation

$$7^x = 0.4$$

Here $a = 7$ and $b = 0.4$, so substituting these values into $x = \dfrac{\log b}{\log a}$, you have

$$x = \frac{\log 0.4}{\log 7} = \frac{-0.3979...}{0.8450...} = -0.4708... = -0.471 \text{ to 3 s.f.}$$

Example 6

Solve the equation

$$e^{-0.3x} = 0.9$$

Here $k = -0.3$ and $b = 0.9$ so substituting these values into $x = \dfrac{\ln b}{k}$, you have

$$x = \frac{\ln 0.9}{-0.3} = \frac{-0.1053...}{-0.3} = 0.3512... = 0.351 \text{ to 3 s.f.}$$

Example 7

Solve the equation

$$e^{-0.15x} = 17$$

Here $k = -0.15$ and $b = 17$ so substituting these values into $x = \dfrac{\ln b}{k}$, you have

$$x = \frac{\ln 17}{-0.15} = \frac{2.833...}{-0.15} = -18.888... = -18.9 \text{ to 3 s.f.}$$

Mathematics in the real world

In 2014, countries in West Africa experienced a serious outbreak of the Ebola virus. Ebola is a highly infectious and very dangerous disease, and the 2014 outbreak spread rapidly and caused many deaths.

The World Health Organization (WHO) in Geneva monitored the number of reported cases of Ebola each week from April 2014.

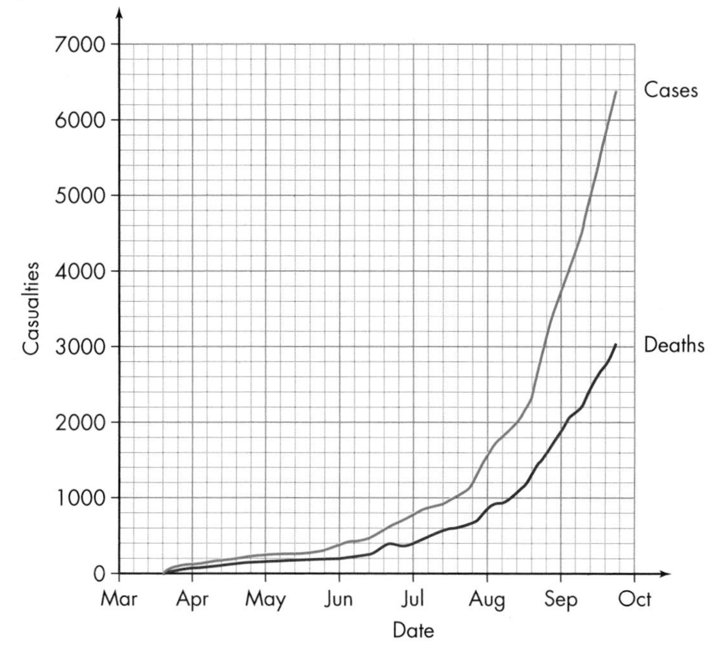

Notice that both these curves are typical of exponential functions. They start fairly flat then suddenly get steeper and steeper.

Mathematical models that included exponential functions were used to predict the spread of the disease and to model the effect of vaccines to halt and reverse the progress of the disease. The models predicted that a programme of vaccinations could prevent hundreds of thousands of deaths. However, at the time, the vaccinations were experimental and untested, and so doctors and scientists had an ethical dilemma about the use of untested drugs.

Fortunately, the spread of Ebola has been contained and Sierra Leone on 7th November 2015 and Guinea on 29th December 2015 were declared Ebola-free. Liberia was declared Ebola-free on 3rd September 2015, but a further case was reported on 20th November 2015, so the WHO still considered Liberia a public health emergency in January 2016.

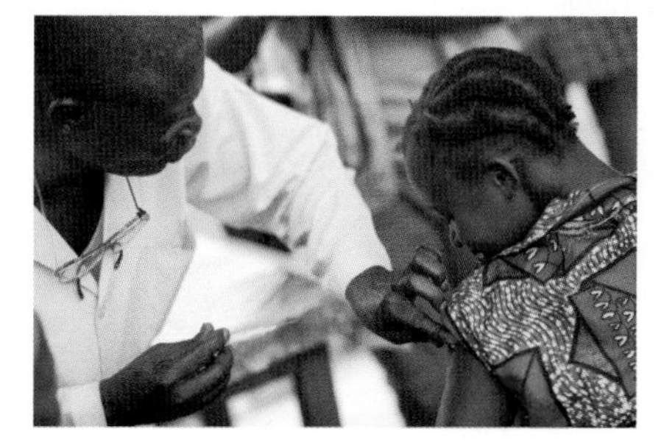

The outbreak was contained by strict restrictions on travel from infected countries, isolation of infected people with medical care and the discovery of a successful vaccine at the end of July 2015.

Exercise 13B

1 A process is modelled by the exponential equation

$a^x = b.$

For each of the pairs of values for a and b below, solve for x.

a $a = 0.5, b = 10$

b $a = 4.7, b = 6$

c $a = 1.5, b = 0.4$

2 A process is modelled by the exponential equation

$e^{kx} = b.$

For each of the pairs of values for b and k below, solve for x.

a $b = 10, k = 1$

b $b = 5, k = -0.2$

c $b = 0.8, k = -0.5$

3 In a biomedical company producing vaccines, a certain type of cell divides into two every day.

This process is modelled by an exponential function with the number of cells, N, after D days, being given by the equation

$N = a^D.$

If the constant $a = 2$, in how many days, D, will the number of cells, N, be 256?

4 The Environment Agency uses an exponential model to predict the growth of a newly-introduced water plant in the rivers it cares for.

The plant grows very quickly. The area, A, in square metres of river it covers in time T measured in days is modelled by the equation

$A = e^{kT}.$

If the constant $k = 0.1$, in what time, T, will the plant cover 100 m^2 of the river?

5 The spread of a disease is recorded and the number of cases in months 5, 6 and 7 after the start of the disease is as follows.

Month number (t)	5	6	7
Number of cases (n)	286	888	2751

This can be modelled by the equation $n = 3.1^t$.

a How many cases were there after 3 months?

b What will be the value of t when there are 10 000 cases?

c What was the value of t when the first case was discovered?

G7: The number e

Learning objectives

You will learn how to:

- Understand that e has been chosen as the standard base for exponential functions.

The exponential functions you have looked at so far have been of the form $y = a^x$ or the form $y = e^x$, where e is a special constant known as Euler's number.

e is an irrational number. (It neither terminates nor repeats.) It is a decimal constant and has a value of 2.718 281 828 459 045 235 ....

The number e is an important mathematical constant that is the base of the **natural logarithm**. It is also the limit of $\left(1+\frac{1}{n}\right)^n$ as n approaches infinity, an expression that arises in the study of compound interest. It can be calculated as the sum of the infinite series:

$$e = \sum_{n=0}^{\infty} \frac{1}{n!} = 1 + \frac{1}{1} + \frac{1}{2\times1} + \frac{1}{3\times2\times1} + \ldots$$

Remember the symbol Σ means 'the sum of' and $n!$ is known as n factorial, and is calculated as

$$n! = n \times (n-1) \times (n-2) \times (n-3) \times \ldots 3 \times 2 \times 1.$$

The constant e can be defined in many other ways. For example, it can be defined as the unique positive number a such that the graph of the function $y = a^x$ has unit slope at $x = 0$.

Exponential functions of the form $y = e^x$, are known as the **natural exponential function**.

In any exponential change with base e, the quantity increases (or decreases) at a rate that is directly proportional to its current value. For example, if a quantity N varies with time,

the rate of change of $N = kN$ (or $-kN$)

where k is a constant that does not change as t changes.

The solution to this equation is

$$N = N_0e^{kx} \text{ (or } N = N_0e^{-kx}).$$

Such functions occur in many natural growth and decay situations, when there is continuous change in a quantity and not just change at fixed time intervals. Note that a positive exponent (e^x) indicates exponential increase (also called exponential growth) and a negative exponent (e^{-x}) indicates exponential decrease (also called exponential decay).

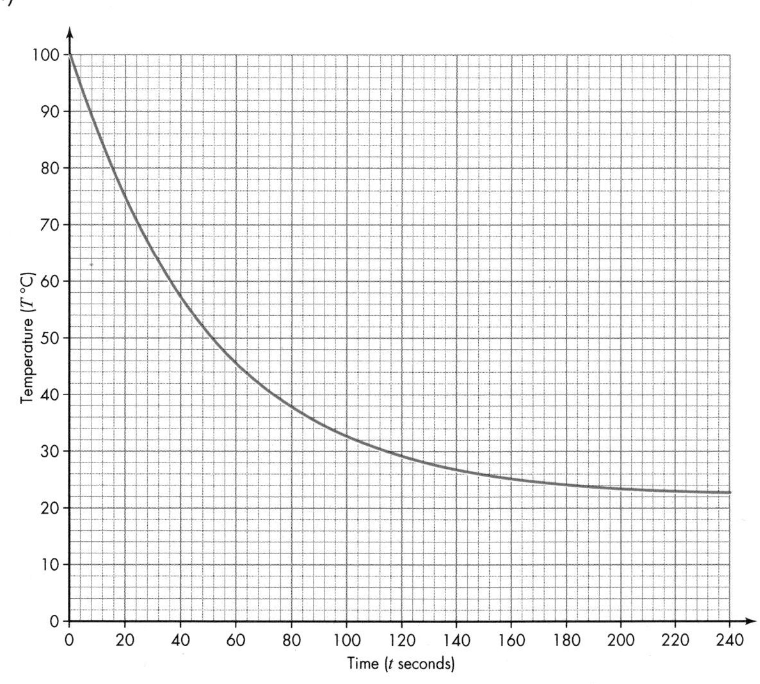

When a hot object such as a cup of tea cools, the temperature decreases rapidly at first but as time passes it cools more slowly. The graph shows how the temperature of boiling water in a kettle changes once the kettle is switched off.

As can be seen from the graph, as time passes and the temperature decreases, the rate at which the temperature decreases also decreases continuously (the gradient becomes less steep). It can be shown that the rate at which the temperature decreases is directly proportional to the difference in temperature between the water in the kettle and its surroundings.

The temperature of the water (T) in degrees Celsius can be modelled as

$$T = 22 + 78e^{-0.02t}$$

where t is the time, in seconds, since the kettle was switched off.

Forensic scientists use similar equations when estimating the time of death. Normal human body temperature is 37°C and the body would cool in a similar way to the graph above. The equation would depend on the ambient temperature: if that was 14°C then the graph would start at the point (0, 37) and would flatten out at 14°C and the equation would be

$$T = 14 + 23e^{-kt}$$

where the value of k affects how quickly the body cools. Therefore if the body temperature is taken at two points after death, the value of k can be found and the time since death can be estimated. In practice, there are many factors that affect the value of k (for example, how the body is lying, whether it is covered, whether the ambient temperature varies), so the model is not accurate.

Radioactive dating also uses a model of exponential decay to estimate the time since an organism was alive. The rate of decay of a sample of radioactive element decreases exponentially with time and the rate of decay depends only the element and not any external factors such as temperature. Carbon dating is a method based on measuring the amount of radioactive carbon-14 present in a sample of once-living material, such as wood or bone.

Bacterial populations grow continuously and the population can double in a fixed time period (such as every hour) if there is no shortage of resources. In these conditions, the rate of growth of the population is exponential with base e. This can be modelled as

$$P = P_0e^{rt}$$

where P_0 is the initial population, r is the relative rate of growth, expressed as a proportion of the population and t is time.

The graph shows exponential growth of bacteria where the population doubles every hour, starting with a population of 10 at $t = 0$. The population, P, is given by the equation

$$P = 10e^{0.693t}.$$

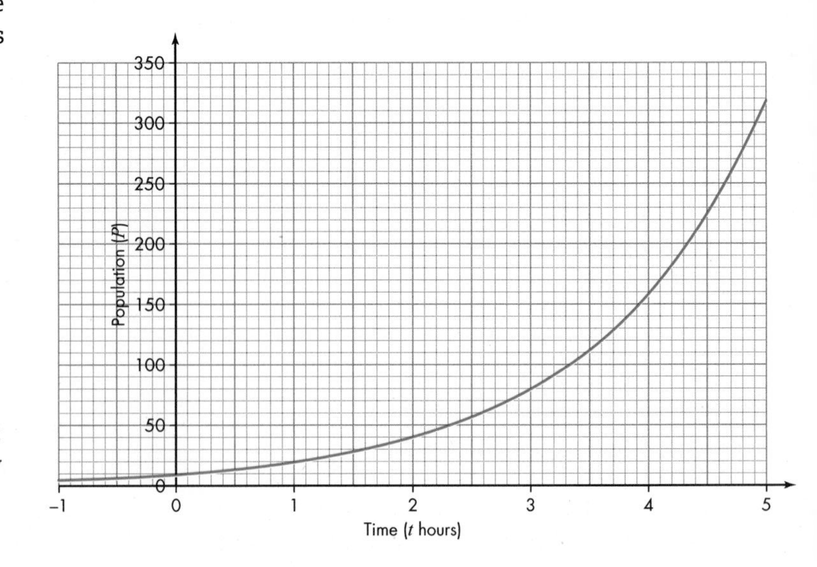

Calculations involving e can be solved using the e^x function on your calculator. For example, if $x = 3$,

$y = e^3 = 20.0855...$

Here is the graph of $y = e^x$, with the coordinates of some points marked on it.

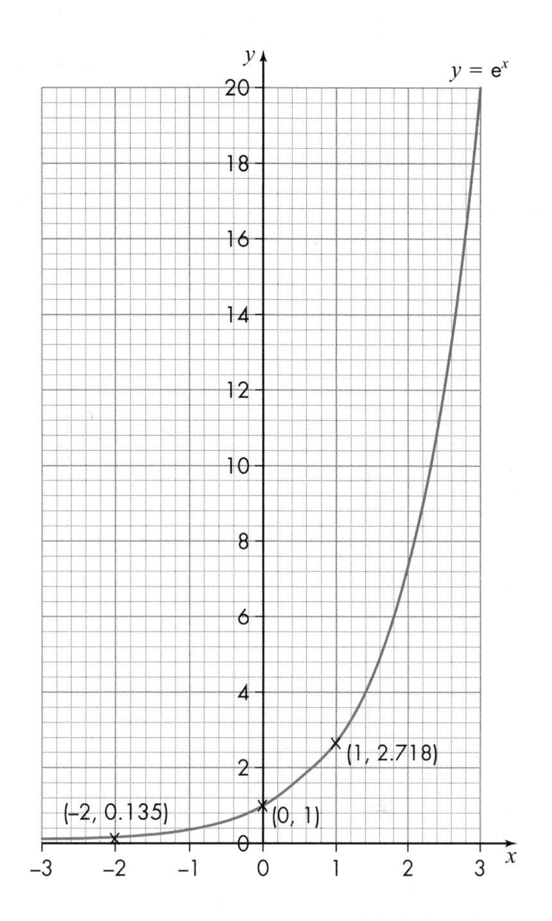

Notice that it passes through the point (0, 1) because when $x = 0$, $y = e^0$ and any number (except 0) to the power 0 is 1.

The graph gets steeper very quickly as x gets larger. If you work out the gradient of the tangent at any point on the graph you will find it is equal to the y-coordinate at that point. It shows **exponential growth**.

y values of the graph are always positive: as x takes larger negative values, the graph gets closer and closer to 0, but it never crosses the y-axis. (The line $y = 0$ is called an asymptote.)

Here is the graph of $y = e^{-x}$, with the coordinates of some points marked on it.

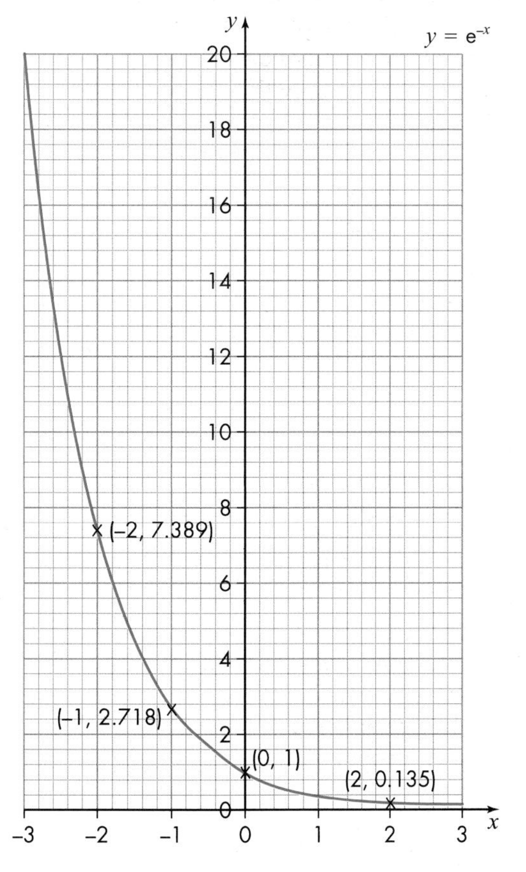

Notice that it passes through the point (0, 1) because when $x = 0$, $y = e^0$ and any number (except 0) to the power 0 is 1.

The graph is steep for large negative values of x. Again, if you work out the gradient of the tangent at any point on the graph you will find it is equal to the y-coordinate at that point. It shows **exponential decay**.

y values of the graph are always positive: as x takes larger positive values, the graph gets closer and closer to 0, but it never crosses the y-axis.

Exercise 13C

For this exercise you can use a graphics calculator and either plot or sketch the graphs. It is important that you have a paper copy of the graphs to remind you of their important features.

1 **a** For the values of x listed below, calculate the value of the natural exponential function, e^x.

i $x = 0$

ii $x = 10$

iii $x = 100$

iv $x = -10$

v $x = -100$

b Copy and complete the table below for the natural exponential function $y = e^x$.

x	0	1	1.5	2	2.5	3	3.5
y			4.481...		12.182...		

Plot the graph of this natural exponential function with x on the horizontal axis and y on the vertical axis.

c Investigate the value of the gradient of the exponential curve at different points.

(See Chapter 12 if you need a reminder about gradients.)

i When $x = 1$, find the gradient of the tangent.
What is the y value when $x = 1$?

ii When $x = 2$, find the gradient of the tangent.
What is the y value when $x = 2$?

2 Copy and complete the table below for the exponential function $y = e^{2x}$.

x	0	1	1.5	2	2.5	3	3.5
y			20.085…				

a Plot the graph of this exponential function with x on the horizontal axis and y on the vertical axis.

b Compare this graph with the graph plotted in question **1**. Are there any points on the graphs that are the same? What happens to the y value as x becomes large and positive?

3 Copy and complete the table below for the exponential decay function $y = e^{-x}$.

x	0	1	1.5	2	2.5	3	3.5
y		0.367…		0.135…			

a Plot the graph of this exponential function with x on the horizontal axis and y on the vertical axis.

b What you think this implies about the value of y when x is either very large and negative or very large and positive? How does this compare with the graphs in question **1**?

An important point

In question **1** if you were to plot an accurate graph of $y = e^x$ and also accurately determine the gradient of the tangent (rather than drawing a tangent by eye) you would find one of the unique features of the natural exponential function – the gradient at any point on an exponential graph is equal to the y value at that point. This is important because it enables you to find the gradient of the tangent at any point without drawing the tangent: you can just work out the y coordinate.

On this graph of $y = e^x$ the tangent has been drawn at the point (1.25, 3.49) and the equation of this tangent is shown below it. Notice that the gradient of that straight line is 3.49, the same as the y coordinate on the curve. (Remember that the formula for a straight line is $y = mx + c$ where m is the gradient of the straight line.)

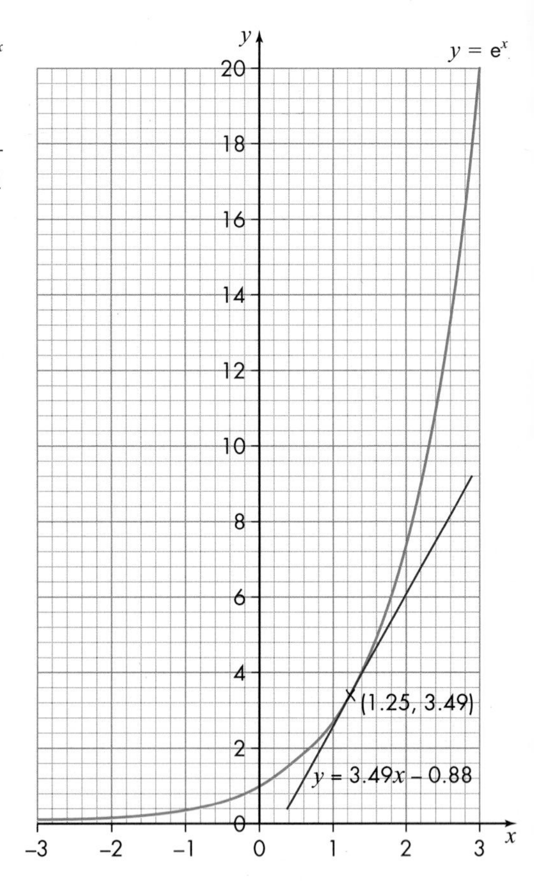

Exercise 13D

This exercise uses the curve $y = e^x$ in all questions.

1 For the values of x listed below, calculate the value of y and the gradient of the tangent at that point.

- **a** $x = 4$
- **b** $x = 2$
- **c** $x = 0$
- **d** $x = -0.1$
- **e** $x = -1000$

Extension: work out the equation of the tangent at the above points.

2 Dara finds that the equation of a tangent on the curve $y = e^x$ is $y = 4.3x - 2.03$.

What is the y coordinate where the curve and tangent touch?

3 Sian finds that the equation of a tangent on the curve $y = e^x$ is $y = 0.48x + 0.83$.

- **a** What is the y coordinate where the curve and tangent touch?
- **b** Is the x coordinate positive or negative at this point?

4 What are the coordinates on the curve where the tangent is $y = x + 1$?

G8: Exponential growth and decay

Learning objectives

You will learn how to:

- Formulate and use exponential equations of the forms $y = Ca^x$ and $y = Ce^{kx}$.
- Use exponential functions to model growth and decay in various contexts.

Most of the real-world examples so far in this chapter have been of exponential growth; the biomedical company producing vaccines by cell division, the Environment Agency predicting growth of an invasive species and the increase in the number of reported Ebola cases.

These models have all been of exponential growth as a result of the size of the constants in the exponential equations used to model the real-world processes.

Formulating and using growth and decay functions

In the table below, x is taken to be greater than or equal to 0.

This is because you are assuming that when you deal with growth or decay, you are measuring from some point in time (here denoted by x), which is taken to be the origin.

	Value of constants	Growth or decay?
$y = a^x$	$a > 1$	growth
$y = e^{kx}$	$k > 0$	growth
$y = a^x$	$0 < a < 1$	decay
$y = a^{-x}$	$a > 1$	decay
$y = e^{kx}$	$k < 0$	decay

Example 8

A small village in Dorset has a population of 1000 people.

Its population increases each year by 2.5%.

What is the exponential function the Parish Council can use to model this exponential growth to enable it to plan for the village expansion?

What will the population be in five years' time?

The equation will be of the form

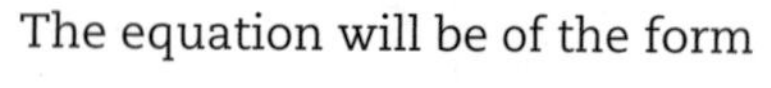

$y = Ca^x$ with a larger than 1 (see table above).

C represents the starting population for the model, here 1000 people.

a represents the **growth factor**.

The population grows by 2.5% each year. As a decimal 2.5% is written as 0.025.

The growth factor a is 1 plus the decimal equivalent of 2.5%, i.e. 1.025.

The equation that models this population growth is

$y = 1000 \times 1.025^x$.

In five years' time the village population will be

$y = 1000 \times 1.025^5 = 1000 \times 1.131... = 1131$ people.

The example above is an example of exponential growth.

What will the village population be in 10 years' time using this exponential model?

What will the village population be in 20 years' time using this exponential model?

The next example is an example of exponential decay.

Example 9

The sales of truffles by a truffle importer are falling by 3% a year.

The sales this year are 1400 truffles.

What is the exponential function the importer can use to predict future sales?

What will the sales be in five years' time?

The equation will be of the form

$y = Ca^x$ with a less than 1.

C represents the starting sales for the model, here 1400.

a represents the **decay factor**.

The sales fall by 3% each year. As a decimal 3% is written as 0.03.

The decay factor a is 1 minus the decimal equivalent of 3%, i.e. $1 - 0.03 = 0.97$.

So the equation that models this decay in sales is

$y = 1400 \times 0.97^x$.

In five years' time the sales will be

$y = 1400 \times 0.97^5 = 1400 \times 0.858... = 1202$ truffles.

What will the sales be in 10 years' time using this exponential model?

What will the sales be in 20 years' time using this exponential model?

Exercise 13E

1 A student attaches a selfie to an email and sends it two friends. Each of her friends then forwards the photograph to two of their friends.

This process is repeated ten times.

How many people would see her selfie? (Never send selfies unless you want the world to see it: think before you press send!)

(Hint: This can be modelled by an exponential growth equation of the form $y = a^x$ with $a = 2$.)

2 There are 100 cells in a test tube. They double in number every day.

How many will there be in the test tube after 10 days?

(Hint: This can be modelled by an exponential growth equation of the form $y = Ca^x$ with $a = 2$ and $C = 100$.)

3 The value of a mobile phone after T years can be modelled as an exponential decay.

It halves in value every year.

If the mobile phone cost £500 when new, how much will it be worth after three years?

4 Kevin's monthly income from the royalties on book sales is £30.

He estimates sales of the books will fall and his income will drop by 5% every month.

What will his income from the sale of books be in one year's time? Does writing books pay?

5 Atmospheric pressure (the pressure of air around you) decreases as you go higher.

It decreases about 12% for every 1000 m.

The pressure at sea level is about 1013 hPa (under normal weather conditions).

Work out the formula that gives the atmospheric pressure, P, in terms of the height, h.

What would the pressure be on the roof of the Shard (309 m) and at the top of Mount Everest (8848 m)?

Using exponential functions to model growth and decay

The first part of this section explains how to solve equations of the other two formulations of exponential models. They include exponential functions and the natural exponential functions (involving e) that can be used to model growth and decay.

Functions written in the form $y = Ca^x$, where C and a are different constants, can be solved for x using your calculator.

$$x = \frac{(\log y - \log C)}{\log a}$$

Example 10

Solve the equation

$5 = 4 \times 2^x$.

Here $y = 5$, $C = 4$ and $a = 2$.

So

$$x = \frac{(\log 5 - \log 4)}{\log 2} = \frac{0.6989... - 0.6020...}{0.3010...} = 0.3219... = 0.322 \text{ to 3 s.f.}$$

Check this using your own calculator and make sure you know the correct keys to use.

Functions written in the form $y = Ce^{kx}$, where C and k are different constants can be solved for x using your calculator. This time, you need to use the ln function, instead of the log function.

$$x = \frac{(\ln y - \ln C)}{k}$$

Example 11

A process can be modelled by the equation $y = Ce^{kx}$.

If $y = 25$, $C = 10$ and $k = 0.1$, find x.

The equation is $25 = 10e^{0.1x}$.

This is solved for x using the ln function on your calculator.

$$x = \frac{(\ln 25 - \ln 10)}{0.1} = \frac{3.2188... - 2.3025...}{0.1} = 9.1629... = 9.16 \text{ to 3 s.f.}$$

You might have to solve a third type of equation that is an extension of the last type since it involves a bit of algebraic manipulation. It is for functions written in the form $y = A + Be^{kx}$, where A, B and k are different constants. To solve this type of equation, you first write it as $y - A = Be^{kx}$, then treat it the same way as Example 11.

Example 12

A process can be modelled by the equation $y = A + Be^{kx}$.

If $y = 18$, $A = 10$, $B = 3$ and $k = -0.2$, find x.

The equation is $18 = 10 + 3e^{-0.2x}$.

Subtracting 10 from each side, the equation becomes $8 = 3e^{-0.2x}$.

Now you can see it is of the same type as in Example 11 and again, this is solved for x using the ln function on your calculator.

$$x = \frac{(\ln 8 - \ln 3)}{-0.2} = \frac{2.0794... - 1.0986...}{-0.2} = -4.9041... = -4.90 \text{ to 3 s.f.}$$

So, here is the flowchart that brings together the procedure for solving these two types of equation as well as the other types of equation you met earlier.

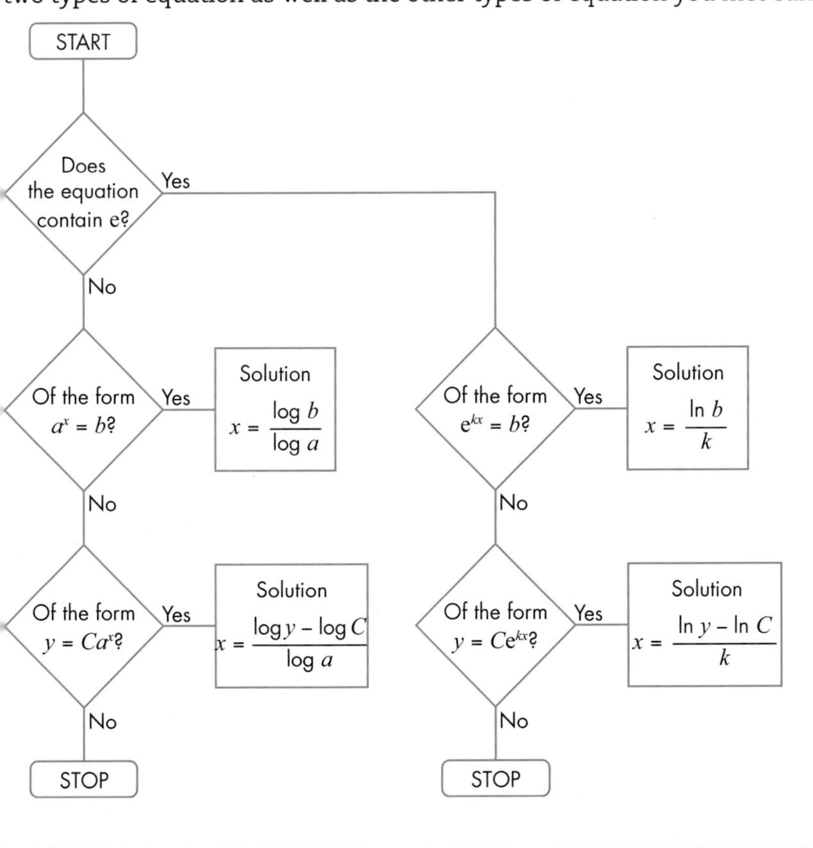

Exercise 13F

1 An auction house uses an exponential model to predict the increase in value of their antiques.

The model for the value, V, of an antique which varies with time T is $V = ka^T$.

With the constant a being 3 and the constant k being 1000, in how many years will the value, V, of the antique be £2000?

2 An archaeological research company uses the radioactive decay of the element radium 226 to measure the age of objects it discovers.

It uses this exponential model to calculate when an object was buried:

$$B = Ae^{-kT}$$

where B is the mass of active radium after time T and A is the mass of active radium when the object was buried.

If $k = 0.000\,44$, the mass $A = 800$ units and the mass $B = 640$ units, how many years, T, has the object been buried?

3 How many years would it take for an initial deposit of £25 000 to grow to £50 000 if the compound interest rate is 10% per annum?

Use $y = Ca^x$ with $y = 50\,000$, $C = 25\,000$ and $a = 1.1$.

If the bank reduced its compound interest rate to 1.5% per annum ($a = 1.015$), for how many years would a new investor have to wait for their £25 000 to increase to £50 000?

4 The number of common blue butterflies on a chalk hill reserve in Dorset is declining.

The numbers of butterflies is currently modelled by the exponential model $y = Ce^{kx}$, where y is the future number of butterflies, C the number of butterflies in the starting year, k a constant equal to -0.9 and x the number of years from the starting year.

This year 1000 butterflies were found.

In how years many will there only be 500 common blue butterflies?

5 Solve the following equations.

a $4 = -3 + 2e^{6x}$

b $7 - 3e^{1.6x} = -5$

c $9 = 7e^{-4x} - 5$

Case study

Hospitals need to manage their resources efficiently to make sure they deliver the best possible healthcare to their patients within an available budget. Managers in hospitals can model bed usage using exponential models which describe the lengths of time patients will spend in hospital.

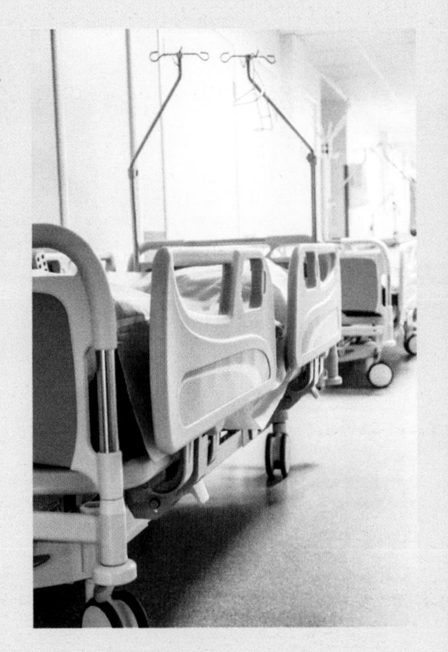

Analysis of statistics of the recovery times for patients shows that there are two major factors which determine the length of hospital stays. These factors are the conditions the patients are being treated for and the age of the patients.

The number of days a patient will stay in a hospital can be modelled by an exponential decay curve.

The graph below shows actual data (blue line) and an associated exponential decay curve (black line) for conditions categorised as general medicine. When dealing with large data sets of real-world cases, you do not expect the fit to be exact: here you see the fit is not very good after just one or two days, but after that the fit improves a lot. Since the cost to the health service of long-term care is quite considerable, the model is acceptable in this instance.

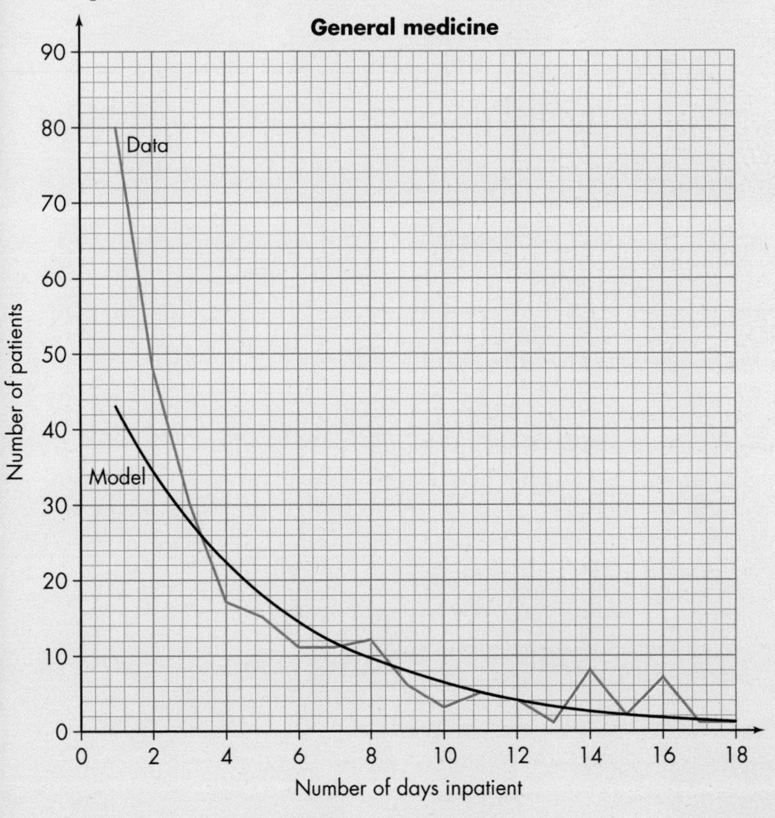

The form of the negative exponential function that models this data is:

$\lambda e^{-\lambda x}$

where the constant λ is known as the treatment rate indicator and x is the length of stay in hospital. Each condition a patient might have, has a different value for the treatment rate indicator constant.

The expectation of time spent as an inpatient is calculated by finding the inverse of this treatment rate indicator.

The treatment rate indicator depends not only on the patient's particular condition but also their age. Remarkably, whatever the condition being modelled the expectation increases by 0.631 days for each decade increase in an patient's age.

The graph below shows the linear trend line increase in expectation with age for different conditions.

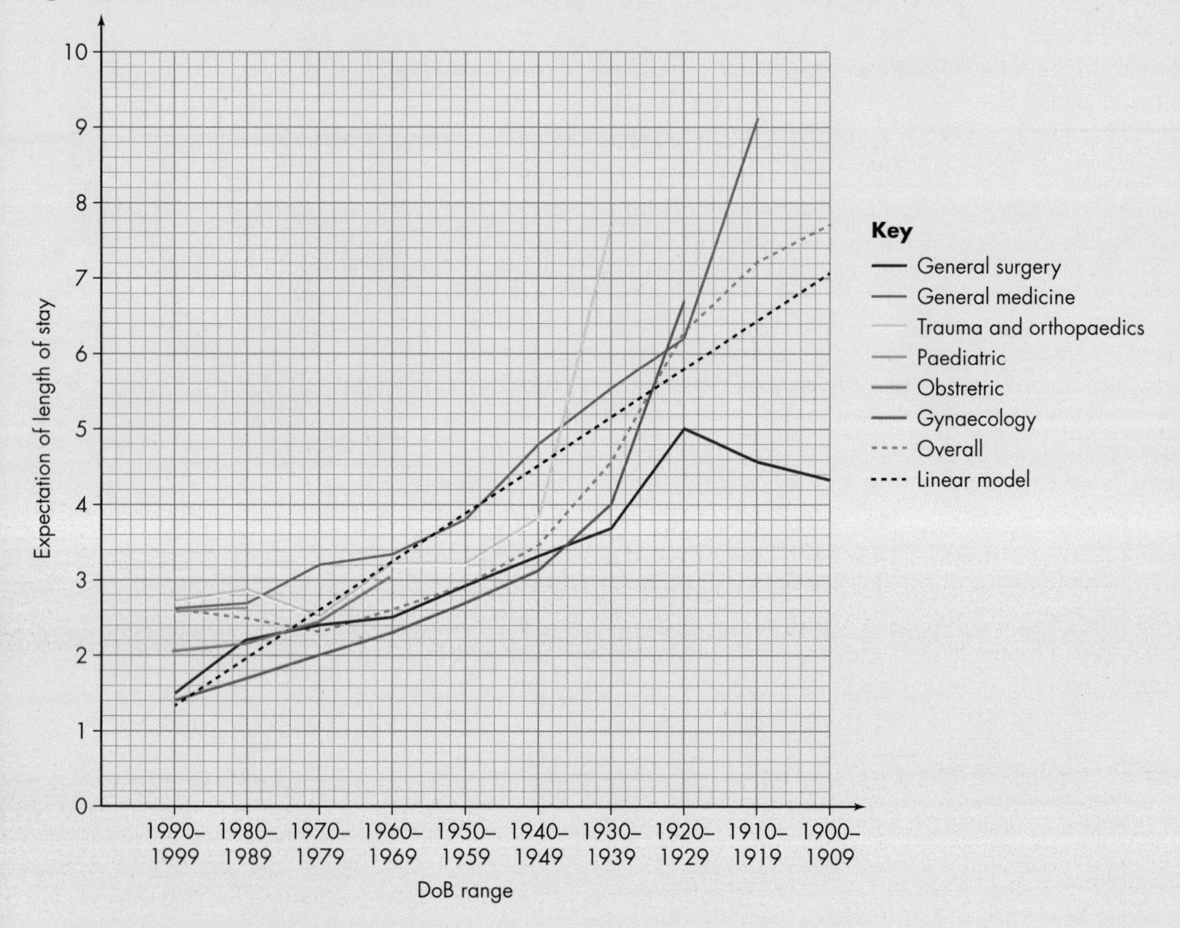

Using this model the managers of hospitals can predict the expectation of time spent in a hospital bed by every patient and so plan bed usage throughout their hospitals in the most efficient way.

Exercise

1. The expectation of the time a patient will spend in a hospital bed is the inverse of the treatment rate indicator.

Using the graph above, what is the difference in the expected length of stay in a hospital bed for someone born in the years 1990 to 1999 and someone born in the years 1930 to 1939?

2. Imagine you are a stressed hospital manager. Discuss with a peer how knowing the value of the treatment rate indicator for different medical conditions would enable you to better plan admissions to your hospital.

Project work

Exponential growth

From Exercise 13A and Section G8 earlier in this chapter, you will have learnt that an exponential growth function $y = a^x$ where $a > 1$ will have a curve of the form shown in blue on the diagram on the right.

An exponential growth function can be modelled by an equation of the form

$y = Ca^x$ where $a > 1$.

In the red graph in the diagram, C has the value 5 for the same value of a as in the blue graph.

If you study the two graphs you will see that the constant C has effectively multiplied each y value by 5 units.

In the blue graph when $x = 0$, the y value is 1.

In the red graph when $x = 0$, the y value is 5.

The value read from the y-axis where the curve intercepts it (the y-intercept) represents the starting value of the process. For example, in Exercise 13F question **1**, the curve for the exponential process would intercept the y-axis at 1000. This represents the starting value of the antique before it increased in value, that is when T was zero.

Exponential decay

From Exercise 13A and Section G8 earlier in this chapter, you will have learnt that an exponential decay function $y = a^{-x}$ where $a > 1$ will have a curve of the form shown in blue on the diagram on the right.

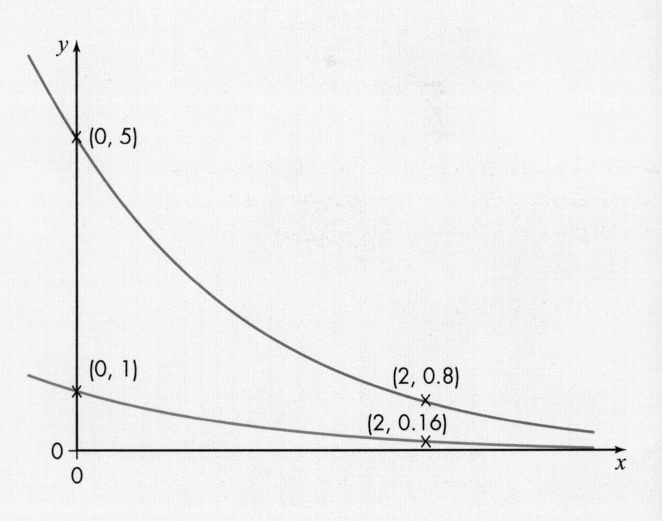

An exponential decay function can be modelled by an equation of the form

$y = Ca^{-x}$ where $a > 1$.

In the red graph in the diagram, C has the value 5 for the same value of a as in the blue graph.

If you study the two graphs you will see that the constant C has effectively multiplied each y value by 5 units.

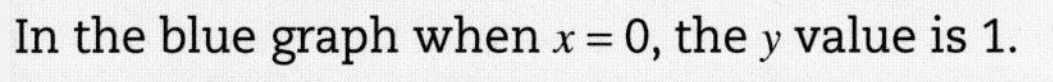

In the blue graph when $x = 0$, the y value is 1.

In the red graph when $x = 0$, the y value is 5.

The value read from the y-axis where the curve intercepts it represents the starting value of the process. For example in Exercise 13F question **2**, the curve for this exponential process would intercept the y-axis at 800 units, which represents the active mass of radium 226 when the object was buried.

Discussion

1. With a peer, discuss how the formulations explained above would look for exponential growth and decay models that include the natural exponential function involving e.

   In what ways, if any, would they differ?

   It will help if you use a graphics calculator to draw some curves of the form $y = e^x$ and $y = Ce^x$.

   Summarise your findings either on a large sheet of paper or as a PowerPoint presentation and deliver a short account of your findings to your peers.

2. With a peer, sketch the shape of the exponential curves for the real-world problems in Exercises 13B, 13E and 13F.

   In particular, decide where the exponential curves will intersect with the vertical axis and what value the curve will be tending towards as the values on the horizontal axis become very large.

Check your progress

How confident are you feeling in your level of knowledge? What do you need to practise more?

Spec reference	Learning objective			
G6.1	Use a calculator to find values of an exponential function			
G6.2	Use a calculator's log and ln functions to solve equations of the forms $a^x = b$ and $e^{kx} = b$			
G7.1	Understand that e has been chosen as the standard base for exponential functions; know that the gradient at any point on the graph of $y = e^x$ is equal to the y value at that point			
G8.1	Formulate and use exponential equations of the form $y = Ca^x$ and $y = Ce^{kx}$			
G8.2	Use exponential functions to model growth and decay in various contexts			

14 Practice questions

Chapter 1: Analysis of data

1 The table shows the number of students studying various mathematics courses in a college.

Course	Core Mathematics	A-level Mathematics	A-level Further Mathematics
Number of students	88	47	15

The principal wants to find out the reasons why students chose to study mathematics.

She chooses three students from each course to ask their opinions.

a Give two reasons why this is not a good sample to take. [2 marks]

b Give a full description of a better sampling method the principal might use. [4 marks]

2 In 2015 the National Trust recorded the number of their staff whose pay and benefits exceeded £60 000.

The results are shown in the table.

Amount, a (£k)	$60 \leqslant a < 70$	$70 \leqslant a < 80$	$80 \leqslant a < 90$	$90 \leqslant a < 120$	$120 \leqslant a < 150$	$150 \leqslant a < 230$
Frequency	43	18	17	9	4	6

a Draw a histogram to represent this information. [3 marks]

b Estimate the mean pay and benefits of staff exceeding £60 000. [3 marks]

3 Didi collected some data from five friends about how many hours they spent watching TV one week.

They all gave their answer to the nearest whole number of hours.

The box and whisker plot shows the results.

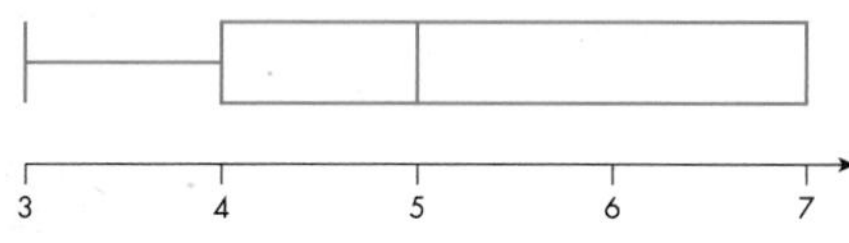

What data did Didi's friends give her? [3 marks]

4 Give one advantage and one disadvantage of the following.

a Range [2 marks]

b Interquartile range [2 marks]

c Standard deviation [2 marks]

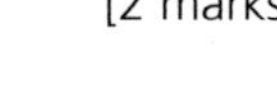

5 This cumulative frequency graph shows the length of time students spent on homework from Chapter 1 of their textbook.

Estimate the following.

a The median number of hours spent on homework [1 mark]

b The interquartile range [2 marks]

c $P_{70} - P_{30}$ [1 mark]

d The number of students who spent over $8\frac{1}{2}$ hours on the homework [1 mark]

e The percentage of students that spent less than six hours on the homework [1 mark]

6 a Explain briefly the difference between a cluster sample and a quota sample. [2 marks]

b A mathematics teacher is asked to provide data on how well the 50 Year 12 and 30 Year 13 students are doing on the course.

To obtain suitable data she decides to give a questionnaire to 16 of the students.

i State a sampling method she should use. [1 mark]

ii Give reasons for you answer to part **i** and explain the process she should use in selecting the sample. [3 marks]

7 The table shows the capacity of a number of different lifts in terms of the maximum mass and the number of people they will hold.

Maximum mass (kg)	3750	450	760	160	1000	1360	2250	1600	1000
Maximum capacity	50	6	10	2	10	18	30	21	13

a Statistically analyse the data numerically, giving reasons for any calculations you make. [4 marks]

b Statistically analyse the data diagrammatically, giving reasons for any diagrams you draw. [4 marks]

Chapter 2: **Maths for personal finance**

1 a Explain briefly the difference between simple interest and compound interest. [2 marks]

b Noah invests £800 in a water-tight bond that guarantees a compound interest rate of 5%. How many years will it take for the value of his investment to exceed £1200? [3 marks]

2 When Indira started work she had a student loan of £27 000.

She started work on 1 January 2016 with an annual salary of £22 000.

Each year she has to pay 9% of all she earns in excess of £21 000 towards repayment of her loan.

Monthly repayments are always rounded down to the nearest £.

Indira was told her salary will increase by £2000 each year.

She sets up a simple spreadsheet as shown:

	A	B	C	D
1	End of year	Salary (£)	Annual repayment (£)	Monthly repayment (£)
2	2016	22,000	90	7
3	2017	24,000	270	22
4	2018	26,000		

a Write a formula for cell C4 to calculate her annual repayment. [1 mark]

b If Indira makes the monthly repayments, how much will she have repaid in 2016? [1 mark]

c What will be her outstanding student loan at the end of 2016? [1 mark]

Simple interest of 3% is added to the outstanding loan at the start of each year, so after the first year the outstanding loan is £27 000 + interest of £810 – repayment of £84.

d Work out what Indira owes at the end of 2019. [4 marks]

3 a State which of the following is the correct interpretation of the abbreviation CPI.

Customer Prevention Index   Consumer Purchase Index   Customer Price Insurance

Consumer Purchase Insurance   Consumer Price Index   Customer Price Index [1 mark]

b Explain briefly the difference between a mortgage and a loan. [2 marks]

c Explain briefly the difference between Income Tax, Value Added Tax and Inheritance Tax. [3 marks]

4 Use the tables on tax and National Insurance for 2015–2016 below to help you answer this question.

Taxable income	£0 to £31 785	£31 786 to £150 000	Over £150 000
Rate	Basic rate of 20% People with the standard personal allowance start paying this on income over £10 600.	Higher rate of 40% People with the standard personal allowance start paying this on income over £42 385.	Additional rate of 45%

The standard personal allowance was £10 600. It was reduced by £1 for every £2 you earned above £100 000.

Your pay	**Class 1 National Insurance rate**
Less than £672 a month	0%
£672 to £3532 a month	12%
Over £3532 a month	2% (on the excess above £3532)

You do not pay National Insurance on the first £8060 of your annual earnings.

A dentist earns £48 000 in a year.

a Work out how much she has to pay in tax that year. [2 marks]

b Work out how much she has to pay in National Insurance that year. [2 marks]

c Work out her monthly take-home pay. [2 marks]

5 **a** What do the initials APR stand for in the context of interest rates? [1 mark]

b Use the formula $C = \sum_{k=1}^{m} \left(\frac{A_k}{(1+i)^{t_k}} \right)$ to calculate the amount of the loan if there are four annual payments of £400 at an APR of 24%. [3 marks]

6 **a** What do the initials AER stand for in the context of interest rates? [1 mark]

b Use the formula $r = \left(1 + \frac{i}{n}\right)^n - 1$ to work out the AER on a loan that charges a nominal rate of 7% interest every quarter. [3 marks]

7 The monthly payment, £P, for a mortgage can be calculated using this formula.

$$P = \frac{i \times A(1+i)^N}{(1+i)^N - 1}$$

where

i is the monthly interest rate expressed as a decimal

A is the amount borrowed in pounds

N is the number of monthly payments.

George takes out a mortgage for £186 000.

The monthly interest rate is 0.4%.

He must pay each month for 15 years.

Work out his monthly payment. You **must** show your working. [4 marks]

Chapter 3: Estimation

1 Alex knits socks for a charity.

He can buy knitting needles in either metric or UK sizes.

The table shows a conversion chart on the knitting pattern he uses.

UK size	14	13	12	11	10	9	8	7	6	5
Metric size (mm)	2.0	2.25	2.75	3.0	3.25	3.75	4.0	4.5	5.0	5.5

Find an expression giving the relationship between the UK and metric sizes that fits the data, remembering to list any assumptions you make. [4 marks]

2 The table shows the altitude of a plane flying from Heathrow to Chicago and the outside temperature at various points of the flight.

Altitude in metres	668	1460	2691	3415	4037	4947	5765	7976	8469	8908	9905
Temperature in °C	12	4	0	−4	−8	−11	−18	−33	−37	−41	−48

a Draw a suitable diagram showing the data, giving one reason why you decided to use that diagram. [3 marks]

b Find a suitable relationship between the altitude and outside temperature, stating clearly any assumptions you make. [3 marks]

3 Jen has to make eight return journeys from Southampton to Leicester on a Saturday each year to attend meetings that start at 10:30 a.m. and finish at 3:30 p.m.

Here is a train timetable and the associated fares.

	Outward journey			**Return journey**		
Departure time	06:00	06:20	06:30	15:48	16:24	16:48
Arrival time	09:29	09:50	10:00	19:41	19:49	20:41
Single fare	£23.50	£28.20	£26.50	£32.60	£26.50	£32.60

The distance from Southampton to Leicester by car is 144 miles and if Jen drives she averages 48 miles per gallon at a speed of 55 miles per hour.

Petrol costs 104.9 pence per litre and 1 gallon = 4.5 litres.

a Work out the cost of the most economical rail journey and the time Jen will be away from the railway station in Southampton. [3 marks]

b Estimate the cost of travelling by car and the time Jen will be away from home. State any assumptions you make clearly. [6 marks]

4 Modern cars contain a computer that gives you data about your journeys such as the average speed (in miles per hour), the distance travelled (in miles) and the average fuel consumption (in miles per gallon).

The table shows the results recorded over a five-day period.

Distance (miles)	50	60	70	80	90
Average speed (mph)	52	33	60	45	65
Average fuel consumption (mpg)	50	61	44	52	42

AJ thinks there may be a relationship between some of this data.

a Explain why a model using the distance as one of the variables is not appropriate. [2 marks]

b Draw an appropriate diagram and suggest a possible model that links the average speed with the average fuel consumption, stating clearly any assumptions you make. [4 marks]

c Using your model estimate the average fuel consumption for an average speed of 40 mph. [1 mark]

5 Estimate the number of A4 sheets that are photocopied in a year at your school or college.

State any assumptions you make.

You **must** show working to support your estimate. [6 marks]

6 Martin is a fashion designer.

He designs a new unisex costume called the U-tube that looks like a tube and is held in place by a belt around the waist and a draw-string around the shoulders.

The material he uses comes in rolls of width 1.40 metres and costs £4.90 per metre length.

Estimate the cost of material used in providing 500 U-tubes to a department store.

State any assumptions you make clearly. [8 marks]

7 Estimate how many words are sent in emails each day from people in the UK.

State any assumptions you make.

You **must** show working to support your estimate. [6 marks]

Chapter 4: Critical analysis of given data and models (including spreadsheets and tabular data)

1 Ahmed has a part-time job in a fast food outlet.

A record of what people buy using their loyalty card is recorded on the till and Ahmed has to analyse the results.

He records the amounts spent, in pounds, in a table as shown.

Customer Id	Visit 1	Visit 2	Visit 2	Visit 4	Median amount
F1	2.95	3.90	5.30	4.00	4.60
F2	1.00	1.00	1.00	1.00	1.00
F3	4.20	4.30	4.50	4.20	4.25
F4	2.50	315	3.20	2.80	3.00
Total	10.65	12.35	14.00	12.00	12.85

a Find one formatting error in Ahmed's table and one data entry error. [2 marks]

b Suggest three improvements he could make to the table. [3 marks]

2 The table shows the amounts of recyclable materials, in kilograms, dumped at a local recycling centre.

Skip number	1	2	3	5	Mixed	Total
Paper	3500	–	–	–	100	3600
Plastics	–	6800	–	–	200	700
Glass	–	–	13267	–	400	13667
Metal	–	–	–	27 000	500	27500
Non-recyclable material	300	700	1306	2 000	1500	5806
Total	3800	7500	14573	27200	2700	

a Find one formatting error in the table and one data entry error. [2 marks]

b Suggest three improvements that could be made to the table. [3 marks]

3 A mountaineer is hoping to climb all the Munros in Scotland.

A Munro is a mountain in Scotland over 3000 feet (914 metres).

She records the mountains she has climbed in the table shown.

Munro name	Ben Nevis	Ben Hope	Lochnagar	Liathach	Total	Mean
Height in metres	1345	927	1155	1055	4481	896.2
Time taken (hours:minutes)	3:12	4:15	3:35	2:50	13:12	3:33
Cost in food and transport (£)	52	29	33	30	144	121.50
Total	1399.12	960.15	1191.35	1087.5	4638	–

a Find one formatting error in the table and one data entry error. [2 marks]

b Suggest three improvements she could make to the table. [3 marks]

4 A newspaper article said "More people realise cashless transactions are easy when paying for low-cost items".

Use the data given below to comment on this headline.

Number of cashless transactions worldwide (in billions)

2010	2011	2012	2013	2014
285	309	333	358	390

[5 marks]

5 Shazza wants to buy a motorbike costing £2500.

The bike shop suggests two purchasing schemes they have on offer.

Scheme 1: A deposit of £100 and 7% interest added on the balance paid off in 12 equal monthly instalments.

Scheme 2: A deposit of £700 and 4% interest added on the balance paid off in 6 equal monthly instalments.

Shazza makes these notes.

2400 × 0.07 = 168
168 ÷ 12 = 14
I can afford £14 each month

1800 × 0.4 = 720
2520 ÷ 6 = 420
I cannot afford £420 each month

Something does not seem right. Have I gone wrong?

Critically analyse Shazza's notes, making corrections where necessary. [3 marks]

6 Colin commutes to college in London using his Oyster card.

Each journey costs him £2.90 in the morning peak period and £2.40 in the afternoon off-peak period.

He attends college for 180 days a year and wants to know how much his travel will cost, on average, each month.

Here are Colin's calculations.

Each day I pay £2.90 + £2.40 = £4.13 in total
180 × £4.13 = £743.04
£743.04 ÷ 12 = £61.92
I'm not sure I'm right. Have I made any mistakes?

Critically analyse Colin's notes, making corrections where necessary. [3 marks]

7 The data below shows the number of annual consumer payments made per UK adult in 2014 and the forecast for 2024.

	Cash	Debit card	Direct debit	Credit card	Cheque	Other
2014	345	172	63	41	7	29
2024	282	225	67	53	2	49

Use the data to write a report making recommendations to an ailing business that currently accepts only cash or cheques.

Your report should include at least one statistical diagram. [7 marks]

Paper 2A Statistical techniques

Chapter 5: The normal distribution

1 A chocolate bar is produced with an advertised mass of 200 g.

A consumer organisation finds that the mass of the chocolate bar is normally distributed with a mean of 200.4 g and a standard deviation of 0.05 g.

a Find the proportion of the chocolate bars produced that have a mass within one standard deviation of the mean. [4 marks]

b 1 000 000 bars of this particular brand and size of chocolate bars are produced in a year.

How many will not be within one standard deviation of the mean? [3 marks]

2 Assume that the number of Facebook friends a Facebook user has is normally distributed with a mean of 176 and a standard deviation of 33.

a What proportion of Facebook users have more than 50 friends? [3 marks]

b Find the number of Facebook friends such that 80% of Facebook users have fewer than this number of Facebook friends. [4 marks]

3 A particular breed of hen produces eggs with a mean mass of 60 grams and a standard deviation of 4 grams and mass is found to be normally distributed.

a Eggs are classified as small if their mass is less than 55 g.

Find the proportion of eggs that are not classified as small. [3 marks]

b Find the range of masses, symmetrical about the mean, such that 50% of the eggs are within this range. [4 marks]

4 A dairy produces cartons of milk with a mean capacity of 1 litre.

Cartons of milk containing less than 960 ml are sold at a reduced price.

The volume of milk in the carton is normally distributed with a standard deviation of 25 ml.

a The dairy refills cartons containing more than 1.05 litres.

What proportion of cartons will not be refilled? [3 marks]

b The production line in the dairy stops for every 1 in a 100 cartons because the cartons cannot hold the amount of milk the machine is trying to put in them.

What is this amount of milk to the nearest ml? [4 marks]

5 Assume the mass of a particular model of mobile phone is normally distributed with a mean of 137 g and a standard deviation of 1.2 g.

a Find the proportion of the mobile phones produced that have a mass below 138.2 g. [2 marks]

b Mobile phones that have a mass greater than three standard deviations from the mean are rejected by the quality assurance process.

What proportion are rejected? [3 marks]

c In a batch of 10 000 phones how many are rejected because they have a mass greater than three standard deviations from the mean? [2 marks]

6 Assume the average UK fixed line residential broadband speed is 15 Mbits/s with a standard deviation of 2 Mbits/s.

a A particular broadband provider claims their comparable broadband speed is 22.5 Mbits/s.

What proportion of broadband speeds are greater than this speed? [2 marks]

b The communications regulator decides to write to any broadband providers whose comparable broadband speed is less than 8 Mbits/s.

What proportion of broadband providers will receive a letter from the communications regulator? [3 marks]

c What broadband speed, to 1 decimal place, is exceeded by 60% of broadband providers? [3 marks]

7 The police regularly monitor the speeds of cars on a section of motorway and it is found that the speeds are normally distributed.

The mean speed is 68.5 mph with a standard deviation of 5 mph.

a What percentage of motorists exceed the 70 mph limit? [2 marks]

b A speed limiter is fitted on a school coach restricting the maximum speed of the vehicle to 55 mph.

What proportion of vehicles travel at a speed greater than this limit? [3 marks]

c While the police are monitoring the speeds on the section of motorway they issue an instant fine to 0.5% of the motorists.

What speed, to the nearest mph, are these motorists exceeding? [4 marks]

8 A manufacturer produces a new fluorescent light bulb and claims that it has an average lifetime of 4000 hours with a standard deviation of 375 hours.

a Find the proportion of light bulbs with a lifetime within two standard deviations of the mean. [4 marks]

b If an office is fitted with 1500 of these light bulbs how many of them, to the nearest whole number, will not have a lifetime within two standard deviations of the mean? [3 marks]

Chapter 6: Probabilities and estimation

1 **a** Would the following situations produce a random sample? If not, why not?

i Conducting a survey of method of transport to work of the visitors to a coffee shop at a railway station. [2 marks]

ii Conducting a survey of favourite genre of music of the visitors to the Last Night of the Proms. [2 marks]

iii Conducting a survey of produce bought by raffle ticket holders at a village fete. [2 marks]

b Describe a method to randomly sample students in a sixth form to determine their lunchtime eating habits. [3 marks]

2 A group of 11 four-year-olds start in the reception class at their local primary school.

Their heights, in cm, are as follows:

91.8 92.5 96.2 98.7 100.0 103.2 103.5 104.1 104.8 109.3 111.0

The teacher of the reception class says that her new pupils must be some of the tallest in the country.

The height of four-year-olds is normally distributed with an unknown mean, μ, and a variance of 0.64 cm.

Calculate the 99% symmetric confidence interval for μ and comment on the teacher's statement. [5 marks]

3 A survey was performed across all the primary schools in a county to study the mode of transport and journey times to and from school.

Would the following situations produce a random sample? If not, why not?

a The data for all of the Class 3 children. (They are the median year group in primary school.) [2 marks]

b The data for all of the rural primary schools. [2 marks]

Various random samples of school journey times were taken and the means (in minutes) of these samples were as follows:

32 45 26 5 34 101

c How should these sample means be used to estimate the population mean?

Give reasons for your answer. [3 marks]

4 A chocolate bar is produced with an advertised mass of 200 g.

The mass of the chocolate bars is normally distributed with an unknown mean and a standard deviation of 2.4 g.

A random sample of 20 chocolate bars was taken by a consumer organisation and was found to have a sample mean of 199.7 g.

Calculate, to 2 decimal places, the 95% symmetric confidence interval for the mean μ and interpret the result. [5 marks]

5 A random sample of 1000 one-litre cartons of milk were found to contain a mean volume of 0.985 litres of milk.

The estimated standard deviation was 25 ml.

a Assuming the volume of milk in a carton is normally distributed, calculate, to 2 decimal places, the 80% confidence interval for μ. [4 marks]

b The 95% confidence interval is calculated to be (983.45, 986.55) and the 99% confidence interval is calculated to be (982.96, 987.04).

Explain the difference in the ranges of these two confidence intervals. [2 marks]

6 A random sample of 30 data values, with sample mean $\overline{x}$, is chosen from a normal distribution with an unknown population mean and variance 16.

Calculate, to 5 decimal places, the 90% confidence interval for μ. [4 marks]

7 A random sample of data values with a mean of 100, is chosen from a normal distribution with unknown mean μ and variance 36.

The 99% confidence interval for μ µ is calculated as (96, 104).

What was the size of the sample? [4 marks]

8 A fuel pump at a filling station is being assessed for accuracy.

The assessor dispenses 1 litre of petrol on ten separate occasions and measures the amount of fuel dispensed.

His findings are as follows:

0.970 0.985 0.950 1.000 0.960 0.940 0.900 1.000 0.960 0.965

The amount of fuel dispensed is normally distributed with an unknown mean and standard deviation of 100 ml.

Calculate, to 2 decimal places, the 99% confidence interval for μ and interpret the result. [5 marks]

Chapter 7: Correlation and regression

1 A number of celebrities have run the London Marathon.

Some of their ages and finish times are detailed in the table.

Age	41	35	43	46	39	31	49	29	32	31
Finish time (mins)	156	172	170	226	234	265	293	209	230	259

a Plot these data on a scatter diagram, labelling the axes clearly. [3 marks]

b Calculate the value of the product moment correlation coefficient, r, between celebrity age and London Marathon finish time. [2 marks]

c Use the scatter diagram and the value of r to comment on the relationship between celebrity age and London Marathon finish time. [2 marks]

2 An egg producer has observed that his chickens lay more eggs in the summer when it the days are longer and it is warmer.

He has recorded the following results.

Month	**Hours of daylight**	**Average temperature (°C)**	**Average number of eggs per week**
June	15	28	19
July	14	30	20
August	12	23	16
September	10	17	11
October	9	12	8
November	8	5	6

The egg producer thinks there is a stronger positive correlation between the number of eggs produced and hours of daylight than between the number of eggs produced and the average temperature.

By calculating product moment correlation coefficients, determine the type and strength of the correlation between the two sets of data and comment on the egg producer's thoughts. [7 marks]

3 The number of users of Twitter and Facebook per quarter are given in the table.

Twitter users by quarter (in millions)	30	40	49	54	68	85	101	117	138	151
Worldwide Facebook users by quarter (in millions)	431	482	550	608	680	739	800	845	901	955

a Find the equation of the regression line of Facebook users on Twitter users, $y = a + bx$, giving the values of a and b to an appropriate level of accuracy. [2 marks]

b Interpret the values of a and b. [2 marks]

c Estimate the number of Facebook users if the number of Twitter users is 200 000 000.

Comment on your estimate. [2 marks]

d Would you say that the number of Facebook users is related to the number of Twitter users?

Provide an argument for your response. [2 marks]

4 An egg producer has asked his daughter to calculate the regression line of hours of daylight on number of eggs produced by his flock of hens.

The equation of the regression line is $y = -10.8 + 2.04x$.

a The egg producer doesn't understand what the values of –10.8 and 2.04 in the equation mean. Interpret these values for him. [2 marks]

b In the data used to calculate the equation of the regression line, the maximum number of hours of daylight was 15.

The egg producer is keen to know how many eggs will be produced if he introduces artificial light throughout the year so that his hens have 18 hours of daylight per day.

Use the equation of the regression line to work out the number of eggs produced for 18 hours of daylight.

What would you say to the egg producer about the reliability of this calculation? [3 marks]

5 The average mathematics test result and number of hours of TV watched per week have been recorded for some GCSE students.

Hours of TV watched per week	11	26	30	23	21	19	12	6	9
Average maths test result (%)	62	54	57	68	73	76	82	85	90

a Plot this data on a scatter diagram, labelling the axes clearly. [3 marks]

b Calculate the value of the product moment correlation coefficient, r, between hours of TV watched per week and mathematics test result. [2 marks]

c Use the scatter diagram and the value of r to comment on the relationship between the number of hours of TV watched per week and mathematics test result. [2 marks]

6 A farmer sells most of his produce wholesale but also has a small farm shop.

The farmer has observed that his shop has more visitors in the summer when it is warmer.

He has recorded the following results.

Month	**Average temperature (°C)**	**Average number of customers per week**
July	27	18
August	29	19
September	22	15
October	16	10
November	11	7
December	4	5

a Plot this data on a scatter diagram, labelling the axes clearly. [3 marks]

b Calculate the value of the product moment correlation coefficient, r, between average temperature and average number of customers per week. [2 marks]

c The farm shop owner finds that in the following January, when the average temperature was 3°C, the average number of customers per week was 9.

Calculate the value of the product moment correlation coefficient, r, between average temperature and average number of customers per week including this additional data.

Comment on this value of the product moment correlation. [4 marks]

7 The number of hits, by quarter, on two neighbouring village websites are detailed in the table.

	Number of website hits									
Village A	30	40	49	54	68	85	101	117	138	151
Village B	431	482	550	608	680	739	800	845	901	955

a Plot this data on a scatter diagram, labelling the axes clearly. [3 marks]

b Calculate the value of the product moment correlation coefficient, r, between the number of hits on the Village A website and the number of hits on the Village B website.

Comment on this value of the product moment correlation coefficient, r. [3 marks]

8 A mathematics teacher has calculated the regression line of the number of hours of TV watched per week on the average mathematics test result for some GCSE students.

The equation of the regression line is $y = 92.36 - 1.17x$.

a Interpret the values of the y-intercept and the gradient of the equation of the regression line. [2 marks]

b In the data used to calculate the equation of the regression line the maximum number of hours of TV watched per week was 30 and the minimum was 9.

The maximum mathematics test result was 90% and the minimum was 54%.

Use the equation of the regression line to work out the mathematics test result if a student watches 1 hour of TV every day of the week.

What would you say to the mathematics teacher about the reliability of this calculation? [3 marks]

Paper 2B Critical path and risk analysis

Chapter 8: Critical path analysis

1 The activities, the immediate predecessors and the durations needed to produce a piece of computer software code are shown in the table.

Activity		Immediate predecessor	Duration (weeks)
A	Write program specification	–	1
B	Write code, first version	A	2
C	Write unit test plan	A	1
D	Unit test first version of code	B, C	2
E	Quality assurance of first version of code	D	3
F	Write code, second version	E	1
G	Unit test second version of code	F	1
H	Quality assurance of second version of code	G	2
I	Write system test plan	–	3
J	System test code	I, H	2

a Draw an activity network for this project. [7 marks]

b List the activities not on the critical path and state the duration of the critical path. [2 marks]

2 The activities, the durations and the start and finish times to produce a piece of history homework are shown in the table.

Activity		Duration (days)	Earliest start time	Latest finish time
A	Set homework	1	0	1
B	Write down homework	1	1	2
C	Research homework	3	2	5
D	Review class notes	1	2	5
E	Cross-reference textbook	1	2	5
F	Write up homework	3	5	8
G	Check homework	1	8	9
H	Hand in homework	1	9	10
I	Mark homework	5	10	15

a Identify the non-critical activities and their associated floats. [2 marks]

b Draw a Gantt chart for this homework project. [6 marks]

c Activity E takes twice as long as planned.
What is the impact on the critical path? [1 mark]

3 The activities, the durations and the start and finish times to write a chapter of a mathematics textbook are detailed in the table.

Activity		Duration (weeks)	Earliest start time	Latest finish time
A	Write author brief	1	0	1
B	Write text and questions, first version	2	1	3
C	Write answers, first version	1	1	3
D	Author review of chapter	2	3	5
E	First editor review of chapter	3	5	8
F	Amend chapter, second version	1	8	9
G	Second author review of chapter	1	9	10
H	Second editor review of chapter	2	10	12
I	Write examination board review checklist	3	0	11
J	Examination board review	2	12	14

a Calculate the float for each activity and list the activities on the critical path. [4 marks]

b Which two activities have no predecessors? [2 marks]

c Which is the only activity without a successor? [1 mark]

d What is the key assumption that has been made in the production of this plan? [1 mark]

4 The activities, the immediate predecessors and the durations to produce a picture of a local landscape are detailed in the table.

Activity		Immediate predecessor	Duration (days)
A	Decide upon medium	–	1
B	Decide upon location	A	1
C	Sketches of different parts of the landscape	B	3
D	Trials of different paint effects	B	1
E	Trials of different types of paper	B	1
F	Paint picture	C, D, E	3
G	Touch up picture	F	1
H	Dry picture	G	1
I	Frame picture	H	5

a Draw the activity network for this project. [7 marks]

b List the activities on the critical path and state the duration of the critical path. [2 marks]

5 The activities, the immediate predecessors and the durations to get ready and prepare for an interview are detailed in the table.

Activity		Immediate predecessor	Duration (minutes)
A	Have a shower	–	10
B	Dry hair	A	5
C	Have breakfast	B	20
D	Listen to travel news on radio	B	5
E	Review interview preparation notes	B	15
F	Iron clothes	C, D, E	10
G	Get dressed	F	15
H	Travel to interview	G	40

a Draw the activity network for this project. [7 marks]

b The interview is at 9:30 a.m. and the interviewee wants to arrive 15 minutes early.

At what time does he need to get into the shower?

Explain your answer. [2 marks]

6 The activities, the durations and the start and finish times to get ready and prepare for an interview are detailed in the table.

Activity		Duration (minutes)	Earliest start time	Latest finish time
A	Have a shower	10	0	10
B	Dry hair	5	10	15
C	Have breakfast	20	15	35
D	Listen to travel news on radio	5	15	35
E	Review interview preparation notes	15	15	35
F	Iron clothes	10	35	45
G	Get dressed	15	45	60
H	Travel to interview	40	60	100

a Identify the non-critical activities and their associated float. [2 marks]

b Draw the Gantt chart to show the activities involved in preparing for an interview. [6 marks]

c Activity E takes 10 minutes longer than planned.

What is the impact on the critical path? [1 mark]

Chapter 9: Expectation

1 A postman has a drawer full of socks containing 10 black socks, 6 blue socks and 2 pairs of yellow football socks.

He takes a sock at random out of the drawer and then takes another sock at random out of the drawer.

a Draw a tree diagram to represent these events. [3 marks]

b What is the probability that both socks will be the same colour? [2 marks]

c What is the probability that he will select a pair of football socks? [1 mark]

d What is the probability he will select a blue sock and a black sock in any order? [1 mark]

e He has a bet with a work colleague that if he goes into work in a pair of matching socks, then his colleague will give him £1 and if he doesn't, then he will give his colleague £1.

How much money is he likely to win? [3 marks]

2 In a sixth-form class of 60 students, 15 are studying maths and 18 are studying chemistry, of whom 10 are also studying maths.

a Represent this information on a Venn diagram. [3 marks]

b What is the probability that a randomly selected sixth former is studying maths? [1 mark]

c What is the probability that a randomly selected sixth former is studying chemistry? [1 mark]

d What is the probability that a randomly selected sixth former is studying maths and chemistry? [1 mark]

e What is the probability that a randomly selected sixth former is studying maths or chemistry? [1 mark]

f What is the probability that a randomly selected sixth former is studying neither maths nor chemistry? [1 mark]

3 Two fair four-sided spinners, with sides numbered 1 to 4, are spun and the sum of their scores is recorded.

a What is the probability of the sum of their scores being 1? [1 mark]

b What is the probability of the sum of their scores being less than 4? [2 marks]

c What is most likely sum of the scores? [2 marks]

d Two friends, Alan and Brian, decide to play a game.

Alan says that Brian has to give him £1 if the sum of the scores is a multiple of 3 and that he will give Brian £1 if the sum of the scores is a multiple of 4.

If the sum of the scores is not a multiple of 3 or 4, then no money will change hands.

Who is likely to win the most? [4 marks]

4 A toddler has been given a small container of sweets.

The pot contains 3 blue sweets, 4 red sweets and 1 green sweet.

He takes a sweet out of the container and eats it.

He then takes another sweet out of the container and eats that too.

a Draw a tree diagram to represent these events. [3 marks]

b What is the probability that the second sweet the toddler eats is green if the first one he ate was green? [1 mark]

c What is the probability that the two sweets he eats will be the same colour? [2 marks]

d What is the probability he eats a red sweet and a blue sweet in any order? [2 marks]

e What is the probability he eats two different coloured sweets? [2 marks]

5 The universal set to be represented on a Venn diagram is the positive integers less than 20.

One subset of the universal set is prime numbers and the other is factors of 20.

a Represent this information on a Venn diagram. [3 marks]

b What is the probability of a randomly selected number being a prime number and a factor of 20? [1 mark]

c What is the probability of a randomly selected number being neither prime nor a factor of 20? [1 mark]

d What is the probability of a randomly selected number being prime but not a factor of 20? [1 mark]

6 The outcomes for the spins of a four-sided spinner and the associated probability distribution are shown in the table.

Outcome	1	2	3	4
Probability	$\frac{1}{3}$	$\frac{1}{6}$	$\frac{1}{6}$	$\frac{1}{3}$

a Calculate E(X). [2 marks]

A student decides to devise a game using a spinner to raise money at the school fair.

He decides that if the spinner lands on 1 or 2, he will give the player (the person who has spun the spinner) 50p.

If the spinner lands on 3 or 4, the player has to give the stall holder 50p.

b How much can the student expect to raise for his school? [2 marks]

The student decides to change his game.

He now decides that if the spinner lands on 2 or 3, he will give the player 50p.

If the spinner lands on 1 or 4, the player has to give the stall holder 50p.

c How much can the student expect to raise for his school now? [2 marks]

7 In a sixth form of 60 students, 24 have a mobile phone, 19 have a tablet and 7 have both a tablet and a mobile phone.

a Represent this information on a Venn diagram. [3 marks]

b What is the probability that a randomly selected sixth former has neither a tablet nor a mobile phone? [1 mark]

c What is the probability that a randomly selected sixth former has only a mobile phone? [1 mark]

d What is the probability that a randomly selected sixth former has a mobile phone or a tablet? [2 marks]

8 A mathematics teacher decides that his class will do extra homework if the sum of the scores on two dice are 6, 7 or 8.

By calculating an expected value, determine how likely is it that the class will do extra homework? [5 marks]

Chapter 10: Cost benefit analysis

1 a What is a control measure? [1 mark]

b Describe a control measure used in a science laboratory. [2 marks]

2 A baker has a £500 000 contract to supply sandwiches to a prestigious publishing firm next year.

The contract depends on not missing the 12:30 delivery time on more than 10 days during the year.

The probability of missing the delivery time on more than 10 days during the year is 0.12.

The penalty for delay is 20% of the value of the contract.

The baker can reduce the risk of delay to 0.04 by employing a courier firm on demand at an annual cost of £4000.

a Calculate the expected penalty if no action is taken to reduce the risk of delay. [3 marks]

b The baker wants to reduce the risk of delay.

State, with justification, whether you recommend the action. [3 marks]

3 A theatre producer wants to make a new theatre production.

He gets financial support from investors (called angels), who provide him with £5 million but insist on a £2 million penalty if the production is late or flops.

The probability the female star will be ill and make the production late is 0.1.

The cost of training a stand-in for the star is £50 000.

The probability the production will be a flop is 0.6.

The cost of insuring against a flop is £400 000.

The table shows information for these control measures.

Control measure	Cost (£ million)	Probability
Stand in	0.05	0.1
Insurance	0.4	0.6
Both	0.45	0

a Work out the expected penalty if no control measures are taken. [5 marks]

b Work out the expected penalties if either or both control measures are taken. [5 marks]

c What control measures, if any, would you recommend are taken?
Justify your recommendation fully. [5 marks]

4 A builder has a £5 million contract to build a new academy.

The builder estimates the probability of delay to be 0.55.

The penalty for delay is 20% of the value of the contract.

The builder can take either of the following actions to reduce the risk of delay.

Employing extra staff at a cost of £200 000 would reduce the probability of delay to 0.2.

Cancelling a contract with a supplier who is known to be unreliable, at a cost of £300 000, would reduce the probability of delay to 0.15.

a Calculate the expected penalty if no action is taken to reduce the risk of delay. [3 marks]

b The builder wants to reduce the risk of delay.
State, with justification, which **one** action you would recommend to the builder. [5 marks]

5 A car manufacturer has carried out a cost benefit analysis for a new device that measures exhaust gas emissions.

There is a penalty of £8 billion if the device is found to give faulty readings.

The device will give faulty readings if one or both of two critical activities (V and W) happen.

The probability of V happening is 0.15.

The cost of implementing a control measure that will prevent V happening is £1 billion.

The probability of W happening is 0.4.

The cost of implementing a control measure that will prevent W happening is £2 billion.

Control measure	Cost (£ billion)	Probability	Expected penalty (£ billion)
V only	1	0.4	3.2
W only	2	0.15	1.2
Both V and W	3	0	0

a Work out what the expected penalty will be if no control measures are taken.
State any assumptions you make. [3 marks]
The table shows information for the control measures V only, W only and both V and W.

b Which control measures, if any, would you recommend the car manufacturer to take?
Fully justify your recommendation. [5 marks]

6 A health centre has a £7 million contract to provide health care to a town next year.

However, if more than 20 people complain about waiting times in a month, the contract will be reduced by 15%.

The probability that more than 20 people complain in a month is 0.4.

Taking either of the following actions would reduce the risk of complaints.

Readjusting the appointment software at a cost of £50 000 would reduce the probability of patient delay complaints to 0.3.

Employing another doctor at a cost of £200 000 would reduce the probability of patient delay complaints to 0.1.

a Calculate the expected penalty if no action is taken. [3 marks]

b The health centre wants to reduce the risk of delay.

State, with justification, which **one** action you would recommend to the health centre. [7 marks]

7 Elton and Suzi play at local gigs and have recently been awarded a regular contract worth £500.

There will be a penalty of £100 if either of them fail to perform to a given standard.

The probability that Elton's guitar fails is 0.12.

The cost of having a spare guitar to hand is £15.

The probability that Suzi's piano is out of tune is 0.28.

The cost of keeping Suzi's piano tuned is £25.

The table shows information for the control measures.

Control measure	Cost (£)	Probability	Expected penalty (£)
Spare guitar	15	0.28	28
Regular tuning	25	0.12	12
Both	40	0	0

a Work out the expected penalty if no control measures are taken.

State any assumptions you make. [5 marks]

b Which control measures, if any, would you recommend Elton and Suzi to take?

Fully justify your recommendation. [5 marks]

Paper 2C Graphical techniques

Chapter 11: Graphical methods

1 A chess board measures 8 squares by 8 squares.

A £1 coin is put on the first square, then two £1 coins are put on the next square, then four £1 coins on the next square, etc.

The amount on any given square is given by the formula 2^{n-1}.

n	1	2	3	4	5	6	7	8	9	10
2^{n-1}										

a Copy and complete the table and draw the graph of 2^{n-1}. [2 marks]

b Describe the type of curve you have drawn in part **a**. [1 mark]

c Using your graph, find the square on which the amount will first exceed £100. [1 mark]

d Using your graph, find the square on which the amount will first exceed £500. [1 mark]

e How many £1 coins will be on the 20th square? [1 mark]

2 A footballer kicks a ball off the ground into the air and the ball lands some distance from the footballer.

The ball follows the path of a parabola given by the equation $y = -x^2 + 5x$.

x (m)	0	1	2	3	4	5
y (m)						

a Copy and complete the table and draw the graph of $y = -x^2 + 5x$. [2 marks]

b If the x-axis is ground level and the ball is kicked from left to right, at what point on the x-axis was the footballer standing when he kicked the ball? [1 mark]

c What is the total horizontal distance travelled by the football? [1 mark]

d When the football is 4 m off the ground, how far has the ball travelled horizontally from the footballer? [2 marks]

3 Solve the simultaneous equations $y = x^2 + 8x + 7$ and $y = x + 7$ graphically by completing tables of values and subsequently plotting the graphs.

Are the equations $y = x^2 + 8x + 7$ and $y = x - 7$ simultaneous?

Use your graphs to support your answer. [8 marks]

4 A swimmer is going to be sponsored by her friends to swim in a swimathon (5 km).

a One friend has agreed to sponsor the swimmer £2.00 per whole kilometre and will give £10 for just starting the swimathon.

Form an equation for this information, make a table and plot the graph for the equation. [2 marks]

b Another friend has agreed to sponsor the swimmer £3.00 per whole kilometre and will give an extra £8 for just starting the swimathon.

Form an equation for this information, make a table and plot the graph for the equation on the same grid as your graph for part **a**. [2 marks]

c At what point in the swimathon will each friend end up sponsoring the swimmer the same amount? [2 marks]

d The swimmer completes the swimathon.

How much will she raise in sponsorship altogether? [2 marks]

5 **a** Draw the graph of $y = x^3 + 2x^2 - x - 2$ for $-4 \leqslant x \leqslant 3$ and indicate clearly the points of intersection with the axes. [4 marks]

b Use your graph to answer the following questions.

i Find the coordinates of the point of intersection of the curve and the line $x = -3$. [1 mark]

ii Find the coordinates of the point of intersection of the curve and the line $y = 4$. [1 mark]

iii Find the coordinates of the point of intersection of the curve and the line $y = 2x - 2$. [2 marks]

6 Dan buys a new kettle for £15.00.

He lives in a hard water area so the kettle needs to be descaled monthly at a cost of £1.50.

a Form an equation to reflect the purchase price and the cost of maintaining the kettle over time, make a table of values and plot the graph. [3 marks]

b Use your graph to answer the following questions.

i How much will the kettle cost to maintain, including the purchase price, in 2 months? [1 mark]

ii How much will the kettle cost to maintain, including the purchase price, in 5 months? [1 mark]

iii At what point in the kettle's life does the cost of descaling equal the price of a new kettle? [2 marks]

7 **a** Make tables of values and subsequently plot the curves of $y = x^3 - x^2$ and $y = 2^x$ on the same axes. [4 marks]

b Describe the two types of curve. [2 marks]

c Explain how the diagram shows that there is one solution to the equation $x^3 - x^2 = 2^x$. [1 mark]

d What is the solution to the equation $x^3 - x^2 = 2^x$? [1 mark]

8 An engineer designs a new section for a roller coaster.

The new section is the shape of a parabola with the equation $y = -x^2 + 6x$.

x (m)	0	1	2	3	4	5	6
y (m)							

a Copy and complete the table and draw the graph of $y = -x^2 + 6x$. [2 marks]

b The start of the parabolic section of the roller coaster is 2 m above ground level.

What is the maximum height the ride will reach on this section of the roller coaster? [1 mark]

c How wide is this section of the roller coaster? [1 mark]

d The engineer decides to put vertical supports every 2 m along this section of the roller coaster.

How many supports will be needed and how tall will they be? [2 marks]

Chapter 12: Rates of change

1 The distance–time graph shows the motion of a car travelling on the M1 from junction 15a (Northampton) to junction 21 (Leicester).

This is a distance of 35 miles.

There is a restricted speed limit on part of this stretch of motorway which is monitored using average speed cameras.

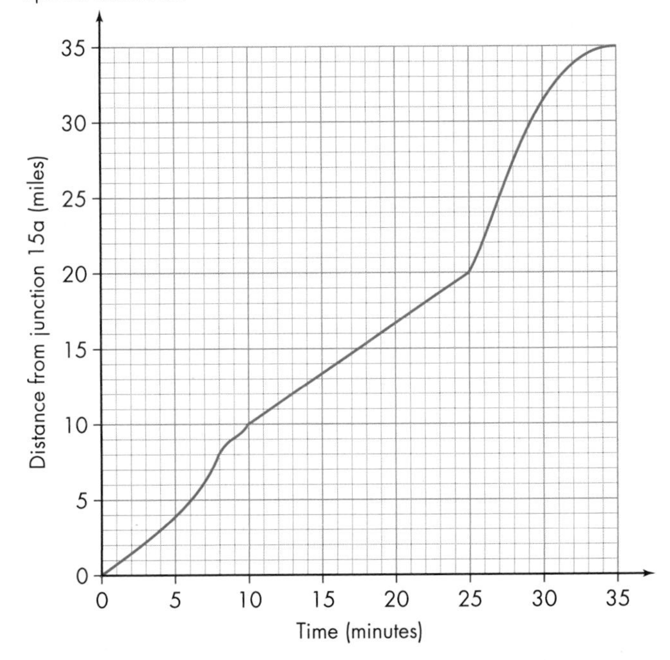

a Work out the average speed of the car in miles per hour for the whole journey. [2 marks]

b The driver was observing the restricted speed limit for part of the journey.

What was the restricted speed limit? [2 marks]

c Did the driver exceed the 70 mph speed limit at any point?

You must show your working and give reasons to justify your answer. [3 marks]

2 The acceleration due to gravity on Mars has been calculated.

If a ball was dropped on Mars, a table of values similar to those shown in the table would be obtained.

Time (t seconds)	0	0.4	0.8	1.2	1.6	2.0
Distance (d metres)	0	0.30	1.12	2.66	4.74	7.40

Write a short report about the acceleration of the ball as it falls, drawing graphs and clearly showing the method and the working you used. [7 marks]

3 The amount of glucose in the blood of an insulin-dependent diabetic is monitored over two hours after a meal. The change in blood glucose levels, in millimoles per litre (mmol/l), is shown in the graph.

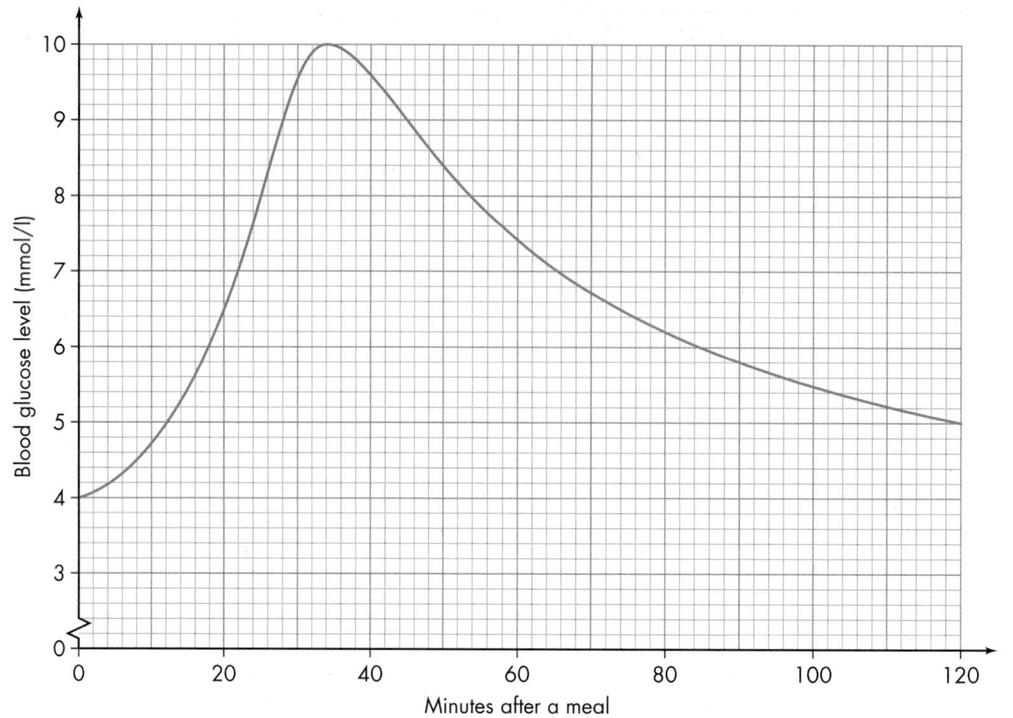

a Work out the average rate of increase in blood glucose level per minute over

  i the first 20 minutes [2 marks]

  ii the last half hour. [2 marks]

b Explain the significance of the gradient of the curve at 34 minutes after the meal. [2 marks]

c Work out the instantaneous rate of change in blood glucose level after 1 hour. [3 marks]

4 The distance–time graph shows the distance between two cyclists who are cycling around a race track. The race ends after 50 seconds.

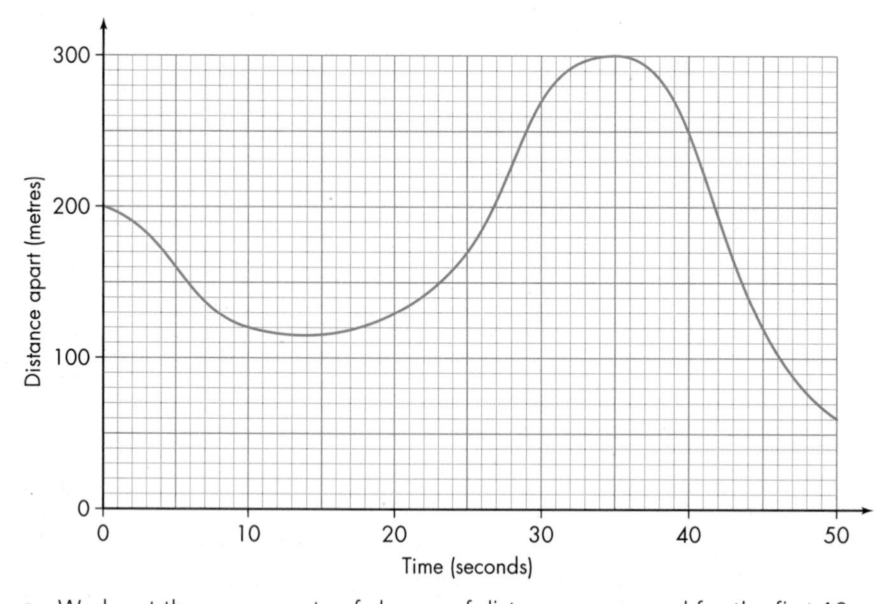

a Work out the average rate of change of distance per second for the first 10 seconds. [2 marks]

b Work out the instantaneous rate of change of distance per second after 25 seconds. [3 marks]

c Work out the instantaneous rate of change of distance per second after 40 seconds. [3 marks]

d State the distance between the two cyclists at the end of the race. [1 mark]

5 The velocity–time graph shows the motion of a skier as she goes down a slope and then jumps into the air.

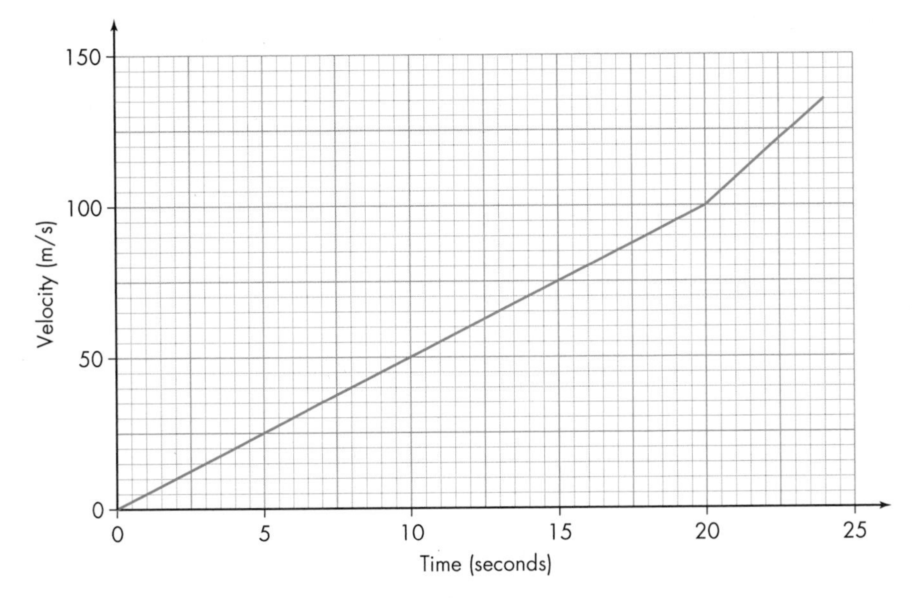

a Work out the average acceleration of the skier for the first 20 seconds. [2 marks]

b Work out the average acceleration of the skier for the last 4 seconds. [2 marks]

6 The vertical distance, d, of a tennis ball t seconds after being hit upwards is modelled by the quadratic equation $d = 2.8 + 8t - t^2$.

a Plot a graph for this quadratic model for $0 \leq t \leq 10$. [2 marks]

b Estimate the time when the velocity is zero. [1 mark]

c Estimate the instantaneous velocity after 7 seconds. [3 marks]

7 The acceleration due to gravity varies slightly according to the height above the centre of the Earth.

If a ball was dropped from a height of 1750 km above the Earth, then the following table would give the time taken for it to drop the given distances.

Distance (*d* metres)	0	1	2	3	4	5	6	7
Time (*t* seconds)	0	0.58	0.82	1	1.15	1.29	1.41	1.53

Write a short report about the acceleration of the ball as it falls, drawing graphs and clearly showing the method and the working you used. [7 marks]

Chapter 13: Exponential functions

1 Solve the following equations.

a $11^x = 0.6$ [2 marks]

b $e^{5x} = 20$ [2 marks]

c $e^{-0.4t} = 100$ [2 marks]

d $16 = e^{0.9x} - 20$ [2 marks]

e $8 - 5e^{-2t} = -3$ [2 marks]

2 When people take medicine the drug is eliminated from the body at an exponential rate.

A new drug is tested and the amount of drug left in the body is given by the formula

$$d = 250(0.8)^t$$

where

d is the mass of the drug, in milligrams, after a time of t hours.

a Sketch the graph of $d = 250(0.8)^t$ for $t \geqslant 0$.

Show the coordinates of any points where the curve crosses an axis. [2 marks]

b Work out the amount of drug in the body after 5 hours. [2 marks]

c Work out how long it takes until only 10 mg of the drug remains in the body. [3 marks]

3 Yorkshire puddings are taken out of the oven at 200 °C.

They are to be served at a temperature of 40 °C.

The temperature, T °C, of the Yorkshire puddings can be modelled as $T = 20 + ae^{-0.008t}$ where a is a constant and t is the time in seconds since the Yorkshire puddings left the oven.

a Work out the value of a. [2 marks]

b Work out the temperature of the Yorkshire puddings after 4 minutes, using your value of a. [2 marks]

c How long does it take for the Yorkshire puddings to cool sufficiently after they are taken out of the oven before they can be served? [3 marks]

4 A snowball is rolled down a hill so that the mass increases proportionally to its mass at a rate of 8% per second.

The initial mass of the snowball is 1 kg

a Calculate its mass after 10 seconds. [2 marks]

b Calculate its mass after one minute. [2 marks]

c Model the mass of the snowball as M and use t as the time in seconds since the snowball started rolling to find an equation that links M and t. [2 marks]

d Sketch the graph of M against t for $t \geqslant 0$. [2 marks]

e State any limitations that might affect your model. [2 marks]

5 A colony of bacteria is growing at a rate of 70% per hour.

The initial population is 200 bacteria.

a Estimate the population after one day. [2 marks]

b Model the growth of the population using P for the population at time t hours after the start to find an equation that links P and t. [2 marks]

c Sketch the graph of P against t for $t \geqslant 0$. [2 marks]

6 DDT is a pesticide that was widely used in the USA until its ban in 1972.

A farmer has some DDT in storage.

DDT decays with time and the farmer has found that the amount of DDT he has left, M litres, after t years since 1972 is given by the formula $M = 6 \times 2^{\frac{-t}{15}}$.

a Copy the axes and sketch the graph of $M = 6 \times 2^{\frac{-t}{15}}$ for $t \geqslant 0$.

Show the coordinates of any points where the curve crosses an axis.

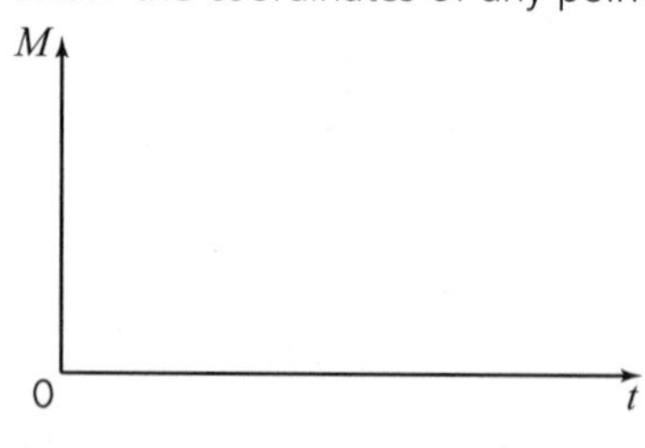

[2 marks]

b Work out the amount of DDT that remained in 2002. [2 marks]

c Work out the year in which the farmer had only one litre of DDT left. [3 marks]

d The farmer says, "It will always take the same time for the amount of DDT to halve from one given value to one that is 50% of that value."

Is the farmer correct? Justify your answer. [3 marks]

7 The spread of a disease is slowly decreasing.

The number of people catching the disease, P, t years after 2000 is given by $P = 10\,000 + 50\,000e^{-0.1t}$.

a Copy the axes and sketch the graph of $P = 10\,000 + 50\,000e^{-0.1t}$ for $t \geqslant 0$.

Show the coordinates of any points where the curve crosses an axis.

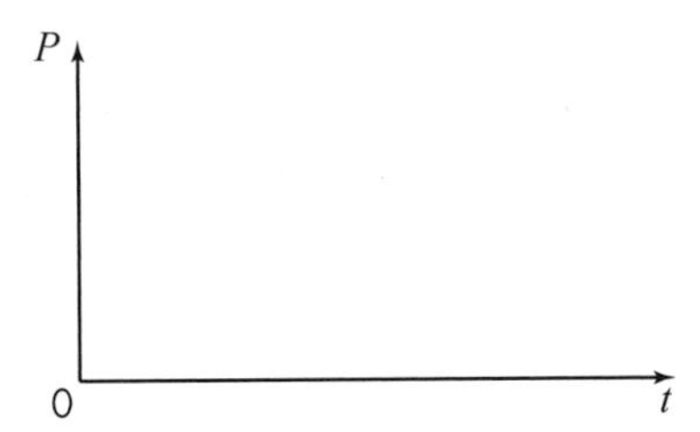

[2 marks]

b Work out the number of people catching the disease in 2020. [2 marks]

c Work out the year in which 11 366 people catch the disease. [3 marks]

d A doctor says, "It will always take the same time for the number of people catching the disease to halve from one given value to one that is 50% of that value."

Is the doctor correct? Justify your answer. [3 marks]

Formulae sheet

These formulae are not required to be learnt. A clean copy of this formulae sheet will be issued to you in the examination.

Volume and surface area

Shape	Volume	Surface area
Cone	$V = \frac{1}{3}\pi r^2 h$	$A = \pi rl + \pi r^2$
Sphere	$V = \frac{4}{3}\pi r^3$	$A = 4\pi r^2$
Pyramid	$V = \frac{1}{3}\text{base} \times h$	

Financial calculation – Annual Equivalent Rate (AER)

The annual equivalent interest rate (AER), r, is given by

$$r = \left(1 + \frac{i}{n}\right)^n - 1$$

where i is the nominal interest rate, and n the number of compounding periods per year.

Note: the values of i and r should be expressed as decimals.

Financial calculation – Annual Percentage Rate (APR)

The annual percentage interest rate (APR) is given by

$$C = \sum_{k=1}^{m} \left(\frac{A_k}{(1+i)^{t_k}} \right)$$

where £C is the amount of the loan, m is the number of repayments, i is the APR expressed as a decimal, £A_k is the amount of the kth repayment, t_k is the interval in years between the start of the loan and the kth repayment.

It may be assumed that there are no arrangement or exit fees.

Glossary

~: means 'is distributed as'.

μ: the Greek letter mu, pronounced 'mew'. It is used to represent the mean of a normal distribution.

σ: the Greek letter sigma. It is used to represent the standard deviation of a normal distribution.

σ^2: the square of the standard deviation and is called the variance.

Activity-on-node network: a diagrammatic representation of a set of activities and their durations, showing the dependencies between activities.

AER: the annual interest rate on savings.

APR: the annual interest rate on loans.

Asymptote: a line (or curve) which approaches a curve closely but never touches it.

Average acceleration: calculated by dividing the change in velocity by the time taken.

The average speed: calculated by dividing the distance travelled by the time taken.

Axes intercept: where a graph meets or crosses one of the coordinate axes.

Backward scan: a backward scan shows the latest finish start time for each activity without extending the length of the critical path.

Base: the base of an exponential function is the number which is raised to a given power.

Bias: a distortion of results.

Braking distance: the distance travelled by a vehicle once the brakes are applied before it stops.

Census: used when every member of the population provides data.

Cluster sample: used to collect data from all members of a randomly selected cluster (or group).

Confidence interval: the range of values within which the population mean will be found, with a given level of confidence.

Consumer Price Index (CPI): a measure used to track changes in prices of goods and commodities.

Complementary probability: if A' represents the event not A then P(A') = 1 – P(A).

Compound interest: the amount paid on the total amount when the interest is calculated.

Continuous data: data that is measured.

Control measure: any measure taken to eliminate or reduce risk

Correlation: a relationship or connection between two or more variables.

Correlation coefficient: a number used to assess the strength of a correlation.

Cost benefit analysis: used to compare the costs of carrying out an action or a project with the benefits delivered by the action or project.

Critical activities: activities where the latest finish minus earliest start minus duration = 0.

Critical path: the longest path (or paths) through a project from the start to the end of the project.

Cubic function: a function with a term in x^3 and the general form $y = ax^3 + bx^2 + cx + d$

Cubic graph: a graph of a cubic function.

Dependent events: events that are affected by previous events.

Discrete data: data that is counted.

Error: the error is the maximum value an amount can differ from a calculated value.

Event: a single result of an experiment.

Exhaustive set: the entire range of possible outcomes.

Expected value: the anticipated value for a given risk or investment

Expected value of X: also known as the mean of X and is written as E(X).

Experiment: an action where the outcome is uncertain.

Experimental probability: the ratio of the number of times an event occurs to the total number of trials.

Exponent: another name for a power or index.

Exponential function: a function where the input variable appears as an exponent OR An **exponential function** has the variable as the power.

Exponential graph: a graph of an exponential function.

Extrapolation: a prediction of a value outside the range of the data.

Fair game: a game in which each player is equally likely to win.

Fermi estimations: rough and ready estimations used to get an idea of the size of the answer.

Float: the amount of time an activity can be delayed without impacting other activities.

Forward scan: a forward scan shows the earliest possible start time for each activity.

Frequency density: the number of items in a given unit.

Gantt chart: a visualisation of a project using a graphical representation which allocates activities and durations to give a project schedule.

Geometric mean: the nth root of a product of n numbers.

Gradient: a measure of how much a line slopes. It is found by dividing the vertical change by the horizontal change.

Income tax: a tax you pay on the money you earn.

Independent events: events that are not affected by other events.

Inflation: the rate of increase in prices for goods and services.

Interest: the amount of money earned on savings or the amount of money added to a loan.

Instantaneous acceleration: calculated by working out the gradient of a tangent to the velocity– time graph at that instant.

Instantaneous rate of change: found by calculating the gradient of the tangent at a particular point.

Instantaneous speed: calculated by working out the gradient of a tangent to the distance–time graph at that instant.

Interpolation: a prediction of a value within the range of the data.

Interpolation: the making of predictions within the range of the known data.

Intersection point: the point at which two or more graphs meet or cross.

Linear graph: a straight line graph.

ln function: the ln on a calculator gives the logarithm of a number to the base e. **e** is the number 2.718 281….

Locus (plural **loci**)**:** the path of an object under certain conditions.

Logarithm or **log:** the logarithm of a number to a given base is the power of that base that gives the number.

Log function: the log function on a calculator gives the logarithm of a number to the base 10.

Lower bound: the minimum value of a rounded amount.

Maximum: the largest value in an interval where the gradient is zero and the gradient changes from positive to negative.

Maximum point: a turning point where the gradient changes from positive to negative as the curve is travelled from left to right.

Minimum: the smallest value in an interval where the gradient is zero and the gradient changes from negative to positive. A horizontal tangent has a gradient of zero.

Minimum point: a turning point where the gradient changes from negative to positive as the curve is travelled from left to right.

Mitigation: measures to control a hazard and to reduce the damage in the event of the hazard happening

Modelling: the process of applying mathematics to a real-world problem to solve it.

Mortgage: a loan taken out to buy property.

National Insurance: a contribution you pay for certain state benefits.

Negative correlation: as one variable increases, the second variable decreases.

Normal distribution: symmetrical about the mean. The area under a normal distribution curve is defined as 1 and represents probability.

Parabola: a special curve shaped like an arch (it can be upside down) with a line of symmetry that goes vertically through the turning point.

Percentile: a value below which a given percentage of the observations fall.

Population: a complete set of items that share a common property.

Population: all the objects or individuals under consideration.

Point estimate: a single estimated value of a parameter of a given population.

Point of inflexion: a point on a curve at which the curve changes from being concave (curving downward)to convex (curving upward), or vice versa.

Positive correlation: as one variable increases, so does the second variable.

Primary data: data that comes directly from firsthand experience.

Probability: the likelihood of an outcome or event.

Probability distribution: the set of all possible values a random variable can have with their associated probabilities.

Quadratic function: a function with a term in x^2 and the general form $y = ax^2 + bx + c$.

Quadratic graph: a graph of a quadratic function.

Qualitative data: non-numerical data that describes a quality.

Quantitative data: numerical data.

Quota sample: used to collect used to collect data chosen by the sampler from a stratum (or layer) where the number of items selected is proportional to the size of the stratum in the population.

Random event: an event with a probability of occurrence determined by some probability distribution.

Random events: events which follow no predictable pattern and whose outcome cannot be predicted.

Random sample: used to collect data from part of a population without bias.

Random sample: a sample in which each member of the population has an equal chance of being selected.

Random variable: the value obtained from an experiment. A random variable must be a numerical value.

Rate of change: the same as the gradient of a graph.

Regression line: a line of best fit for a given set of values, using the equation of a straight line $y = a + bx$.

Retail Price Index (RPI): is used to track changes in prices of goods and commodities.

Risk analysis: assessment of potential risks and management of those risks.

Sample: a subset of a population which can be used to represent the whole population.

Sample mean: the mean of a sample of a population, denoted by x.

Sample space diagram: shows all the possible outcomes for an experiment.

Sampling: used to collect data from part of a population.

Secondary data: data that already exists or has been processed.

Simple interest: the amount paid only on the original amount invested or borrowed.

Stratified sample: used to collect random data from a stratum (or layer) where the number of items selected is proportional to the size of the stratum in the population.

Standard deviation: a measure of the spread of a distribution that uses all the data.

Standardising: a mathematical process to transform any normal distribution to the standardised normal distribution with mean = 0 and standard deviation 1.

Subset: a group of things within the defined universal set.

Tangent: a straight line that just touches a curve at a given point.

Theoretical probability: the number of ways that the event can occur, divided by the total number of outcomes.

Thinking distance: the distance travelled by a vehicle in the time it takes the brain to see the hazard to the application of the brakes.

Truncate: to chop off without rounding.

Turning point: a point on a curve is where the gradient is zero.

Universal set: the collection of all things in a particular context. All other sets within the universal set are subsets.

Upper bound: the maximum value of a rounded amount.

Validating: checking that the solution found is realistic.

VAT (Value Added Tax): a tax on most goods and services.

Velocity: the velocity of an object is its speed in a stated direction.

Weighting: weighted values are adjusted to reflect their relative importance.

Answers

Chapter 1: Analysis of data

There are many possible answers for some questions in this chapter. It will be useful to discuss your answers with others as a means of checking them.

Exercise 1A

1 i Qualitative
ii Quantitative, continuous
iii Quantitative, discrete
iv Qualitative
v Qualitative
vi Quantitative, continuous

2 Answers may vary

3 Answers may vary

4 Qualitative because even though numbers are involved they represent qualities.

5 Answers may vary

Exercise 1B

1 a Hospital data about the time spent in the ward, the facilities used etc.
b Data from surveys or questionnaires given to hospital staff or patients etc.

2 a Primary data
b Secondary data

3 Primary: could have sent a questionnaire or done a survey asking householder opinion etc.
Secondary: could have compared data from previous years etc.

4 Secondary data might be hard to obtain due to privacy laws.
Primary data might be unreliable as people would consider such information personal and might be quite likely to refuse giving information or might lie. Many answers are possible.

5 a Whether people would buy them, what price they might pay, etc.
b Previous sales in various areas at various times of year etc.

Exercise 1C

1 Answers may vary

2 Many answers are possible such as primary data: daily mass of babies, mass and time of food intake, time spent with mother etc. Independent variables are such as the time when the measures are taken, dependent variables are mainly the other quantities.

3 a Student answers such as the daily number of spoons used/lost
b Might be no need or more expensive to change etc.
c Survey or questionnaire for customer or staff
d Number of customers, day of week etc.

4 Answers may vary

Exercise 1D

1 a Sample: too many items to do a census
b Census: probably small enough to survey them all
c Depends on how large the estate agent is
d Sample: too many items to process

2 All those who are eligible to use the bars in the House of Commons

3 All those who use, or might wish to use, the facilities on a Sunday

4 a Census: group small enough to ask all parents and possibly toddlers.
b All the toddlers in the group and their parents. (Could also include the venue etc.)

Exercise 1E

1 It only surveys those who are able to attend on a Friday. Some might not use the catering facilities etc. Better to survey people who use the catering and do the survey at various times on various days during the week.

2 Only takes place on one month, which is not representative of the year. Data should be taken each month, or at least every three months to work out the yearly average.

3 a Car owners in County Durham, plus all those who have driven/parked in the county.
b Varies: cluster, stratified or quota could be used.
c Answers suitable for your chosen method in b

4 a Probably quota
b Easily identifiable groups: bankers and non-bankers. Difficult to divide bankers into strata when their bonuses or salaries are probably not known.

5 a All the galls on the tree
b Lighter ones might have been blown elsewhere, Might just be the old ones that have fallen etc.
c Collect samples from other oak trees.

6 Answers may vary

Exercise 1F

All answers should round to those given as there may be some small acceptable differences when reading the chart.

1 a 66cm
b 82cm

2 31.25 – 29.75 = 1.5 inches

3 32.75 inches or 83 cm

4 about 2%

5 any answer between 25% and 10%

6 6 months to 13 months

7 90.5 – 82.5 = 8 cm or 35.5 – 32.3 = 3.2 inches

8 12 months

Exercise 1G

1 a mean = 9, s.d. = 4.80

b mean = 90, s.d. = 48.0

c mean = 15, s.d. = 4.80

2 a mean = 15.4 cm, s.d. = 2.2 cm

b median and modal length are both 14.0 cm, so the mean is greater than both

3 Here we need to make a reasonable estimate of the unlimited group. Since the maximum length of a tweet is 140 characters, we take it as 101 to 140. Using mid-point values the answer given here is based on the following table: mean = 81.9 characters, s.d. = 32.3 characters

No of characters	20	55	85	120
Frequency	18	39	58	52

4 a mean = $\sqrt{44} = 6.63$, s.d. = $\sqrt{69-44} = \sqrt{25} = 5$

b mean = $\frac{72}{8} = 9$, s.d. = $\sqrt{16} = 4$

c mean = $\frac{35}{10} = 3.5$, s.d. = $\sqrt{\frac{612.5}{10} - 3.5^2} = \sqrt{49} = 7$

5 a $\frac{56-50}{3} = 2$ and $\frac{45-50}{3} = -1.67$ so the first result is better.

b $\frac{23-30}{4} = -1.75$ and $\frac{57-65}{3.7} = -2.16$ so the first result is better.

c $\frac{4.5-3.9}{0.45} = -1.33$ and $\frac{3.9-2.1}{1.2} = -1.5$ so the second result is better.

6 Answers may vary

Exercise 1H

1 Sam: median = 25, LQ = 8, UQ = 40, IQR = 32, range = 59. Ella: median = 25, LQ = 19.5, UQ = 49, IQR = 29.5, range = 45

Although their medians are the same, Sam's scores are more spread out indicating he is less consistent than Ella.

2 a median = 70 g, LQ = 57 g, UQ = 88 g

b frequency densities are 0.9, 3, 1.9, 1.6, 2.2

3 a 17 years 6 months

b 2 years 11 months

c 1 year 2 months

d 17 years 7 months

e 25%

4 a, c, d and e are all true. It is not possible to determine if g is true.

5 a 2.6 minutes

b 5 minutes

c 1.6 minutes

d 2.3 minutes

e 50 calls

6 Frequency densities are 1.5, 2.5, 3, 5, 0.2. Missing frequency is 6

7 a Experiment 3 as it has the smallest IQR, OR experiment 5 as it has the next smallest IQR and its median is closest to the true speed.

b Experiment 1 as it has the largest IQR, the largest range and the median is furthest away from the true speed.

8 Answers may vary

9 Answers may vary

10 Answers may vary

Chapter 2: Maths for personal finance

Opener

1 They are weekly intervals

2 Weekends and bank holidays

3 Take the mean of the highest and lowest for example.

Exercise 2A

1 The amount you want to borrow might vary

2 It is the decimal equivalent of 2%

3 To find the interest of 1%

4 Multiplication is commutative: the order does not matter

5 The previous amount minus the repayment plus the interest

6 In theory, it will never be repaid, but the 1 % interest and the 2% repayments eventually become unrealistically small. That is why credit cards have the £5 minimum.

Exercise 2B

1 a £2440

b Expenditure is more than Income

c =A3 – B3

d i £15 067 ii £6187 iii –£4455 iv £2366

2 a Good – it means what you owe is less than your income

b 0.881

c 0.85

d 0.78 If paid weekly, divide the total liability by 52 and the monthly debts by 4 (or 52/12)

Exercise 2C

1 a	$345 \leqslant C < 355$	£10
b	$325 \leqslant V < 335$	10 ml
c	$1.55 \leqslant s < 1.65$	0.1 Mbs
d	$0.545 \leqslant s < 0.555$	0.01 Mbs
e	$11.995 \leqslant m < 12.005$	0.01 g
f	$28.395 \leqslant d < 28.405$	0.01 mm

2 a £37

b £68

3 a 99p

b 99p

4 a

Capacity	$233.5 \leqslant c < 234.5$	1 cc
Compression ratio	$9.15 \leqslant r < 9.25$	0.1
Maximum power	$14.55 \leqslant p < 14.65$ at $8750 \leqslant rev < 9250$	0.1 kW, 500 rpm
Dry weight	$131.5 \leqslant w < 132.5$	1 kg
Fuel capacity	$15.5 \leqslant f < 16.5$	1 l

b 60 mpg, error interval 1 mpg

c Upper bound = $16.5 \times 0.2205 \times 60.5 = 220.11$ miles
Lower bound = $15.5 \times 0.2195 \times 59.5 = 202.43$ miles

5 Upper bound = 384 403.5 + 43 592.5 = 427 996 km
Lower bound = 384 402.5 − 43 592.5 = 340 810 km
Not possible to give the distance to 1 s.f. because the upper and lower bounds disagree to 1 s.f.

6 a $487.5 \leqslant e < 492.5$

b $24.5 \leqslant p < 25.5$

c $1911.76 \leqslant GDA < 2010.2$

d 2000 kcal

Exercise 2D

All answers in this exercise are estimates, so answers can vary quite considerably according to the assumptions made.

1 195 college days ≈ 20 weeks so she would have spent £200 on bus.
Bike plus 6 weeks of bus costs £95 + £60 = £155, so she would save about £45.

2 Cheese ≈ £5 per kilo, so weekly cost is about £1.50 plus £5 × 0.6, or £4.50.
5x £1.85 ≈ £9, so he saves about £4.50 each week.
195 college days ≈ 20 weeks so would save ≈ £4.50 × 20 = £90

3 Earnings = (5 + 3 × 1.25) × £6.80 × 52 ≈ (5 + 4) × £7 × 50 = £3150

4 Profit = (289 × £16 + 58 × £8 + 164 × £12)/4 ≈ £(300 × 16 + 60 × 8 + 200 × 10)/4 ≈ £1600

Exercise 2E

1 a 0.91 b 1.47 c 0.03 d 0.007

2 a 38% b 4% c 0.8% d 234%

3 a £31.59 b £25.44 c £178.20 d £7.71
e £42.63 f £1.63

4 a 66 g $65.625 \leqslant f < 66.875$

b 20 g $19.5833 \leqslant sf < 20.4167$

c 92 $91.5 \leqslant sf < 92.5$

d 6.67 $6.5 \leqslant sf < 6.833$

5 a 1.44% b 3.79% c 8.83% d 0.0321%

6 a 38.7% b 42.5% c 43.2% d 34.9% e 36.7% Lovers

7 Double shot (19.23 mg per fluid ounce)

Exercise 2F

1 a $2\frac{3}{4}$, 2.75 b $\frac{467}{100}$, 467% c 6.25, 625%

d $\frac{1009}{100}$, 1009% e $\frac{607}{100}$, 6.07

2 a 105.3 m b £24.07 c 1546.3 kg
d 2.726 cl e 1890.72 tonnes

3 a £990 b £990 c £997.50 d £960
e £999.90

All answers are less than £1000 and the order of applying the percentages does not matter.

4 a increase of 25.44% b decrease of 22.56%
c decrease of 4.45% d increase of 18.8%
e no change

5 a 34p b £311 c £413.93 d 22.6g

6 Those with loans benefit when interest rates drop.

a Interest received would have dropped: £15, £13.50, £9, £6, £4.50, £3, £1.50

b Interest added to the loan would have dropped: £100, £90, £60, £40, £30, £20, £10

Exercise 2G

1 a £1065.75 b £852.60

2 Formula in cell D3 should be =A3*(B3/100)*5

3 a 18 b 7.2% c =72/A3 d =72/B3

4 a So that the reference to cell E1 does not change when the formula is copied down

b It means that there is an inconsistent error. This does not necessarily mean that the formula is wrong: just that it does not match the other formulas near it. In this case the other formulas near it change as they are copied down, so the triangle has appeared. It is not wrong because we always want to use cell E1.

Exercise 2H

1 a £60 b £28 c £328.32

2 a £649.51 b 1.69 years c 0.91%

3 a £729.99, £129.99 b £616.96, £84.96
c £424.09, £26.09

4 a 4% b 6% c 0.352%

5 a 14.8% b 74.9% c 36.5% d 788%

6 a 2.21% b 0.506%

Exercise 2I

Since the RPI and interest rates vary from time to time, the answers will vary and so it is not appropriate to give fixed answers here. The website used for the example given can be found at: http://www.studentloanrepayment.co.uk

Exercise 2J

1 £82.50

2 £30

3 £105, £118.50, £132

4 £60, £71.25, £82.50

5 £26 280.00, £25 389.60, £24 313.39

6 £29 880.00, £29 625.60, £29 234.11

Exercise 2K

1 £238.59

2 £329.46

4 a £1137.36, £1800 b £244.83, £400
c £262.42, £400

Exercise 2L

1 Two line graphs on the same chart, or a diagram like the one on page 63.
Total amount repaid = £66 560

2 Line graph probably best. Drop = 32.1%

3 NHS spending as a percentage of GDP increased over the period shown. During periods of recession there is a bigger increase in the percentage of GDP spent on the NHS than in other periods.
At the end of a period of recession the percentage drops. The periods of recession are of different lengths. The periods of recession do not occur at equal intervals.

4 a The sale of iPhones is generally increasing. The sales of iPhones drop after a peak in sales. The sales of iPhones start to increase about two quarters after a peak.
b The significant peaks probably indicate the quarter when a new iPhone is released.

Exercise 2M

1 a £5379 b £25 403 c £1153

2 a £9.36 b £99.36 c £350.06

3 a £9.40 b £17.25

Exercise 2N

1 a £15 787.20 b £15.60 c £238.36
d Better off by £158.88 per year

2 a i €1628.64 ii £440.83
b i TL 1964.10 ii £174.06

Exercise 2O

1 £1410

2 a £18680 b £192.86 per month c £2934.32

3 a Answers vary. Assume Jitesh buys 2 goals and 3 footballs for the answers that follow.
b 15 weeks c £18

Chapter 3: Estimation

Exercise 3A

1 $2\pi r^2 + 2\pi rh$ or $2\pi r(r + h)$

2 $V = \pi r^2 h$

3 $\pi r^2 h = 330$

4 $\pi r^2 h = 500$

5 3.8 cm

Exercise 3B

There are many possible answers for this section. The following answers have been obtained by plotting a graph of time against date and assuming that there is a linear relationship between them. The line of best fit used to obtain the answers was $y = -0.007263x + 4.37276$, obtained by using graphing software. It will be useful for you to discuss your answers with others as a means of checking them.

1 Sometime around April 1990

2 The assumptions made could include the following: that the time will continue to reduce linearly; that humans will always get faster without a limit.

3 Sometime around February 2020

Exercise 3C

There are many possible answers for this section. One of the first assumptions to make is which data to use from that given. The sprint races are the 100 m, 200 m and 400 m. The long distance races are those of 3000 m or more. The following answers have been obtained by plotting a graph of time against distance and assuming that there is a linear relationship between them. The line of best fit used to obtain the answers was obtained by using graphing software. It will be useful for you to discuss your answers with others as a means of checking them.

1 About 22 minutes 56 seconds (20 to 25 minutes acceptable)

2 The assumptions made could include the following: only the long distance results are used; that the time will continue to increase linearly for the long distance races of 3000 m or more; that people will have the stamina to keep going; that the conditions were the same for these races. The equation used to obtain the answer in Q1 was *time* = 0.003614*distance* – 4.17598.

3 Sprint data: *time* = 0.12*distance* – 2.2, where the time is in seconds and distance in metres.
Long distance: *time* = 3.6*distance* – 4.2, where the time is in minutes and distance in km.

4 Road distance: *time* = 4.1*distance* – 19, where the time is in minutes and distance in km
It is interesting to note that although the intermediate distance times are quicker, the outlier of the 100 km race means the gradient is more for the road race, probably because this long distance is so tiring that a slower pace results.

Exercise 3D

1 Remember to convert the speed in km/h to m/s!

2 $d_t = 0.1875u$

3 $d_b = 0.006u^2$

4 $d_s = 0.1875u + 0.006u^2$

5 A variety of answers are possible, such as: it assumes people are always paying attention to road traffic, it assumes road conditions and tyre conditions are good, it assumes all people have the same reaction times etc.

Exercise 3E

There are many possible answers for some questions in this section. It will be useful for you to discuss your answers with others as a means of checking them.

Exercise 3F

The answers here depend on your original diagrams. The use of dynamic geometry software is invaluable.

1–4 Students own diagrams

5 If there is only one lion all the den belongs to the lion so there is no safe locus for Daniel.

Exercise 3G

1 For example: The total distance travelled needs to be a minimum; it needs to be sited near major roads for ease of transportation, etc.

2 At the same place as the supermarket

3 a It can go anywhere on a straight line between the two supermarkets, assuming no problem with planning permission. It is probably most practical if it is sited at the same place as one of the supermarkets.

b Put it at the same place as the supermarket with the largest demand.

4 In the centre of the triangle formed by the three supermarkets, such that the lines from the warehouse to the supermarkets are all at 120°.

5 Imagine the four supermarkets to be the vertices of a quadrilateral. The warehouse should be situated at the intersection of the diagonals.

Exercise 3H

The answers here depend on your assumptions. Check with another student that your assumptions are reasonable. The answers given here are based on the author's assumptions.

1 0.5 h; 2 h; geometric mean 1 hour

2 5 h; 20 h; geometric mean 10 hours

3 10 g; 100 g; geometric mean 30 grams

4 1 million; 4 million; geometric mean 2 million

5 100 000; 500 000; geometric mean 200 000

Exercise 3I

The answers here depend on your estimates. Check with another student they are reasonable. The answers given here are based on the author's assumptions.

1 $C = \pi d$, 70 cm diameter, $C \approx 2$ m

2 You will probably use area of rectangles at some point.

3 $V = \pi r^2 h$, 5 mm diameter, 12 cm long, $V \approx 3000$ mm^3

4 You will probably use volume of a cuboid

Exercise 3J

The answers here depend on your assumptions. Check with another student that your assumptions are reasonable. The answers given here are based on the author's assumptions.

1 500 million to 1000 million

2 1 000 to 10 000

3 10 million to 100 million

4 4 m^2 to 12 m^2

5 1 000 to 2,000

6 100 000 to 1 000 000

7 30 million to 300 million

8 £100 000 000 to £10 000 000 000

9 1 to 4 £thousand million

10 £10^9 to £10^{10}

Chapter 4: Critical analysis of given data and models (including spreadsheets and tabular data)

Exercise 4A

1 Descriptive: states what happened, says when something occurred, gives information

2 Critical: structures information, draws conclusions

3 Descriptive: explains how something works, gives information

4 Critical: identifies the significance of events, argues a case according to the evidence, indicates why something might work

5 Descriptive: states what happened, lists in any order, gives information

Exercise 4B

1 Depends on your results

2 The time is not proportional to the strikes. There are 3 gaps (seconds) when the clock strikes four and 7 gaps or seconds when it strikes eight.

3 Richard. The first mile took 2 minutes, the second mile took 1 minute so he travelled 2 miles in 3 minutes, which is 40 mph

4 For example: Not every person is a single child of two parents

Exercise 4C

This exercise focusses on your reports and critical analysis from your peers.

Exercise 4D

1 You cannot square root money! Equality is not maintained when the units differ. Also $x = 6$ or -6.

2 a $\frac{12}{24} = \frac{1}{2}$

b Fred has just cancelled the same digits rather than factorising the numerators and denominators and then cancelling out the same factors.

3 a Although the diameters of the balls are in proportion to the price, the area of the balls are not in proportion and that is what has most impact on your eye.

b A better diagram could have all the balls the same size, representing £100 and then using a pictogram.

4 The Brie sector sticks out giving an inappropriate exaggeration. It would be good to know how many were surveyed and the angles at the centre of the pie chart.

5 a The lengths are in the correct proportion, but the visual impression of the object is misleading because your brain interprets each diagram as a solid 3D object.

b A better diagram could be a pictogram using books the same size representing the numbers or a simple bar chart.

6 a There is no vertical scale so the 2015 result appears about five times the 2014 result.

b A bar chart (a horizontal one would use less space on the page) with a scale from 0 to 350 000 (or a clearly marked broken scale) would be fine.

7 It does not take into account the number of people in each age group. There are many more 20–24 year olds than those aged 80+ in the country, so we would expect fewer fatalities in the older groups. A better diagram would show the percentages of drivers involved in fatal crashes in each age group.

Exercise 4E

1 She has added 20% per month, not 2% per year! 2% compound interest annually corresponds to 0.165% compound interest per month because $1.00165^{12} = 1.02$. So the amount at the end of the year is £121.30. The correct results are shown in the spreadsheet below

	A	B	C	D	E	F	G
1	Month number	1	2	3	4	5	6
2	Amount in account at start of the month	£10.00	£20.02	£30.05	£40.10	£50.17	£60.25
3	Interest	£0.02	£0.03	£0.05	£0.07	£0.08	£0.10
4	Amount in account at end of month	£10.02	£20.05	£30.10	£40.17	£50.25	£60.35

	A	H	I	J	K	L	M
1	Month number	7	8	9	10	11	12
2	Amount in account at start of the month	£70.35	£80.46	£90.60	£100.75	£110.91	£121.10
3	Interest	£0.12	£0.13	£0.15	£0.17	£0.18	£0.20
4	Amount in account at end of month	£70.46	£80.60	£90.75	£100.91	£111.10	£121.30

2 He should divide the £40 by 1.15, then by 0.9 (or vice versa) to give £38.65 as the original price of each cartridge.

3 a Dick adds to 35, not 45. Tweets read add to 74 not 84. 'Total' is not a person.

b He should use formulas in the spreadsheet. 'Total' would be best in cell E6

4 a The vertical scale does not increase in equal intervals. The width of each column looks the same but represents different amounts. A histogram using frequency density would be best.

b She has divided by the number of categories instead of 39, the number of students. It would be better to use 50 for the mid–point of the last group (usually when the last group is unbounded we use the same width as the one next to it). A better estimate would be £17.31

c She probably took the mid–point of the middle interval rather than taking into account the distribution. With 39 items the median amount is the 20th value – this occurs at the end of the second interval, so £15 is a better estimate of the median, assuming the amounts are equally distributed in the interval.

5 a Test 3 and Test 4 would be better arranged in order. It is not clear whether column F is the median or mean amount. The mean of a question number (in cell A6) is nonsense. His mean for Test 3 is inappropriate – would be better if he recorded a 0 in cell E3.

b In Test 3 he scored a total of 21 marks – his lowest score. The mean should be 5.25. We don't know how many marks were available in total for each question or each test, so without that information we cannot make any valid conclusions.

Chapter 5: The normal distribution

Exercise 5A

1 a The distribution is symmetrical about the mean.

b The mode, median and mean are all equal.

c The total area under the curve is 1.

d The distribution is defined by two parameters (defining characteristics), the mean and the variance.

2 Student free to create any appropriate crossword. Teacher assessment required.

3 The two distributions have different location parameters but the same dispersion.

4 The two distributions have the same location parameters but different dispersions.

5 a In the middle of the line

b 50%

c $\frac{1}{3}$

d 47.5%

e $\frac{1}{3} + \frac{95}{200} = \frac{485}{600} = \frac{97}{120}$

Exercise 5B

1 a $X \sim N(3, 4)$

b $X \sim N(2, 3^2)$ or $x \sim N(2, 9)$

2 a $X \sim N(0, 5)$

b $X \sim N(-1, 2^2)$ or $x \sim N(-1, 4)$

3 a mean = 5, variance = 7^2 or 49 and standard deviation = 7

b mean = 6, variance = 9^2 or 81 and standard deviation = 9

4 The student confused the mean with standard deviation. It should be $X \sim N(7, 11^2)$ or $X \sim N(7, 121)$

5 The student confused the mean with standard deviation. It should be $Z \sim N(0, 1)$ or $Z \sim N(0, 1^2)$

Exercise 5C

1

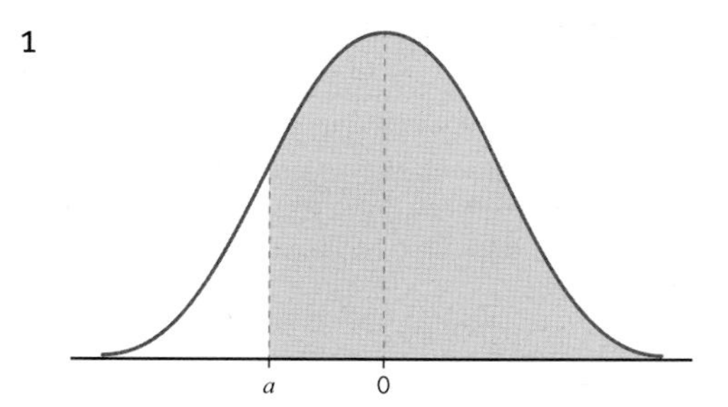

2

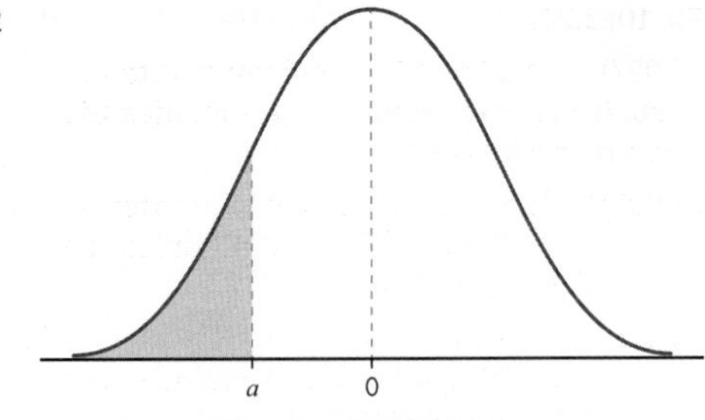

3

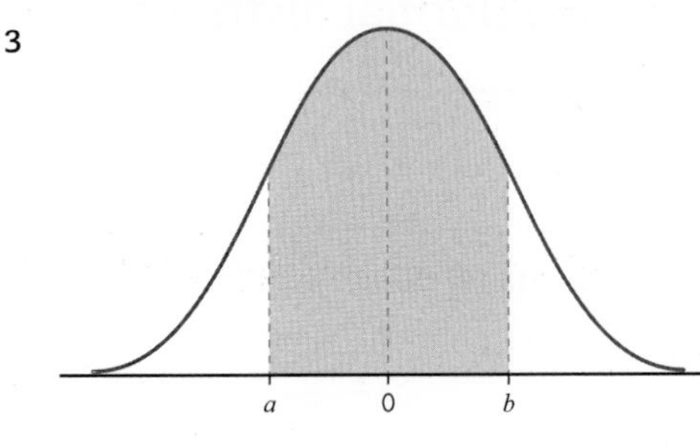

4

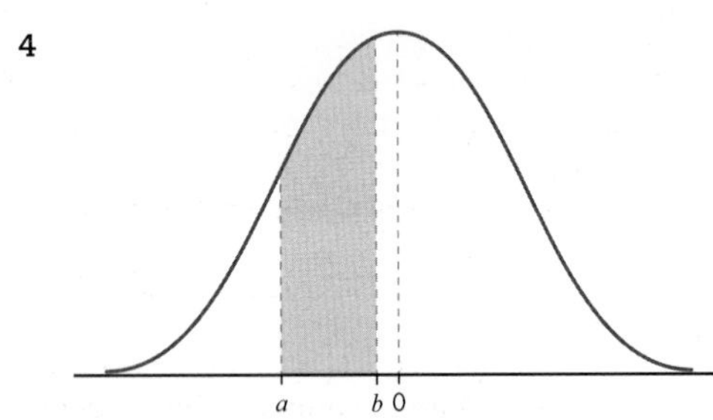

5

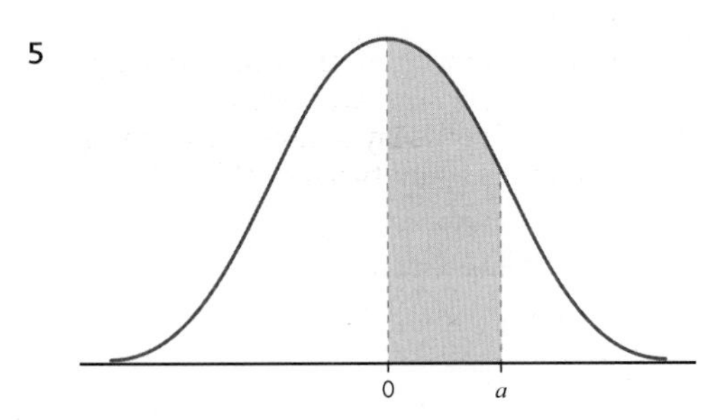

Exercise 5D

1 0.93319

2 0.71566

3 0.0197

4 0.15866

5 0.84134

6 0.13591

7 0.68268

8 0.13591

Exercise 5E

1 a 0.69146
 b 0.19489
 c 0.84134
 d 0.14988
 e 0.51245
 f 0.03887

2 0.00379 or 0.379%

3 0.2843 or 28.43%

4 a 0.10565
 b 0.7887
 c 0.10565
 d 0.0829

5 a 0.38209
 b 0.04457
 c 0.00048 or 0.048%

Exercise 5F

1 a 53.263
 b 50.5
 c 38.1
 d 36.745
 e 6.737
 f 46.449

2 4480.6 hours

3 1.859712 m or 1.860 m

4 57.302 g

5 62.092 to 74.908

6 a 5.48%
 b £1 036.25

Chapter 6: Probabilities and estimation

Exercise 6A

1 Student free to create any appropriate diagram. Teacher assessment required.

2 a No. Crowds as football matches tend to be segregated by team to avoid difficulties Consequently home team might have walked whereas away supporters will, more than likely have travelled some distance using car, bus or train.

2 b No. A&E is just one department in a hospital. Either just concentrate survey on A&E or develop a method so that all departments, where appropriate, are considered.

3 Student free to list any reasonable methods. Teacher assessment required.

4 Student free to use any reasonable method. Teacher assessment required.

5 Student free to use any reasonable method. Teacher assessment required.

Exercise 6B

1 Student free to list any reasonable improvements including take more sample sections from pillow case and count number of mites then find mean across samples. Teacher assessment required.

2 Not possible to say. Instead find the mean of the sample means and use this as the estimator for the population mean.

3 Student free to complete any reasonable methods. Teacher assessment required.

4 Student free to complete any reasonable methods. Teacher assessment required.

5 Student free to complete any reasonable methods. Teacher assessment required.

Exercise 6C

1

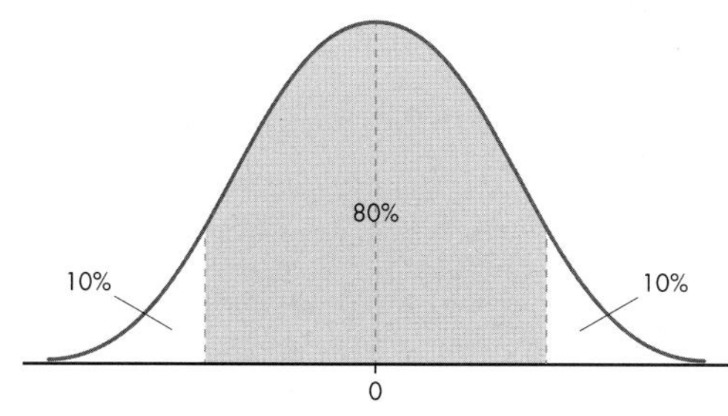

2 upper limit = 1.28 ; lower limit = −1.28

3

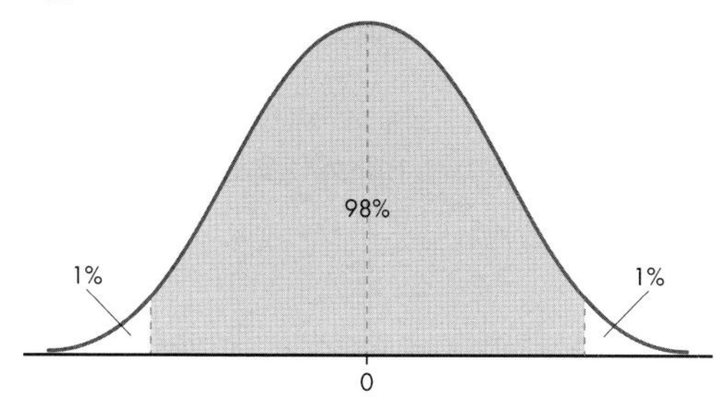

4 upper limit = 2.33 ; lower limit = −2.33

5

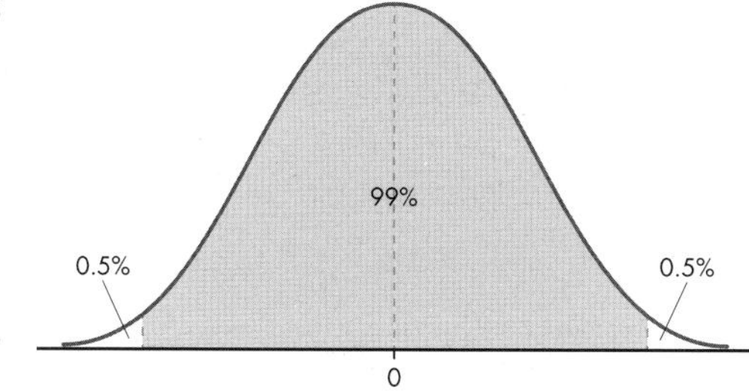

6 upper limit = 2.58 ; lower limit = −2.58

Exercise 6D

Confidence level	Confidence interval
80%	$\left(\bar{x}-1.28\frac{\sigma}{\sqrt{n}}, \bar{x}+1.28\frac{\sigma}{\sqrt{n}}\right)$
90%	$\left(\bar{x}-1.645\frac{\sigma}{\sqrt{n}}, \bar{x}+1.645\frac{\sigma}{\sqrt{n}}\right)$
95%	$\left(\bar{x}-1.96\frac{\sigma}{\sqrt{n}}, \bar{x}+1.96\frac{\sigma}{\sqrt{n}}\right)$
98%	$\left(\bar{x}-2.33\frac{\sigma}{\sqrt{n}}, \bar{x}+2.33\frac{\sigma}{\sqrt{n}}\right)$
99%	$\left(\bar{x}-2.58\frac{\sigma}{\sqrt{n}}, \bar{x}+2.58\frac{\sigma}{\sqrt{n}}\right)$

Exercise 6E

1 (1.76, 2.87)

2 (96.0, 104.0)

3 (249.70, 252.30)

4 (1016.78, 1023.22)

5 (0.983, 0.987) This is the 98% confidence interval (the interval such that the probability it includes the population mean, μ, is 98%).

6 (7.658, 8.009) This is the 99% confidence interval (the interval such that the probability it includes the population mean, μ, is 99%).

7 (5.065, 8.685) This is the 80% confidence interval (the interval such that the probability it includes the population mean, μ, is 80%).

8 $\left(\bar{x}-1.645\frac{5}{\sqrt{10}}, \bar{x}+1.645\frac{5}{\sqrt{10}}\right)$

Chapter 7: Correlation and regression

Exercise 7A

1 Observations:

- chickens are divided into three groups: high rate of lay, medium rate of lay and low rate of lay.
- high rate of lay chickens lay in excess of 240 eggs in the first year.
- medium rate of lay chickens lay in excess of 160 eggs in the first year.
- low rate of lay chickens lay in excess of 100 eggs in the first year.
- the rate at which high rate of lay chickens produce eggs is much faster than medium and low rate of lay chickens.
- it seems that low rate of lay chickens live longer than the other two faster rate layers
- low rate of lay chickens lay relatively more eggs for longer than the other two rate of lay chickens.

2

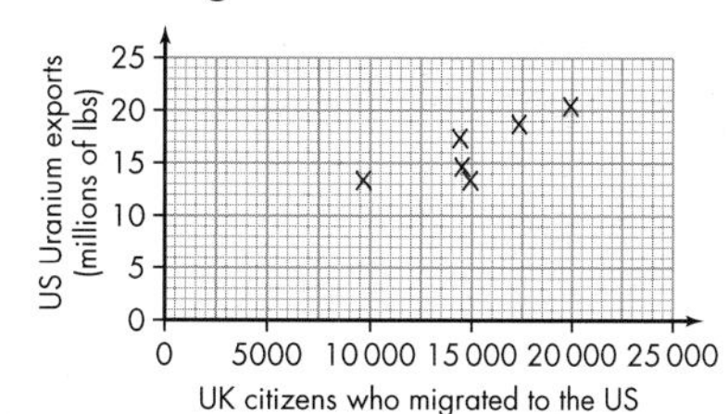

The scatter diagram suggests that there is a positive correlation between UK citizens who migrated to the US and US Uranium exports (millions of lbs) but in reality this is not the case (correlation does not mean causation).

3 a A strong positive correlation. Example hours of sunshine and ice cream sales.

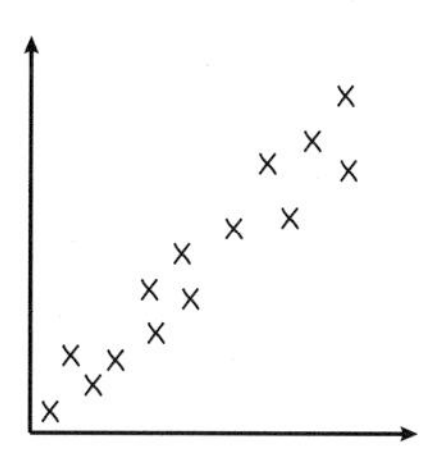

3 b A weak negative correlation. Example hours of sunshine and umbrella sales (some people still might purchase umbrellas to keep the sun off!).

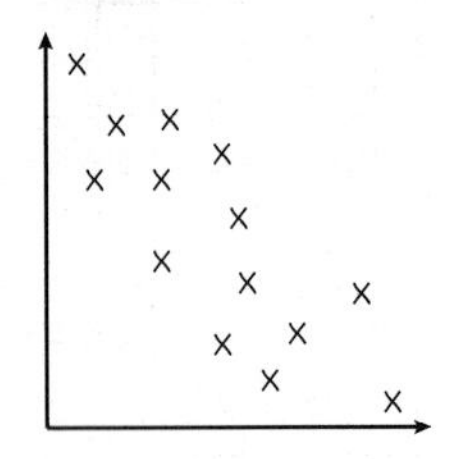

3 c No correlation. Example ipod sales and apple consumption.

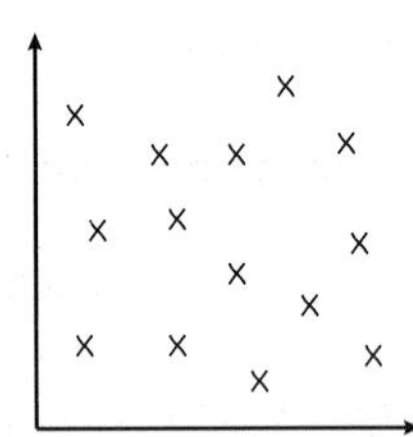

3 d A nonsense correlation. Example UK citizens who migrated to the US and US Uranium exports (millions of lbs)!

4

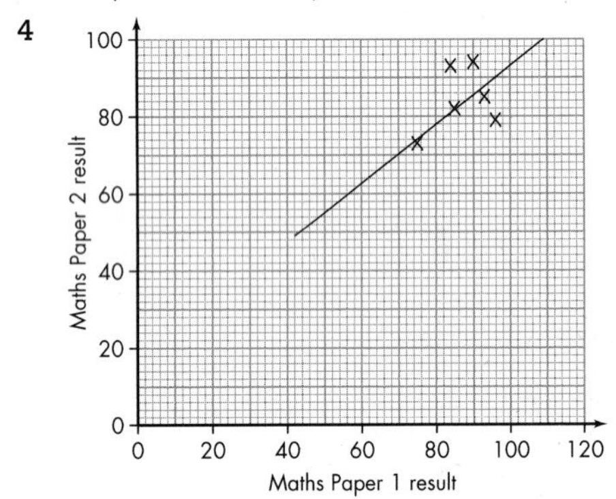

5 a

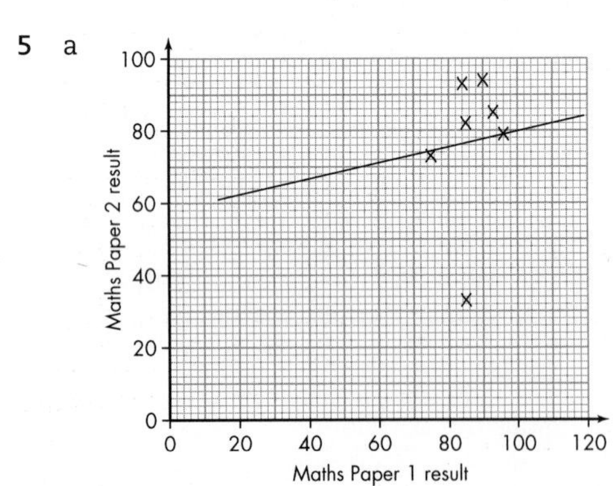

5 b Poor maths paper 2 result possible causes: personal problem, ill, not revised, etc..

Exercise 7B

1 For a positive linear correlation the range of value is $0 < \text{pmcc} \leqslant 1$.

2 a 0.98 strong positive

2 b –0.36 weak negative

2 c 1.00 perfect positive

2 d 0.02 very close to no correlation

3 pmcc = 0.5194; positive correlation (neither very strong nor very weak)

4 –0.85 strongest correlation (the fact that negative doesn't affect strength).
–0.05 weakest correlation (the fact that negative doesn't affect strength).

5 pmcc = –0.0282; virtually no correlation

Exercise 7C

1 $y = 41.42 + 0.4875x$; If the pupil scores 0 of maths paper 1 then they will get 41.42 on maths paper 2. For every extra mark on maths paper 1 this correlates to almost a half mark increase on maths paper 2.

2 79.45 ($x = 78$). A mark of 78 on maths paper 1 correlates to a similar mark on maths paper 2.

3 57.51 ($x = 33$). A mark of 33 on maths paper 1 correlates to a much higher mark on maths paper 2.

4 $y = 47.10 - 0.067x$; If the pupil scores 0 of maths paper 1 then they will get 47.10 on art paper 1. For every extra mark on maths paper 1 this correlates to small mark decrease on art paper 1.

5 42.00 ($x = 76$). A mark of 76 on maths paper 1 correlates to a low mark on art paper 1.

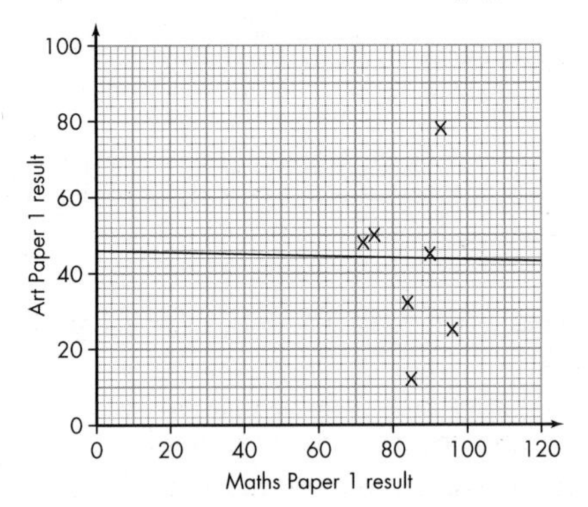

6 a

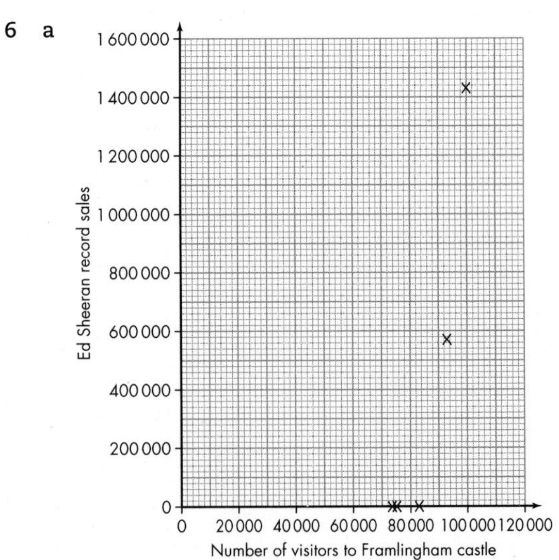

6 b pmcc = 0.909944

6 c a = –3,831,544, b = 49.88404, $y = -3{,}831{,}544 + 49.88404x$ The equation suggests that when there are no visitors at Framlingham castle Ed Sheeran will have nearly 4,000,000 negative record sales and that Ed will sell 49 more records for every additional visitor to Framlingham castle.

6 **d** It is highly unlikely that the proximity of Ed Sheeran's private house to Framlingham castle is affecting the number of visitors to this tourist attraction. If his house was opened to the public and managed by English Heritage then the relationship might be very different!

Chapter 8: Critical path analysis

Exercise 8A

1

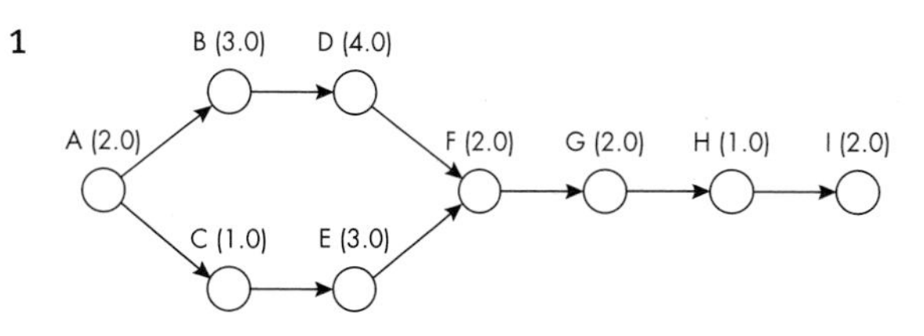

2 Any appropriate diagram matching precedence table. Teacher assessment required.

Exercise 8B

1 a

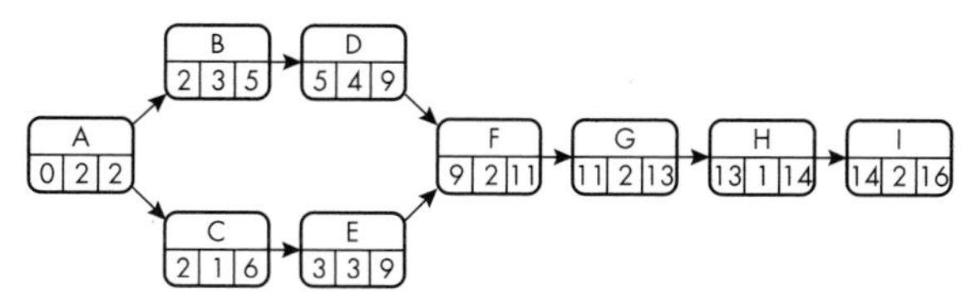

1 **b** Critical activities: A, B, D, F, G, H, I. Critical path length: 16

1 **c** A creation of or amendment to any critical activity. Teacher assessment required.

2 Any appropriate diagram matching precedence table. Teacher assessment required.

Exercise 8C

1 a

Activity	Duration (in days)	Earliest time	Latest time	Float
A	2.0	0	2	0
B	3.0	2	5	0
C	1.0	2	6	3
D	4.0	5	9	0
E	3.0	3	9	3
F	2.0	9	11	0
G	2.0	11	13	0
H	1.0	13	14	0
I	2.0	14	16	0

1 **b**

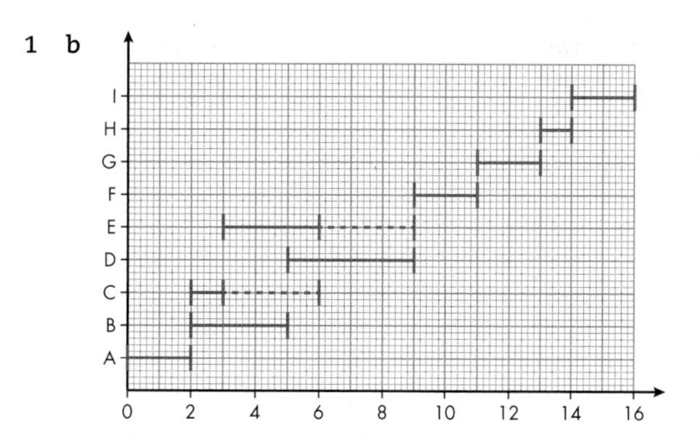

1 **c** Yes, the zero float activities do correspond with the critical activities.

2 **a** Tabulate the information from your answer to Exercise 8B question 2.

2 **b** Construct a Gantt chart to reflect your tabulated data.

2 **c** The activities with non–zero float should correspond with non–critical activities.
Teacher assessment required.

3 Student project. Teacher assessment required.

Chapter 9: Expectation

Exercise 9A

1 Students own experiment

2 0.64

3 Students own experiment

4 **a** $\frac{1}{3}$ **b** $\frac{2}{3}$

5 0.03

6 Sample space diagram:

+	1	2	3	4	5	6
1	2	3	4	5	6	7
2	3	4	5	6	7	8
3	4	5	6	7	8	9
4	5	6	7	8	9	10
5	6	7	8	9	10	11
6	7	8	9	10	11	12

The student who said 7 is correct. It is likely to come up $\frac{1}{6}\left(\frac{6}{36}\text{ simplified}\right)$ times.

7. **a** $\frac{18}{36}\left(\frac{1}{2}\right)$

b $\frac{7}{36}$

c $\frac{29}{36}$

8. **a** 50 times **b** 550 times

Exercise 9B

1 a
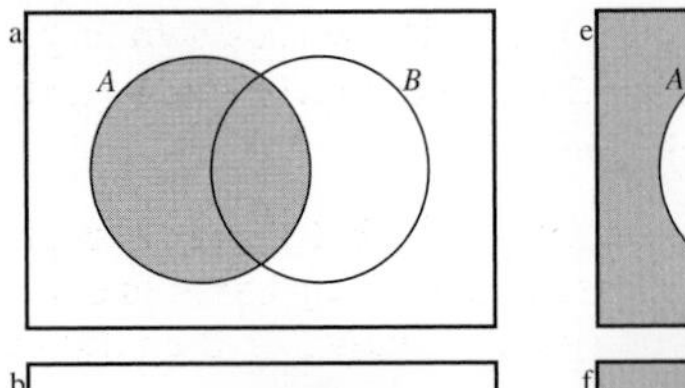

b
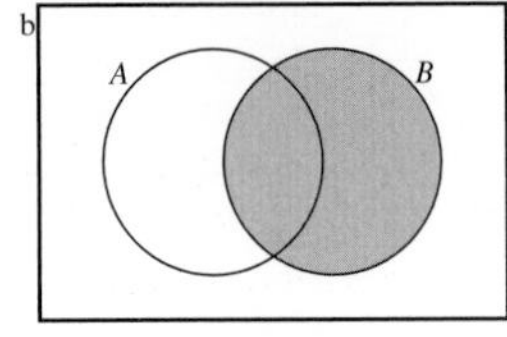

c
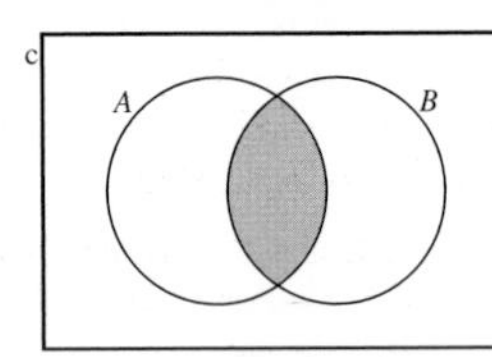

d
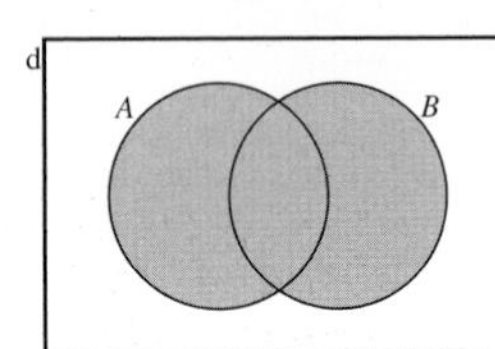

e
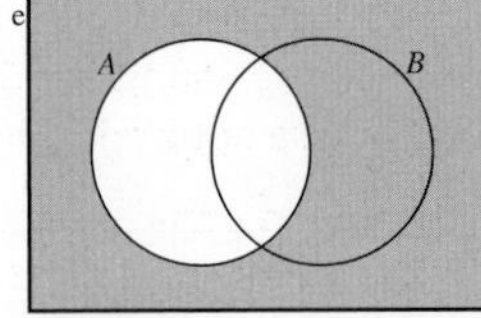

f
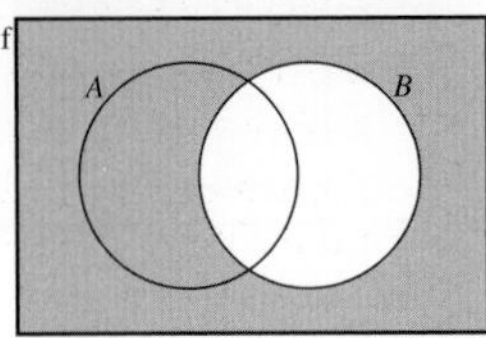

g
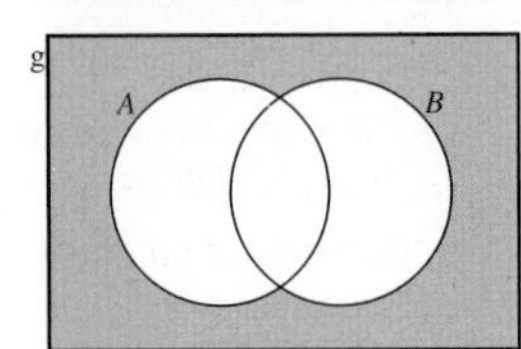

h
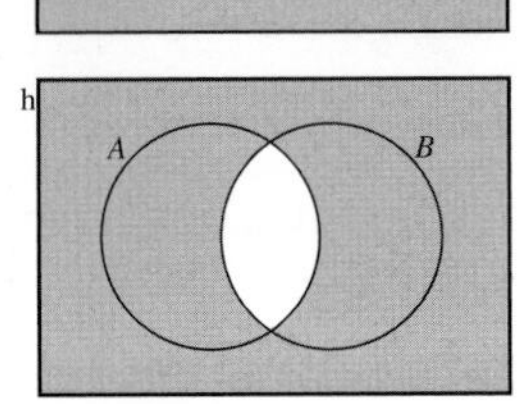

2 a Moulin Rouge, Paddington, Grace of Monaco, Days of Thunder, Eyes Wide Shut

b Mission Impossible II, Top Gun, Valkyrie, Oblivion, Days of Thunder, Eyes Wide Shut

c Nicole Kidman and Tom Cruise films: Days of Thunder, Eyes Wide Shut

d Nicole Kidman or Tom Cruise films: Moulin Rouge, Paddington, Grace of Monaco, Days of Thunder, Eyes Wide Shut, Mission Impossible II, Top Gun, Valkyrie, Oblivion

e None of these: Spectre

3.
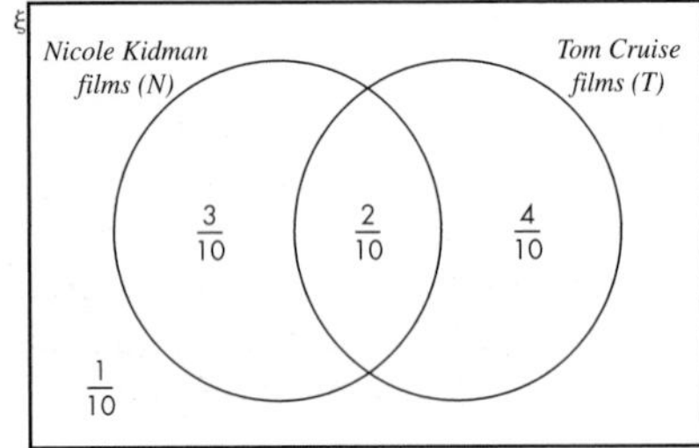

a $P(N) = \frac{5}{10}$

b $P(T) = \frac{6}{10}$

c $P(N \cap T) = \frac{2}{10}$

d $P(N \cup T) = \frac{9}{10}$

e $P(N') = \frac{5}{10}$

f $P(N' \cap T') = \frac{1}{10}$

g $\frac{2}{6}$

4 a
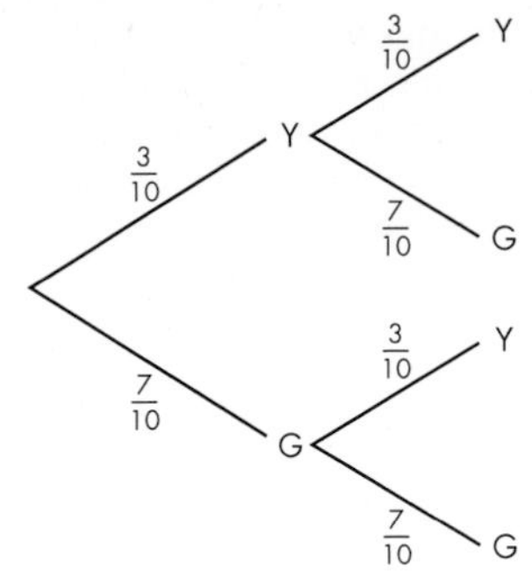

b $\frac{9}{100}$

c $\frac{42}{100}$

d $\frac{9}{100}$

e $\frac{58}{100}$

5 a
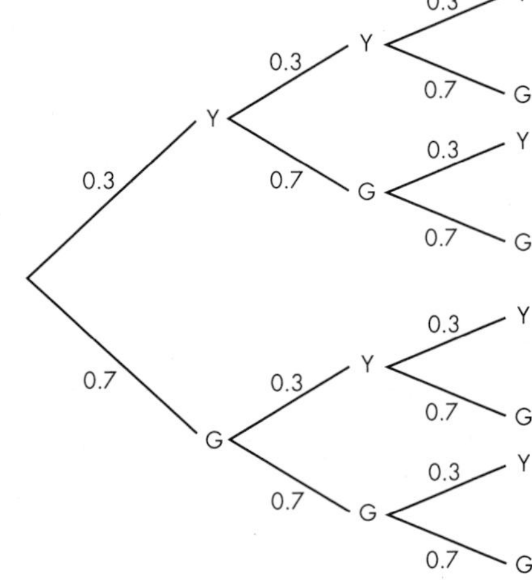

b $\frac{27}{1000}$

c 0

d $\frac{630}{1000}$

e $\frac{370}{1000}$

6 a
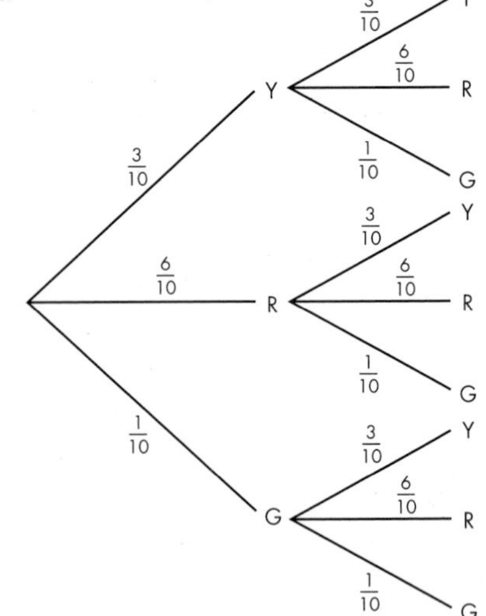

b $\frac{9}{100}$

c $\frac{54}{100}$

d $\frac{81}{100}$

e $\frac{46}{100}$

Exercise 9C

1 a $\frac{2}{132}$ b $\frac{48}{132}$ c $\frac{44}{132}$

2 They are independent events. The throwing of a die has no affect on the throwing of a second die.

3 a 0.12 b 0.58 c 0.42

4 They are independent events. The winning with one raffle ticket has no affect on winning with the other raffle ticket. However by buying two raffle tickets you have doubled your chance of winning.

5 a $\frac{6}{90}$ b $\frac{42}{90}$ c $\frac{6}{90}$ d $\frac{48}{90}$

6 0.000 000 0222 (3 significant figures)

Exercise 9D

1 $E(X) = \frac{15}{6}$

2 $E(X) = -0.3$

3 a

O	2	3	4	5	6	7	8	9	10	11	12
P	$\frac{1}{36}$	$\frac{2}{36}$	$\frac{3}{36}$	$\frac{4}{36}$	$\frac{5}{36}$	$\frac{6}{36}$	$\frac{5}{36}$	$\frac{4}{36}$	$\frac{3}{36}$	$\frac{2}{36}$	$\frac{1}{36}$

b $E(X) = 7$

4

Outcome	0	1	2	3
Probability	$\frac{1}{8}$	$\frac{3}{8}$	$\frac{3}{8}$	$\frac{1}{8}$

$E(X) = \frac{12}{8} = £1.50$

5 Dependent on student response.

6 Dependent on student response.

Chapter 10: Cost benefit analysis

Exercise 10A

Answers may vary

Exercise 10B

Answers may vary

Exercise 10C

Answers may vary

Exercise 10D

In this exercise you are expected to write down any assumptions you make. Some assumptions are given, but they are not exclusive.

1 Expected penalty is £100, so probably best to replace thermostat in both cases, assuming baker is concerned about reputation and wants to keep the order each month.

2 a £300

b £3, assuming the lap top will be replaced under guarantee. Otherwise the loss is £403, in which case it is worth taking the control measure.

3 The expected loss is £21.60, assuming you will try and recover it from lost property. Assuming you can pay the insurance it is worth it to avoid any fraudulent use of the phone and loss of personal data.

4 Expected loss if third party insurance taken is £210. Difference in insurance is £600. Assuming the value of the car is not much, it is probably not worth the extra as usually there is a compulsory excess that you don't get back.

5 Assuming that the probability of each fitter being sick is independent, then the probability that none of them is sick is 0.9025. Therefore the probability the job is delayed is 0.0975. Expected loss is £1950, so the control measure is worth it.

6 Answers may vary

Exercise 10E

In this exercise you are expected to write down any assumptions you make. One assumption you should mention in each question is that the probabilities are independent: i.e. the outcome of one event does not influence the other. Some assumptions are given, but they are not exclusive. Your justifications may also differ, but the important thing is that they support your numerical results in some way.

1 a £560

b Cost of A only £500, B only = £550, both = £425, so worth using both control measures

2 a £9600

b Cost of computer back up only £8000, dongles only = £7500, both = £5000, so worth using both control measures. Expected profit £45 000

3 a £228000

b Cost of code only £90 000, test only = £220 000, both = £40000, so worth using both control measures. Expected profit £460000

4 a £14 100

b Cost of wrapping only £19 000, guards only = £26 000, both = £30 000, so not worth using any control measures. Expected profit £25 900

5 a, b Use of spreadsheet

c Cost of control measure A is always more than cost of control measure B

d 0.1275, 0.255

Chapter 11: Graphical methods

Exercise 11A

1. $c = 10 + 0.5m$

m	c
1	10.5
2	11
3	11.5
4	12
5	12.5
6	13
7	13.5
8	14
9	14.5
10	15
11	15.5
12	16
13	16.5
14	17
15	17.5
16	18
17	18.5
18	19
19	19.5
20	20
21	20.5
22	21
23	21.5
24	22
25	22.5
26	23

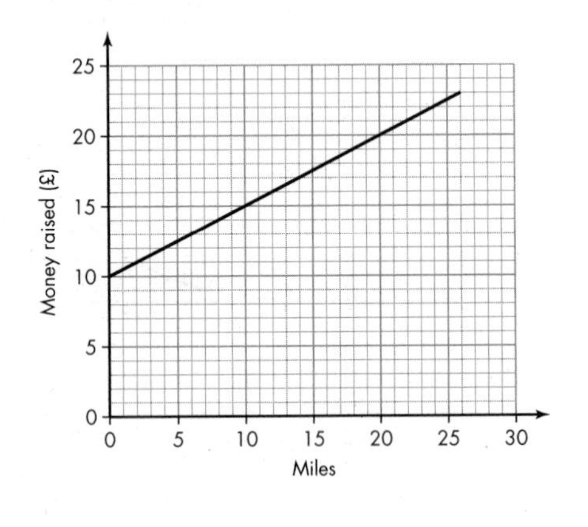

2 $y = 3x - 2$

x	−5	−4	−3	−2	−1	0	1	2	3	4	5
y	−17	−14	−11	−8	−5	−2	1	4	7	10	13

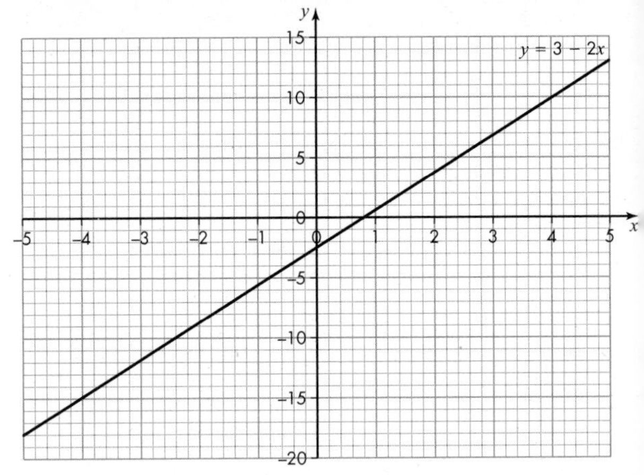

3 Plumber 1: $y = 30x + 50$
Plumber 2: $y = 50x$
$y = 30x + 50$

x	2	4	6	8	10	12	14
y	110	170	230	290	350	410	470

$y=50x$

x	2	4	6	8	10	12	14
y	100	200	300	400	500	600	700

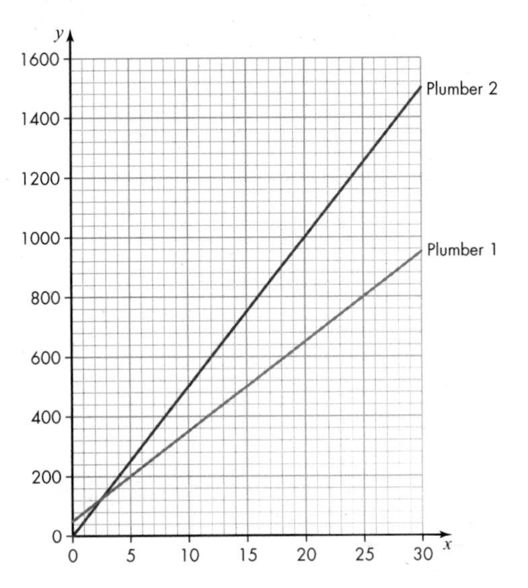

Plumber 2 who charges £50 per hour is cheaper on smaller jobs (< 2.5 hours) but as soon at the time takes more than 2.5 hours (where the graphs intersect) then plumber 1 is cheaper. You would use plumber 2 to fix a leaky tap (small job) but plumber 1 to fit a bathroom (a longer job).

4 $y = \dfrac{4-x}{2}$

x	−5	−4	−3	−2	−1	0	1	2	3	4	5
y	$\frac{9}{2}$	4	$\frac{7}{2}$	3	$\frac{5}{2}$	2	$\frac{3}{2}$	1	$\frac{1}{2}$	0	$-\frac{1}{2}$

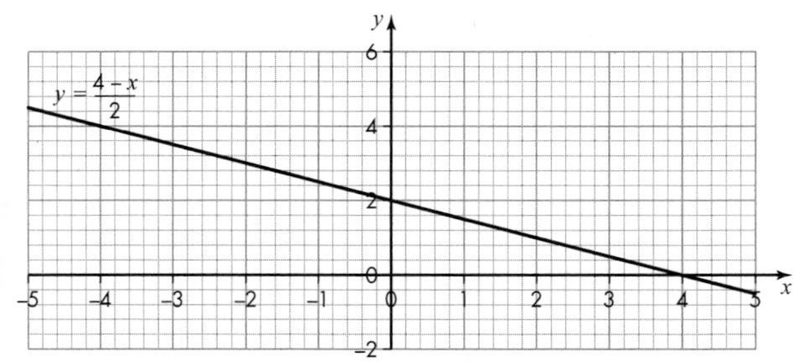

This graph has a negative gradient whereas the graph in question two has a positive gradient.

5 Mistakes corrected and highlighted in table below.

x	−4	−3	−2	−1	0	1	2	3	4
y	11	9	7	5	3	1	−1	−3	−5

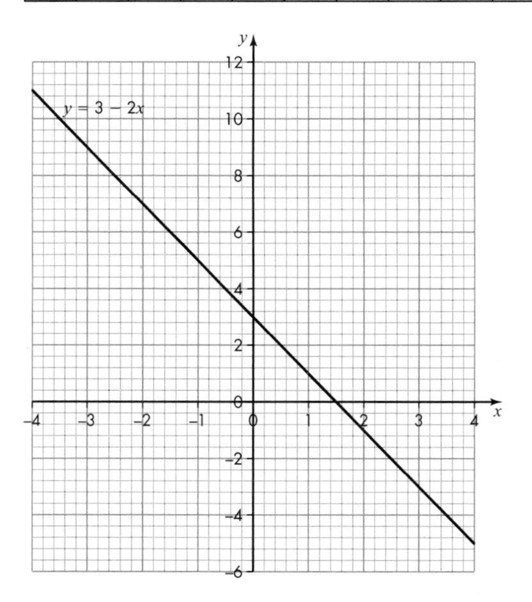

Exercise 11B

1 $y = x^2 - 5x + 6$

x	−1	0	1	2	3	4	5	6
y	12	6	2	0	0	2	6	12

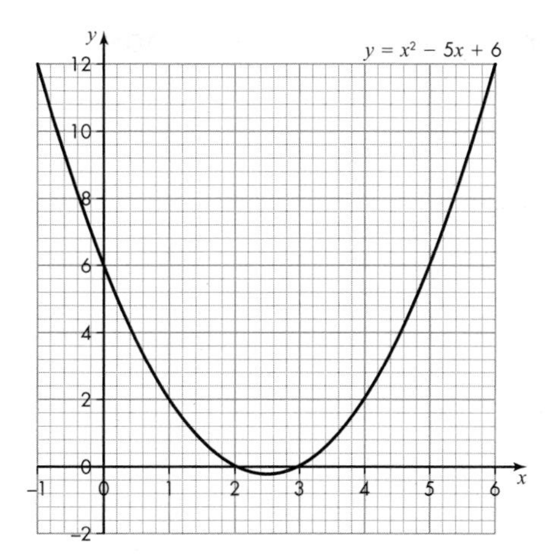

2 $y = -x^2 + 5x - 6$

x	−1	0	1	2	3	4	5	6
y	−12	−6	−2	0	0	−2	−6	−12

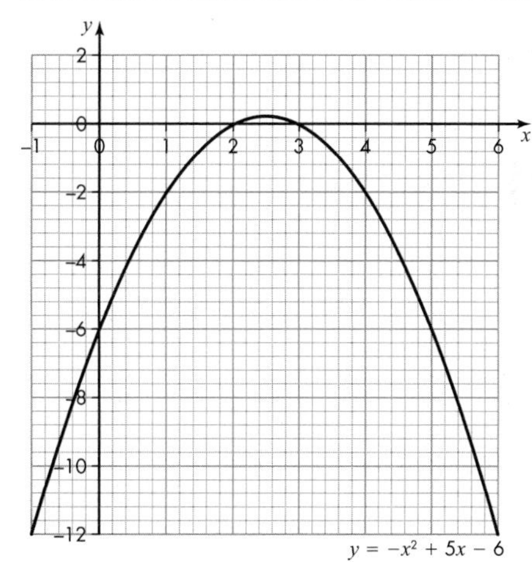

Maximum height ball reaches is 0.25m. This graph is a reflection of the graph in question 1.

3 $y = x^3$

x	−5	−4	−3	−2	−1	0	1	2	3	4	5
y	−125	−64	−27	−8	−1	0	1	8	27	64	125

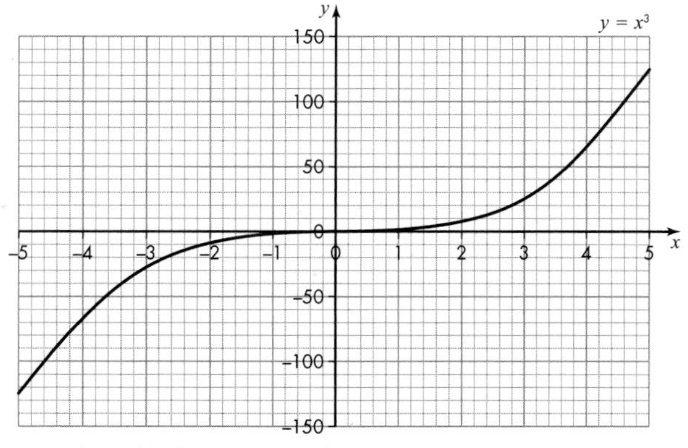

Point of inflexion at (0,0)

4 $y = x^3 - x$

x	−5	−4	−3	−2	−1	0	1	2	3	4	5
y	−120	−60	−24	−6	0	0	0	6	24	60	120

5 $y = x - x^3$

x	−5	−4	−3	−2	−1	0	1	2	3	4	5
y	120	60	24	6	0	0	0	−6	−24	−60	−120

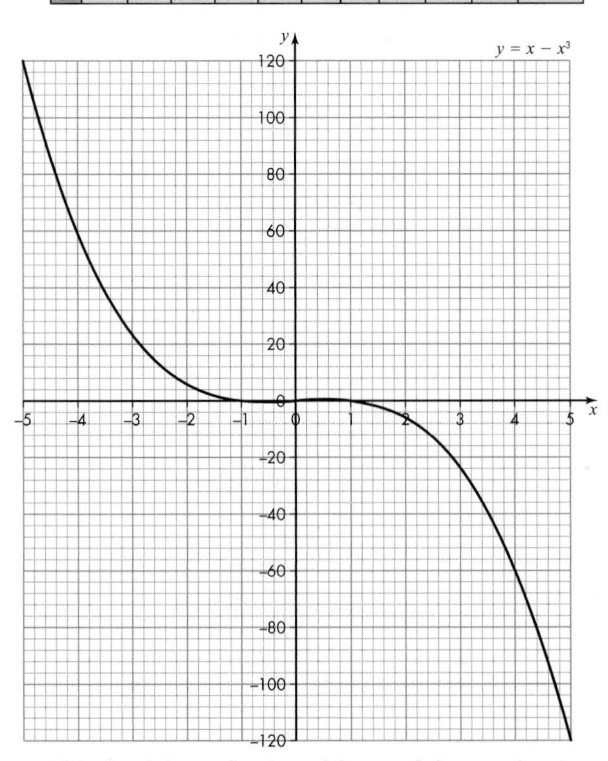

This graph is a reflection of the graph in question 4

Exercise 11C

1 a ii

b v

c iv

d i

e iii

2 An exponential growth graph.

3 $y = 2^{-x}$

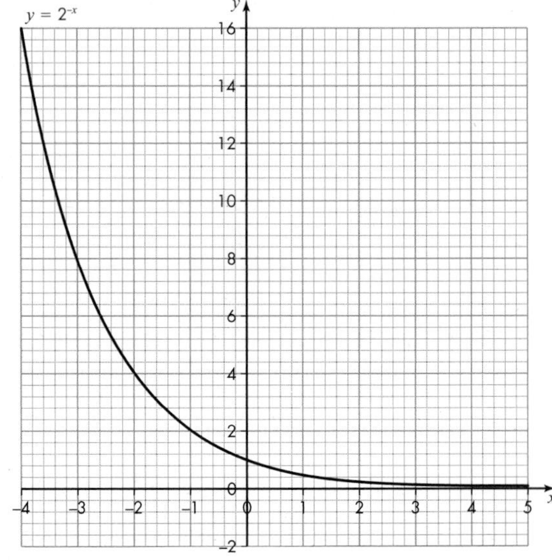

$y = 2^{-x}$ is a reflection of $y = 2^x$

4 $y = 2^x + 1$

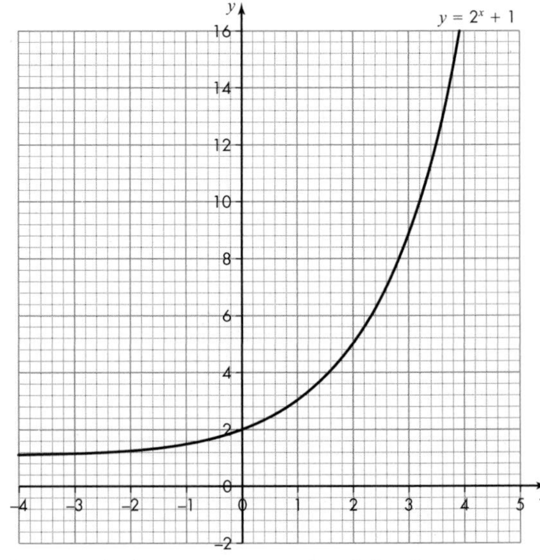

$y = 2^x + 1$ is the same $y = 2^x$ as but it has been translated 1 unit upwards (in the positive y direction)

5 a £7

b $y = 2x + 3$

c The graph is shown. The asymptote is $y = 3$.

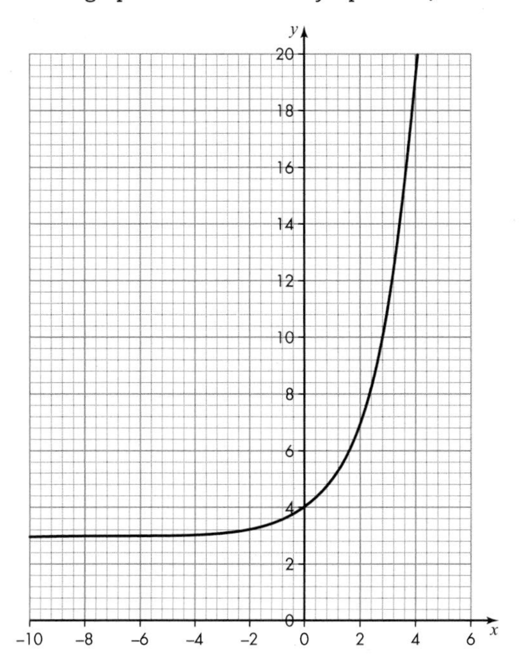

6 Any fraction to the power of x e.g. $\left(\frac{1}{2}\right)^x$ in this case x is positive but the graph looks like exponential decay.

Exercise 11D

1 $y = 180x + 150$

x	1	2	3	4	5	6	7	8	9	10
y	330	510	690	870	1050	1230	1410	1590	1770	1950

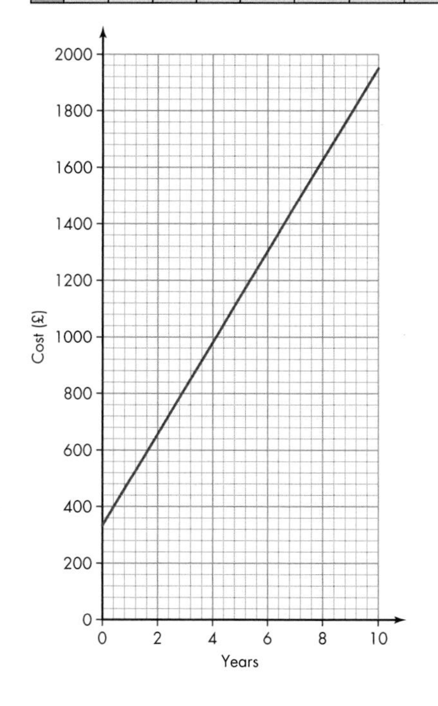

1 a Using the graph when x = 2 y = £510

b Using the graph when x = 5 y = £1050

c Using the graph when y = 500 x = 1.8 years (approximately)

2 $y = -x^2 + 4x$

x	0	1	2	3	4
y	0	3	4	3	0

$y = -x^2 - 4x$

x	–4	–3	–2	–1	0
y	0	3	4	3	0

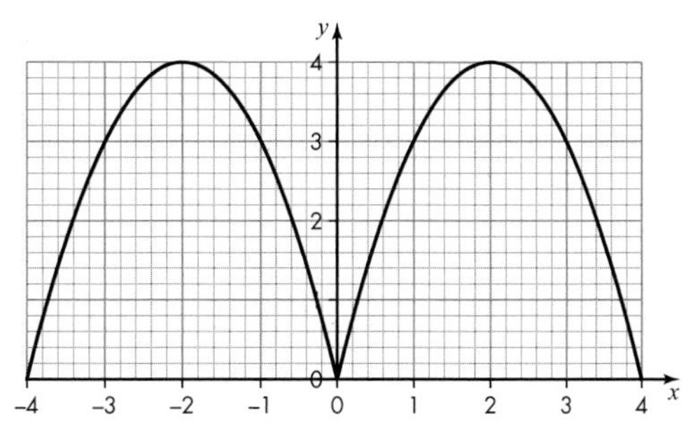

The new logo will be 8 units wide and 4 units tall.

3 $y = -x^3 + 4x$

x	–3	–2	–1	0	1	2	3
y	15	0	–3	0	3	0	–15

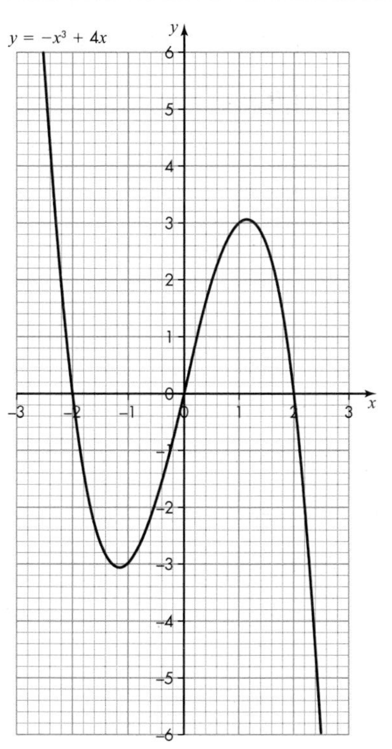

The shape will need to be trimmed above $y = 3$ and below $y = -3$.

4 The graph is $y = 1\,367\,000\,000 \times 0.995x$ where $x = 1$ for 2015 and so on.

x	y
2014	1 367 000 000
2015	1 360 165 000
2016	1 353 364 175
2017	1 346 597 354
2018	1 339 864 367
2019	1 333 165 046
2020	1 326 499 220
2021	1 319 866 724
2022	1 313 267 391
2023	1 306 701 054
2024	1 300 167 548
2025	1 293 666 711
2026	1 287 198 377
2027	1 280 762 385
2028	1 274 358 573
2029	1 267 986 780
2030	1 261 646 846
2031	1 255 338 612
2032	1 249 061 919
2033	1 242 816 610
2034	1 236 602 527
2035	1 230 419 514
2036	1 224 267 416
2037	1 218 146 079
2038	1 212 055 349
2039	1 205 995 072
2040	1 199 965 097
2041	1 193 965 271
2042	1 187 995 445
2043	1 182 055 468
2044	1 176 145 190
2045	1 170 264 464
2046	1 164 413 142
2047	1 158 591 076
2048	1 152 798 121
2049	1 147 034 130
2050	1 141 298 960

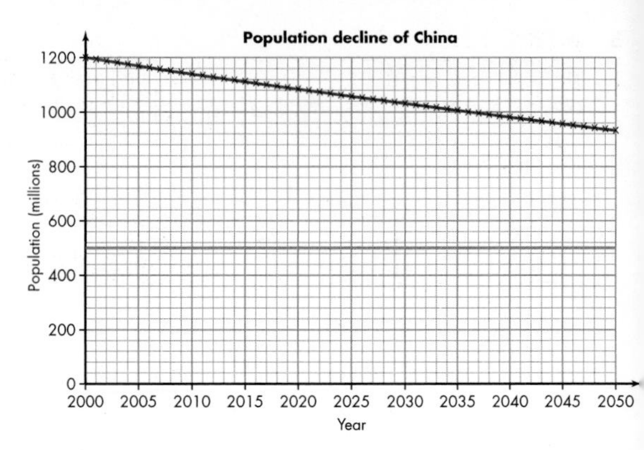

No China will not reach its goal of a population of 700 000 000 by 2050 (especially now it has relaxed its one child policy!).
The graph appears linear because we have not plotted enough of the graph to see the full exponential curve.

5 Answers may vary. Check against explanation of interpolation and extrapolation earlier in this section.

Exercise 11E

1 Both companies charge £40 at 100 minutes. If the total usage is less than 100 minutes then the cheaper company is A however if the total usage is greater than 100 minutes then the cheaper company is B.

2 $y = 2x + 4$

x	−4	−3	−2	−1	0	1	2	3	4
y	−4	−2	0	2	4	6	8	10	12

$y = -x + 1$

x	−4	−3	−2	−1	0	1	2	3	4
y	5	4	3	2	1	0	−1	−2	−3

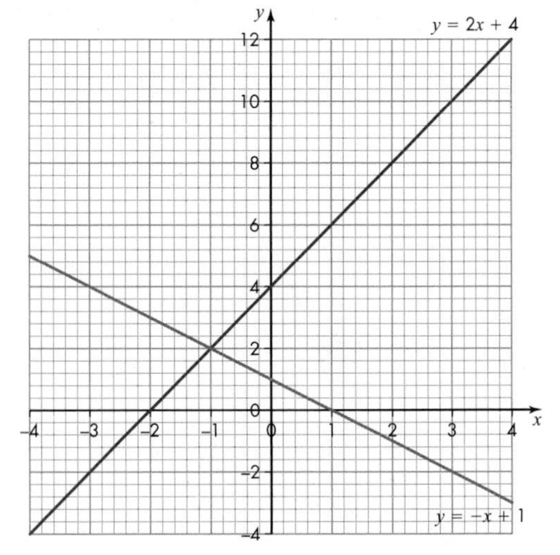

x and y have same values at the intersection point at (−1, 2)

3 $y = 10 + 0.5x$

x	2	4	6	8	10	12	14	16	18	20	22	24	26
y	11	12	13	14	15	16	17	18	19	20	21	22	23

$y = 5 + 0.75x$

x	2	4	6	8	10	12	14	16	18	20	22	24	26
y	6.5	8	9.5	11	12.5	14	15.5	17	18.5	20	21.5	23	24.5

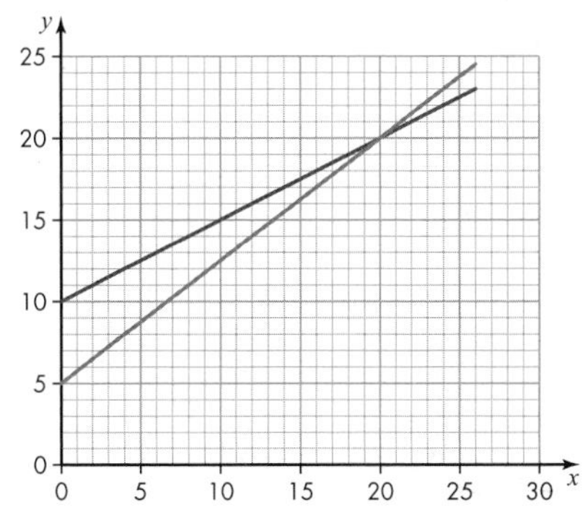

Sponsorship money raised from both friends will be the same at the intersection point of the two graphs when the runner has run 20 miles and raised £20.

4 $y = x + 1$

x	−3	−2	−1	0	1	2	3
y	−2	−1	0	1	2	3	4

$y = x^2 - 1$

x	−3	−2	−1	0	1	2	3
y	8	3	0	−1	0	3	8

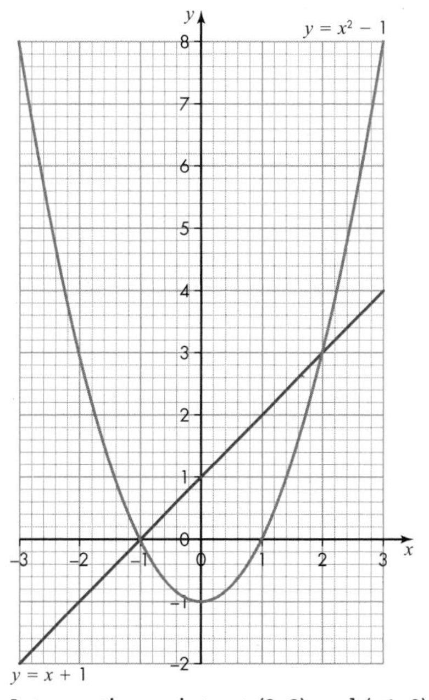

Intersection points at (2, 3) and (−1, 0).

5 These equations cannot be solved simultaneously; there are no solutions. The graphs do not intersect.

Chapter 12: Rates of change

Exercise 12A

1 a Sketch graph, straight line through the origin, gradient 60
  b Yes. Gradient represents the speed of the car.

2 a One graph is a straight line through (0, 20), gradient 6 (£ per hour). Other graph is a straight line through the origin, gradient 8.4 (£ per hour).
  b The cost of calls in £ per hour (though this depends on the units used in the graph)
  c The second option is cheaper if you expect to spend less than 8 hours 20 minutes on calls per month.

3 a Anton's graph is a straight line through (0, 600), gradient −60 (€ per day). Becky's graph is a straight line through (0, 500), gradient −40 (€ per day).
  b The amount they spend each day.
  c The gradient is negative, showing the graph decreases as the days go by.
  d At the end of the 5th day.

4 Bristol: F = 15 + 20W Manchester: F = −20 + 30W Newcastle: F = −5 + 10W
  a The number of people entering the shops each week.
  b The one with the largest gradient (i.e. the steepest slope)

5 a Graphs to be drawn: all should pass through the origin.
  b Group A 100 mm per newton; Group B 140 mm per newton; Group C 0 mm per newton; Group D 75 mm per newton
  c Group C had the stiffest spring since the gradient was least and the graph was flat. Group B had the slackest spring because the gradient was largest and the graph was steepest.

6 Students own answers.

Exercise 12B

These answers are exact. Allow some tolerance for tangents being slightly incorrectly.

1 a 12 m/s
  b −8 m/s
  c 0 m/s
  d Between 1.5 and 2.5 seconds

2 a −4 million$/year, 21 million$/year
  b The first is negative, showing the profits are decreasing, the second is positive, showing the profits are increasing
  c The company makes more and more of a loss until after about 1.7 years the losses start to decrease. After about 2.8 years the profit is 0 and then the profits increase rapidly.

3 a Student's graph
b $3\ m^3 s^{-2}$
c $6\ m^3 s^{-2}$, $15\ m^3 s^{-2}$
d $3\ m^3 s^{-2}$, after 1 hour
e Because the formula uses a cubed term, it increases very quickly after 4 hours and if the storm has stopped then there will be no more rain entering the rivers.

4 a Student's graph
b $0.8\ km\ h^{-2}$
c The braking distance is not directly proportional to the speed but increases far faster than that

5 a Student's graph
b 60, 100, 120
c It implies the rate of registration increases by 20 people per day

6 a Student's graph
b 22.5 m when the horizontal distance is 150 metres
c Perhaps it will, as the height of the ball when the distance is 100 is 20 m and the ball is still rising at this point

7 Since the temperature is stable, the patient could be asleep, resting, stable or could have died a while ago and the body temperature is at room temperature.

8 It depends on what has happened before that. If the rate of change had been increasing previously it might be a time to advise the client to sell as the rate of change could be starting to drop. If the rate of change had been decreasing previously it might be a time to advise the client to buy as the rate of change could be starting to increase. Buy when shares are low, sell when they are high!

9 Answers may vary

Exercise 12C

1 a 14 m/s
b 0.06 m/s
c 1200 m/s
d 90 000 m/s
e 3 m/s
f 2.8 m/s
g 5.3 m/s
h 6.7 m/s
i 0.006 m/s

In order: i) snail; b) blink; f) hat; e) pan; g) wildfire; h) snappy; a) bat; c) bullet; d) lightning

2 36 mph

3 a All work out to be 30 mph
b 30 mph

4 51.4 mph

5 Jane's speed is $22.5\ km\ h^{-1}$, Ben's is $15\ km\ h^{-1}$, so Jane is $7.5\ km\ h^{-1}$ faster.

6 a 08:35
b 15:58 so she could just make it, but if setting up time and dismantling time is taken into account, it is not possible. It would also depend on the time of the masterclass.

7 a If she uses the shortest distance she would arrive at 15:36.
b If she uses the distances in the chart (which use motorways) she would arrive at 15:25, so that is the earliest time she could arrive back.

Exercise 12D

1 a –125 m/min; 125 m/min; the speed she started walking home
b 333 m/min; 333 m/min; the speed she walked back to college
c –250 m/min; 250 m/min; the speed she finally walked home

2 a $3\ m\ s^{-1}$; $3.5\ m\ s^{-1}$
b $5.5\ m\ s^{-1}$; $3\ m\ s^{-1}$
c Comments such as 'because the instantaneous speed is slowing down, after 2 seconds it is less than the average speed'

3 a 25 mph; 0 mph; 50 mph
b 38 mph; 19 mph; 24 mph
c Comments such as 'The cyclist's instantaneous speed was very fast at 8 hours but because of the rest involved the average speed was less – a total distance travelled of 190 miles in 8 hours'
d He started the ride fast and his speed decreased steadily through the first part until the rest period at 6 hours. After the rest, he again started fast and steadily rode slower towards the end.

4 a Graph approximates to a straight line graph with gradient 0.35
b about $0.35\ m\ s^{-1}$
c about $0.35\ m\ s^{-1}$
d Between 23 and 33 seconds; $0.5\ ms^{-1}$

5 a Student's graph
b $20\ m\ s^{-1}$
c $55\ m\ s^{-1}$
d Student's graph
e $1.7\ m\ min^{-1}$
f $1.3\ m\ min^{-1}$

Exercise 12E

1 a $0.33\ m\ s^{-2}$
b $0\ m\ s^{-2}$
c $-1.83\ m\ s^{-2}$

2 a $-2.5\ m\ s^{-2}$
b $0.45\ m\ s^{-2}$
c $0.25\ m\ s^{-2}$

Exercise 12F

1 a $5\ m\ s^{-2}$; $3\ m\ s^{-2}$
b $1\ m\ s^{-1}$ and $18.3\ m\ s^{-1}$ (at 0.2 s and 4.8 s)
c $t > 4.8$ s

2 a $0.55\ m\ m^{-2}$
b $-2.2\ m\ m^{-2}$; $1.2\ m\ m^{-2}$
c $15.5\ m\ m^{-1}$; $43\ m\ m^{-1}$
d The acceleration is negative and approaches zero until at t = 20 it is zero. It the increases until at t = 40 it starts to reduce, returning to zero at t = 58. After that it is negative.

3 a A straight line graph passing through (0, 25) and (5, 0)
  b -5 m s^{-2}
4 a Part of a quadratic graph passing through (0, 15), maximum at (3.75, 17.8) ending at (12, 4.2)
  b $0 \leqslant t < 3.75$
  c $t = 3.75$ s
  d -2.1 m s^{-2}
5 a Student's graph
  b Gradients at the various points are 0, 1.6, 3.2, 4.8, 6.4, 8.0, 9.6. 11.2, 12.8
  c Student's graph using the gradients in part b as the velocities.
  d It is a straight line given by $v = 3.2t$
  e The acceleration is constant at 3.2 m s^{-2}

Chapter 13: Exponential functions

Exercise 13A

1 a 625
  b 1
  c 49
  d 32 768
2 a 1
  b 1
  c 1
  d 1
  Any number, except 0, to the power 0 is 1
3 a 0.015625
  b 1.55×10^{-7}
  c 3.81×10^{-6}
  d 0.100
  All less than 1
4 a Missing values are 0.25, 1, 2, 8
  b Missing values are 0.579, 0.833, 1, 1.44, 1.728
  c Missing values are 0.296, 0.444, 0.667, 1.5, 2.25, 3.375
  d (0, 1)
  e y becomes very large
5 a Missing values are 4, 1, 0.5, 0.125
  b Missing values are 1.95, 1.25, 1, 0.64, 0.512
  c Missing values are 15.625, 6.25, 2.5, 0.4, 0.16, 0.064
  d (0, 1)
  e y approaches 0

Exercise 13B

1 a −3.32
  b 1.16
  c −2.26
2 a 2.30
  b −8.05
  c 0.446
3 8
4 46.1 days
5 a 30
  b 8.1
  c 0

Exercise 13C

1 a i 1 ii 22 026 iii 2.69×10^{43} iv 0.000 045 4
    v 3.72×10^{-44}
  b Missing values are 1, 2.718, 7.39, 20.1, 33.1 followed by student's graph
  c i 2.718, 2.718 ii 7.39, 7.39
2 Missing values are 1, 7.39, 54.6, 148, 403, 1097
  a Student's graph
  b This graph is steeper. (0, 1) is common to both. y becomes very large
3 Missing values are 1, 0.223, 0.0821, 0.0498, 0.0302
  a Student's graph
  b If x is very small, y approaches 1; if x is very large, y approaches 0. (If x is very large and negative, y is very large)
    It is the reflection of $y = e^x$ in the y–axis

Exercise 13D

1 a 54.6, 54.6
  b 7.39, 7.39
  c 1, 1
  d 0.905, 0.905
  e 0, 0 (as close as possible)
2 gradient = 4.3, so y–coordinate = 4.3
3 a 0.48
  b negative, since the y–coordinate is less than 1
4 gradient = 1, so y–coordinate = 1, (0, 1)

Exercise 13E

1 2048 (excluding herself)
2 102 400
3 £62.50
4 £16.21
5 $P = 1013 \times 0.88^{0.001h}$, Shard 974 hPa, Everest 327 hPa

Exercise 13F

1 0.631
2 507
3 7.27
4 0.770
5 a 0.209
  b 0.866
  c −0.173

Chapter 14: Practice questions

Chapter 1

1 a Sample too small, not representative of each group

b Total 150 so a better sample size is $\sqrt{150} \approx 12$, so should take $\frac{12}{150} \times 88 = 7; \frac{12}{150} \times 47 = 4; \frac{12}{150} \times 15 = 1$ from the respective groups

2 a Histogram with frequency densities of 4.3, 1.8. 1.7. 0.3. 0.13 and 0.075

b $\frac{8215\text{k}}{97}$ = £84.7k

3 3, 4, 5, 7, 7

4 a Easy to calculate, only uses the end two values

b Not influenced by extreme values, only half the data is used

c All data used, can take a long time to calculate

5 a 6.7 hours

b 7.2 – 6.2 = 1 hour

c 7.1 – 6.3 = 0.8 hour

d 3

e 20%

6 a A cluster (or group) is selected at random and data is collected from all members of that cluster. When quota sampling is used, the person doing the sampling selects a specific number of members from a stratum.

b i Stratified random sample

ii Needs to ensure each of the strata (Y12 and Y13) are properly represented. $\frac{16}{80}$ = 0.2, so she should select $50 \times 0.2 = 10$ Y12 students and $30 \times 0.2 = 6$ Y13 students.

7 a Find the mean mass and capacity: uses all the data with a small sample so accessible. Mean mass = 1370 kg. Mean capacity = 18 (17.78 rounded) people. Average mass of a person = $\frac{1370}{17.78}$ = 77 kg.

b A scatter graph is best. This will see if there is any correlation between the mass and capacity. Data lies on a line of best fit approximating to capacity = $\frac{\text{mass}}{74}$

Chapter 2

1 a With simple interest the interest is only calculated on the principal amount. With compound interest the interest is calculated on the principal amount plus any accrued interest.

b 9 years

2 a =(B4–21000)*0.09

b £84

c £26 916

d £28 935.55

3 a Consumer Price Index

b A mortgage is a loan used to buy a house. A loan is money that you borrow for other purposes.

c Income tax is a tax you pay on the money you earn, VAT is a tax on most goods and services. Inheritance tax is a tax on the value of your estate when you die.

4 a $(48000 - 42385) \times 0.4 + (42385 - 10600) \times 0.2 = £8603$

b $\frac{(48000 - 8060)}{12} \times 0.12 = £399.40$

c $\frac{(48000 - 8603 - 399.40)}{12} = £3249.80$

5 a Annual Percentage rate

b $\frac{400}{1.24} + \frac{400}{1.24^2} + \frac{400}{1.24^3} + \frac{400}{1.24^4} = £961.71$

6 a Annual equivalent Rate

b $r = \left(1 + \frac{0.28}{4}\right)^4 - 1 = 0.31$ i.e. 31%

7 $P = \frac{0.004 \times 186000(1.004)^{15\times12}}{(1.004)^{15\times12} - 1} = £1451.57$

Chapter 3

1 Metric size = 7 – 0.4 × UK size.
This assumes i) there is a linear relationship between them and ii) that they are measured accurately.

2 a Your own graph to see if a linear relationship exists.

b temperature = 17 – 0.006 × altitude

This assumes there is a linear relationship between the two variables and the data was not subject to adverse weather conditions.

3 a £50.00, 13 hours and 49 minutes

b Assumptions can include parking at the venue at 10:30, leaving at 3:30, return journey is same distance etc. Time away from home is about 10 hours (5 hours meeting, 5 hours travelling). Uses about 6 gallons or 27 litres, cost about £28

4 a Since the consumption goes up, down, up, down as the distance increases there is no obvious relationship there. Consumption depends more on the speed and the way a driver drives the car.

b Assumptions made could be that these data are all from the same car; the sample is small and we should have more data; the driver is driving in the same manner each time. The scatter graph shows a reasonable model would be consumption = 80 – 0.6 × speed

c This depends on the your model. Using part b a reasonable estimate would be 56 mpg

5 Your own estimate: depends on the assumptions made.

6 Assumptions such as modelling the U-tube as a cylinder, 250 costumes at height 140 cm diameter 40 cm, 250 costumes at height 100 cm, diameter 30 cm – answers will vary, but at least two sensible dimensions. Need to show use of circumference formula or estimate. Final cost can vary.

7 Your own estimate: depends on the assumptions made.

Chapter 4

1 a Formatting errors examples: two 'Visit 2's; 'Total' in the Customer id column

Data entry errors examples: Median for F1, F4 are incorrect; missing decimal point for F4 Visit 2 (or the wrong total for Visit 2 or wrong median for F4)

b E.g. Move the word 'Total' to the cell next to the 12.85; use cell formulas to add up each column; use cell formulas to calculate the medians; put the units in the table somewhere

2 a Formatting errors examples: no skip number 4; not clear if 'Mixed' refers to a skip number or not; don't use a dash for a number as they cannot be added

Data entry errors examples: totals wrong for skip 5, plastic total wrong

b E.g. number all the skips in order; use cell formulas to add up each column; use cell formulas to sum each row; put the units in the table somewhere; use a 0 rather than a dash in the table

3 a Formatting errors example: spurious adding of the data in the columns – cannot add heights, time taken and costs!

Data entry errors examples: all the means are incorrect, total height and time taken are incorrect (there are 60 minutes in an hour, not 100!)

b E.g. delete the 'Total' row; use cell formulas to add up each row; format the 'Time taken' row to be in single units of time (hours or minutes); use the 'average' formula to find the mean for each row

4 Answers may vary, but need to mention that the data doesn't mean the transactions are easy, just that they are increasing. Comments such as the first three increases were about the same, larger increase in the final year. Assuming world population is about 7 billion an estimate for the number of transactions made by one person can be made, provided mention that not all the world's population will have cashless transactions.

5 For the first scheme Shazza has forgotten to add on the loan to the interest so she should have calculated 2568 ÷ 12 = £214 each month. For the second scheme she should have multiplied by 0.04 as that is the decimal equivalent of 4%. This gives 1872 ÷ 6 = £312 each month.

6 Colin adds incorrectly. £2.90 + £2.40 = £5.30. When he did 180 × £4.13 he should get £743.40 – he interpreted the calculator display incorrectly. He should have done £5.30 × 180 = £954, £954 ÷ 12 = £79.50 each month

7 Your own report, which should have a statistical diagram, perhaps a dual percentage bar chart, or proportional pie chart etc.

Chapter 5

1 a 68.268%

b 317 320 bars

2 a 0.99993

b 204 to the nearest Facebook friend

3 a 89.435%

b 57.302 g – 62.298 g

4 a 97.725%

b 1058 ml

5 a 84. 134%

b 0.135%

c 14

6 a 0.009%

b 0.023%

c 14.5 Mbits/s

7 a 38.2%

b 99.653%

c 81 mph

8 a 95.45%

b 68

Chapter 6

1 a i No. Most people who go to a railway station will be using the train as the mode of transport consequently the sample questioned will be biased.

ii No. Although people who go to the Last Night of the Proms are likely to appreciate lots of different types of music the proms is essentially a classical music event consequently the sample questioned will be biased.

iii If only interested in the produce bought at the village fete then choosing raffle ticket holders would seem reasonable. However some people do not like any form of gambling and consequently this is an example of a small group of the population that will not have been included in the sample.

b For example:

Obtain a list of student names from the Head of the Sixth Form.

Number the students.

Use a random number generator on a calculator to select students from the list.

Subsequently question these students about their lunchtime eating habits.

2 (100.750, 101.995)

This means that we can be 99% confident that we will find the mean of the heights of all four years olds between these limits. Some of the individual children in the class may well be some of the tallest in the country but the sample mean height is less than the lower limit of the 99% confidence interval suggesting that on average this class is shorter than the average four year old height, based on this confidence interval.

3 a No. The fact that class 3 are the median year group at primary school is not relevant. A random sample across all year groups would need to be taken.

b No. All counties have a mix of both rural and urban schools. A random sample across all schools would need to be taken.

Find the mean of the sample means and use this as the estimator for the population mean. The means of the samples are reflective of the data in the samples. The best estimator for the population mean is the one that is reflective of most data.

4 (198.65, 200.75)

This means that we can be 95% confident that we will find the mean of the mass of all these chocolate bars between these limits.

5 a (983.99, 986.01)

b The confidence intervals mean that we can be 99% confident that we will find the mean of the all cartons between (983.45, 986.55) and we can be 95% confident that we will find the mean of all the cartons between (982.96, 987.04). The 99% confidence interval is wider than the 95% interval because there is a greater range of values in which the mean can be found.

6 $(\bar{x} - 1.20134, \ \bar{x} + 1.20134)$

7 15

8 (0.88, 1.04) litres or (881.41, 1044.59) ml

This means that we can be 99% confident that we will find the mean of all the litres of fuel dispensed by this pump between these limits.

Chapter 7

1 a

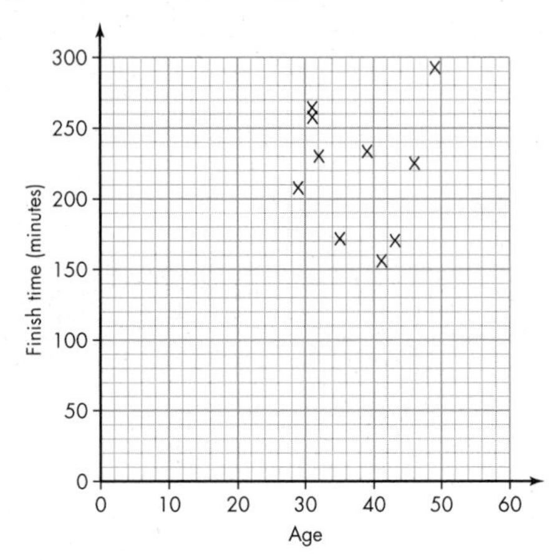

b pmcc = 0.008315

c No correlation. Age cannot be used as a predictor of London Marathon finish time.

2 Pmcc between eggs produced and hours of daylight = 0.97858, a strong positive correlation.

Pmcc between eggs produced and average temperature = 0.98887, a strong positive correlation.

From the data provided there is a slightly stronger correlation between eggs produced and average temperature than eggs produced and hours of daylight.

3 a $y = 351.04 + 4.178x$ where x is the number of Twitter users by quarter and y is the number of Facebook users by quarter.

b a – before there were any Twitter users there were over 351 million Facebook users, b – for every one million extra Twitter users there will be over 4 million new Facebook users.

c If the number of Twitter users was 200 million then the regression line would suggest in excess of 1.1 billion Facebook users. This is extrapolation (estimating outside the range of given data) and therefore unreliable.

d From the data is seems that as the number of Twitter users increases then so does the number of Facebook users and vice versa in fact it seems to be quite a strong correlation. However social media is relatively new with many people wanting to sign up to all forms of social media and others just selecting particular ones. From this data alone it is difficult to draw conclusions.

4 a The value of –10.8 represents the number of eggs produced when there are no hours of daylight. This is not a true value in real-life as a hen cannot produce a negative number of eggs!
The value of 2.04 represents the number of eggs produced for every hour of daylight i.e. 2 eggs per hour of daylight.

b. When $x = 18$, the number of hours of daylight, then $y = 25.92$ eggs. However, $x = 18$ is extrapolation as it is a value outside the known range of data and consequently cannot be relied upon.

5 a

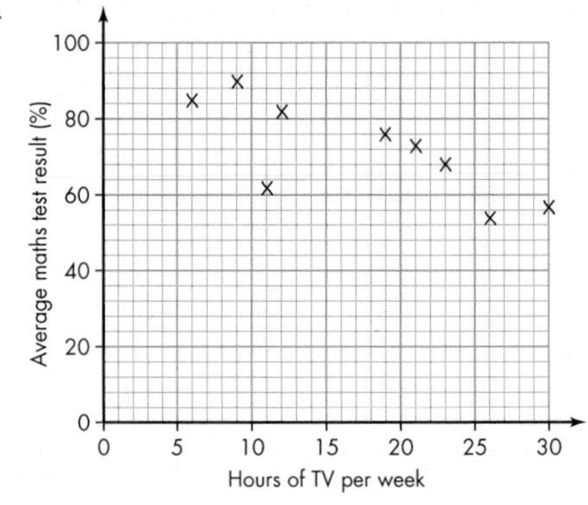

b pmcc = –0.77078

c Negative correlation. As the number of hours of TV watched increases so the mathematics test result decreases. Time spent watching TV is time not spent doing maths!

6 a

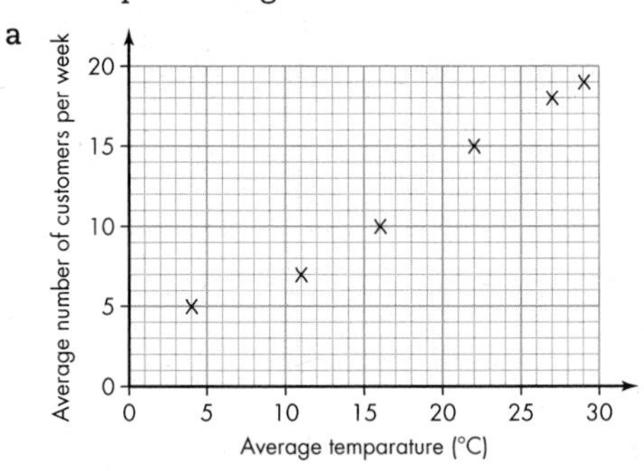

b Pmcc = 0.988869187

c Pmcc = 0.932075. Although the correlation remains a strong positive one it is not as strong as in part b.

7 a

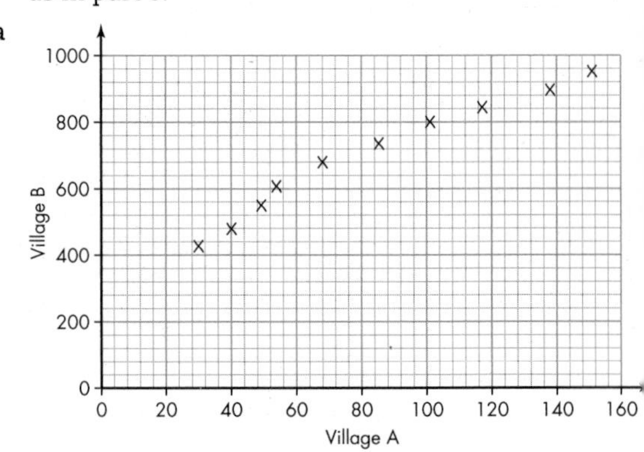

b. pmcc = 0.98316. There value of the pmcc suggests a strong positive correlation between the number of websites hits for villages A and B. Both villages may have only just developed and publicised their websites with both having a steadily increasing number of hits consequently it cannot be concluded that there is a relationship between the number of websites hits for each village.

8 a The value of 92.36 represents the Maths test score when no TV is watched. The value of 1.17 represents the increase in test score for every hour of TV not watched.

b When $x = 7$, the Maths test score $y = 84.17$. $x = 7$ is interpolation as it is a value within the known range of data and consequently could be relied upon.

Chapter 8

1 a

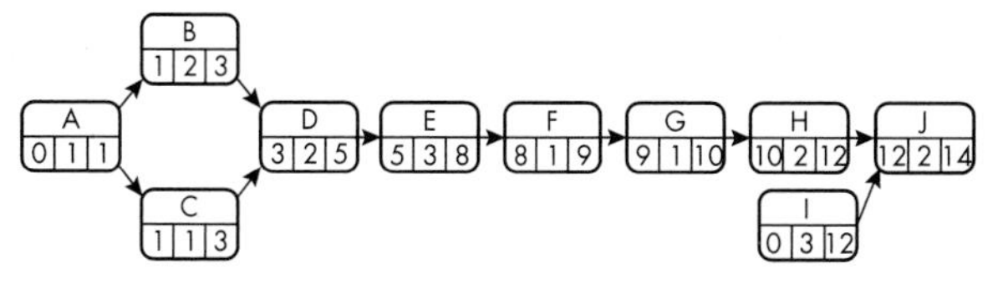

b Activities C (Write unit test plan) & I (Write system test plan) are not on the critical path. The critical path is 14 weeks long.

2

Activity	Duration (in days)	Earliest start time	Latest finish time	Float
A Set homework	1.0	0	1	0
B Write down homework	1.0	1	2	0
C Research homework	3.0	2	5	0
D Review class notes	1.0	2	5	2
E Cross reference text book	1.0	2	5	2
F Write up homework	3.0	5	8	0
G Check homework	1.0	8	9	0
H Hand in homework	1.0	9	10	0
I Mark homework	5.0	10	15	0

a The non-critical activities are D (Review class notes) and E (Cross reference text book). They both have a float of 2 days.

b

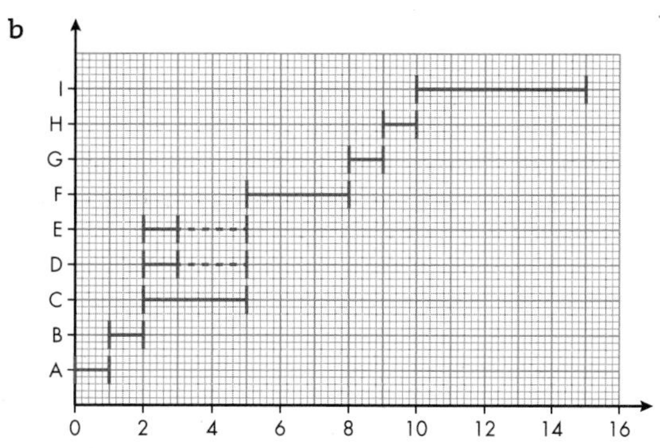

c No impact on the critical path at activity as E has a float of 2 days. If it takes twice as long it will only have a float of 1 day.

3

Activity	Duration (weeks)	Earliest start time	Latest finish time	Float
A Write author brief	1	0	1	0
B Write text and questions, first version	2	1	3	0
C Write answers first version	1	1	3	1
D Author review of chapter	2	3	5	0
E First editor review of chapter	3	5	8	0
F Amend chapter, second version	1	8	9	0
G Second author review of chapter	1	9	10	0
H Second editor review of chapter	2	10	12	0
I Write examination board review checklist	3	0	11	8
J Examination board review	2	12	14	0

a The activities on the critical path are A (Write author brief), B (Write text and questions, first version), D (Author review of chapter), E (First editor review of chapter), F (Amend chapter, second version), G (Second author review of chapter), H (Second editor review of chapter) and J (Examination board review).

b A (Write author brief) and I (Write examination board review checklist)

c J (Examination board review)

d Key assumption made is that after two iterations of writing and reviewing the chapter will be ready to be sent to the examination board.

4 a

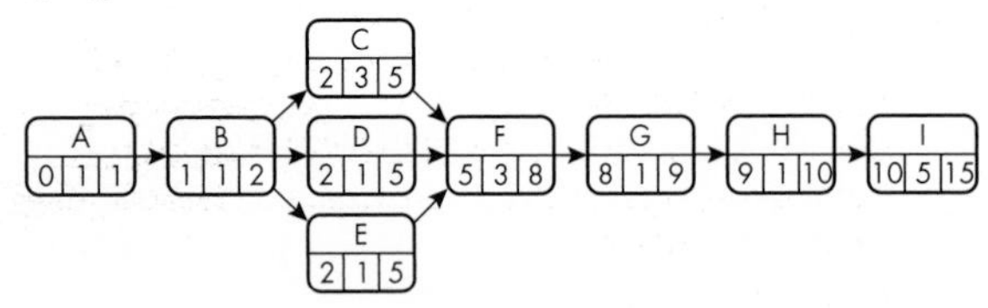

b The activities on the critical path are A (Decide upon medium), B (Decide upon location), C (Sketches of different parts of the landscape), F (Paint picture), G (Touch up picture), H (Dry picture) and I (Frame picture). The critical path is 15 days long.

5 a

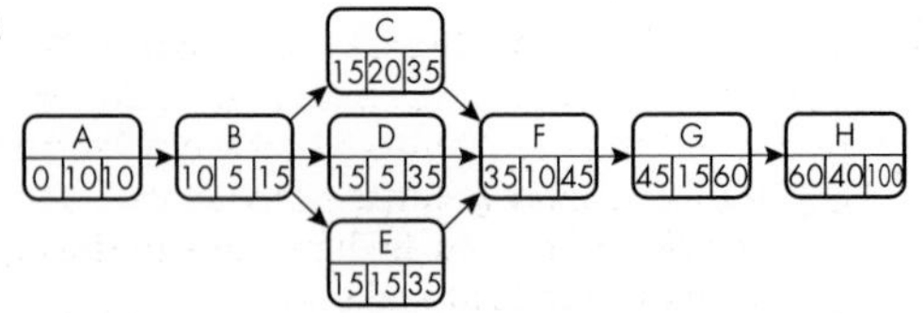

b If he wants to arrive 15 minutes early then he needs to arrive by 9.15am. The critical path is 100 minutes long which means he needs to get in the shower at 7.35am.

6 a The non-critical activities are D (Listen to travel news on radio) with a float of 15 minutes and E (Review interview preparation notes) with a float of 5 minutes.

b

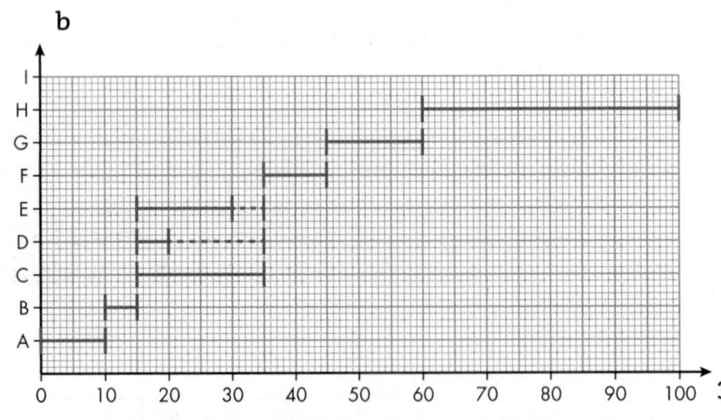

c The critical path will be extended by 5 minutes.

Chapter 9

1 a

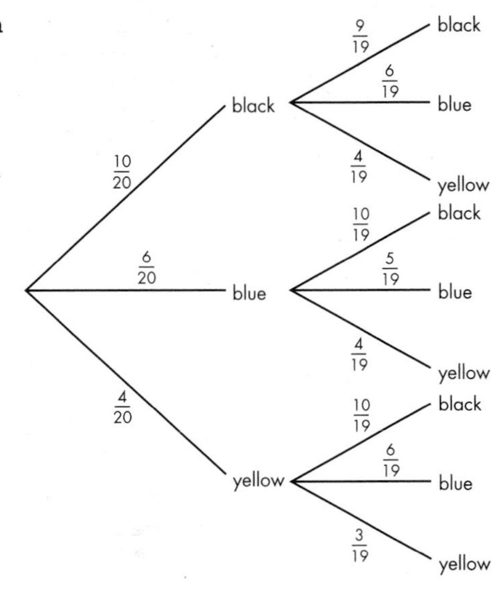

b $\frac{132}{380}$

c $\frac{12}{380}$

d $\frac{120}{380}$

e $E(X) = \left(\frac{1 \times 132}{380}\right) + \left(-\frac{1 \times 248}{380}\right) = -\frac{116}{380}$!

(he will lose 31p)

2 In a sixth form of 60 students 15 are studying maths, 18 are studying chemistry of which 10 are also studying maths.

a

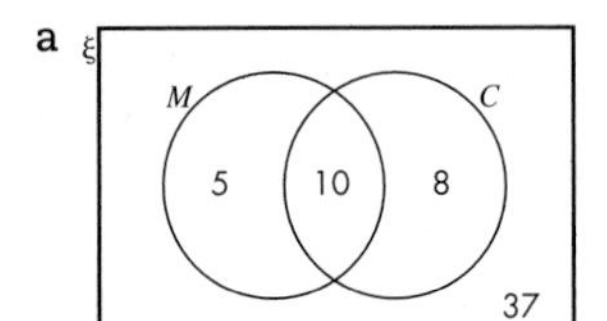

b $\frac{15}{60}$

c $\frac{18}{60}$

d $\frac{10}{60}$

e $\frac{23}{60}$

f $\frac{37}{60}$

3 a 0

b $\frac{3}{16}$

c 5

d Alan winning = $(\frac{5}{16} \times 1) + (4 \times 16 \times -1) + (\frac{7}{16} \times 0)$

$= \frac{1}{16}$. Alan most likely to win.

4 a

blue $\frac{3}{8}$: blue $\frac{2}{7}$, red $\frac{4}{7}$, green $\frac{1}{7}$

red $\frac{4}{8}$: blue $\frac{3}{7}$, red $\frac{3}{7}$, green $\frac{1}{7}$

green $\frac{1}{8}$: blue $\frac{3}{7}$, red $\frac{4}{7}$, green $\frac{0}{7}$

b 0

c $\frac{18}{56} = \frac{9}{28}$

d $\frac{24}{56} = \frac{12}{28} = \frac{6}{14} = \frac{3}{7}$

e $1 - \frac{9}{28} = \frac{19}{28}$

5 a

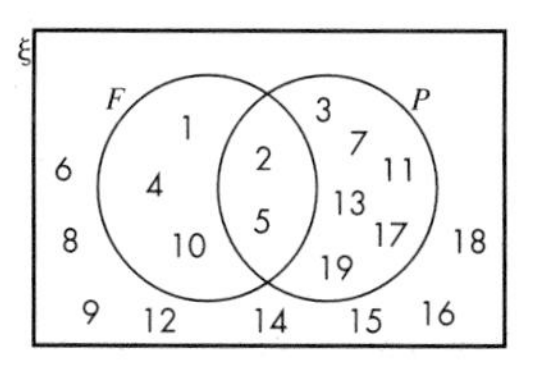

b $\frac{2}{19}$

c $\frac{8}{19}$

d $\frac{6}{19}$

6 a $\frac{15}{6}$

b 0

c $\frac{50}{3}$ or £16.67

7 a

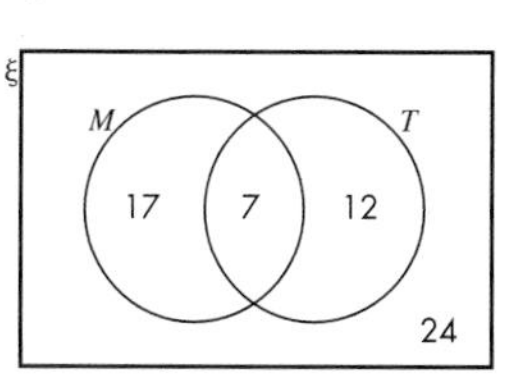

b $\frac{24}{60} = \frac{12}{30} = \frac{6}{15} = \frac{2}{5}$

c $\frac{17}{60}$

d $\frac{36}{60}$

8 If students assign −1 to probabilities for sum score of 2, 3, 4, 5, 9, 10, 11 and 12 and assign 1 to probability for sum score of 6, 7 and 8 then $E(X) = \frac{-4}{36} = \frac{-1}{9}$.

Chapter 10

1 a Any measure taken to eliminate or reduce risk

b e.g. a fume cupboard to control noxious smells; safety goggles to stop harmful substances entering the eyes; a fire extinguisher to stop or control fire etc.

2 a £12 000

b £8 000, so probably worth taking the control measure as it would save about £4 000

3 a £1 280 000

b Stand in only - £1 250 000; insurance only - £600 000; both - £450 000

c Worth taking out both control measures as there is a considerable potential saving. Stand in only would save £30 000, insurance only saves £680 000, but both saves £830 000. Not worth having just one control measure as with both the saving is considerable.

4 a £550 000

b Cost of extra staff - £400 000; cancelling contract cost - £450 000, so employing extra staff is the better option as it is both the cheaper control measure and saves a potential £150 000.

5 a £3.92 billion - assumption is that the probabilities are independent

b V only cost is £4.2 billion; W only cost is £3.2 billion; cost of both is £3 billion. Recommendation is not to take V only as that would result in a greater loss than not having any control. Probably best to take both as little difference in the cost and a reputation is at stake.

6 a £420 000

b Software only cost £365 000, extra doctor cost £305 000. Recommend having the extra doctor as it would save £115 000 which is £60 000 more than the software option and having an extra doctor is a more visible to the patients.

7 a £36.64 - assumption is that the probabilities are independent

b Guitar only costs £43; tuning only costs £37; both costs £40.

Not much to choose between all three options: taking one or both costs more than the expected penalty, but not by much. If they are building up a reputation it would be advisable to use both control measures as then the show would always be able to go on and would enhance their reputation.

Chapter 11

1 a Table and graph

n	1	2	3	4	5	6	7	8	9	10
2^{n-1}	1	2	4	8	16	32	64	128	256	512

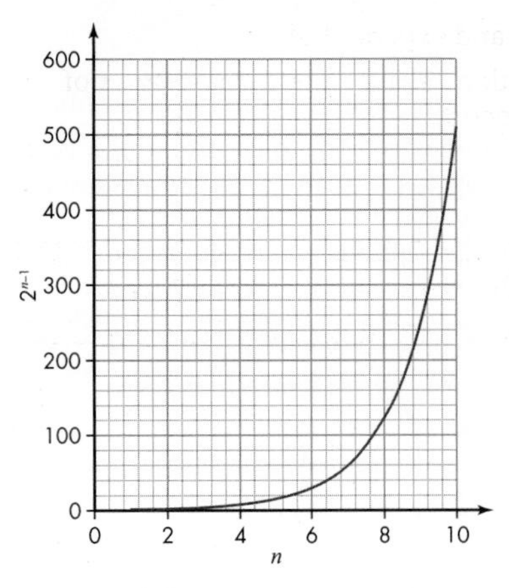

b Exponential.

c Square 8.

d Square 10.

e £524 288

2 a. Table and graph.

x (m)	0	1	2	3	4	5
y (m)	0	4	6	6	4	0

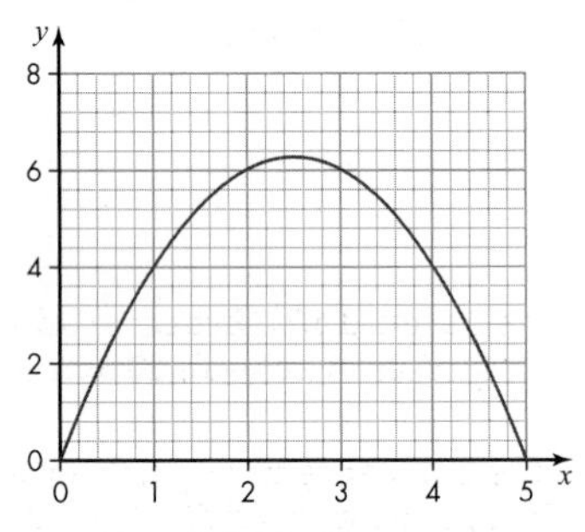

b At the origin (0, 0)

c 5 m

d 1 m and 4 m

3

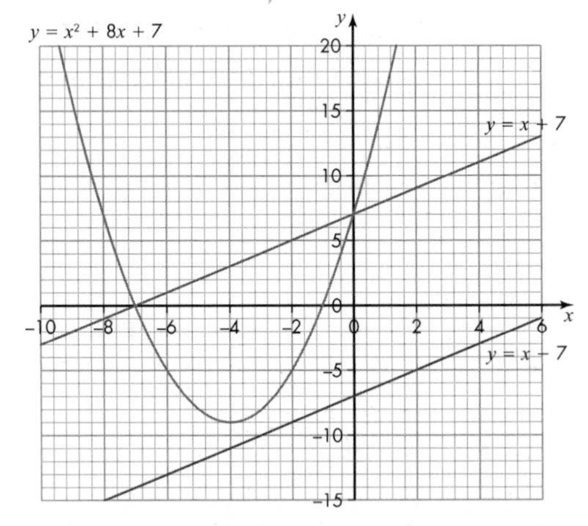

$x = -7$ and $y = 0$ and $x = 0$ and $y = 7$

The equations $y = x^2 + 8x + 7$ and $y = x - 7$ are not simultaneous because there are no points of intersection.

4 a $y = 2x + 10$

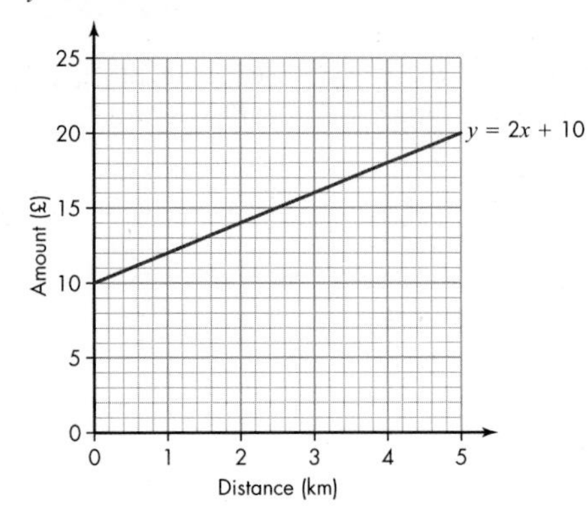

b $y = 3x + 8$

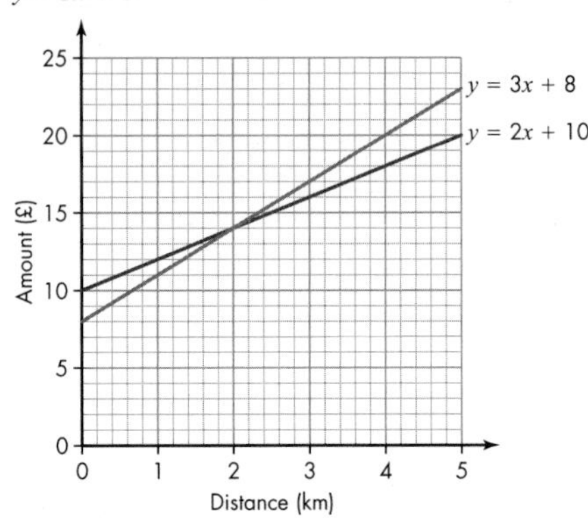

c Both pay the same amount (£14.00) at 2km – the point of intersection.

d The friend who sponsored the swimmer for £3.00 per mile plus £8 pays the most.

e £43.00

5 a

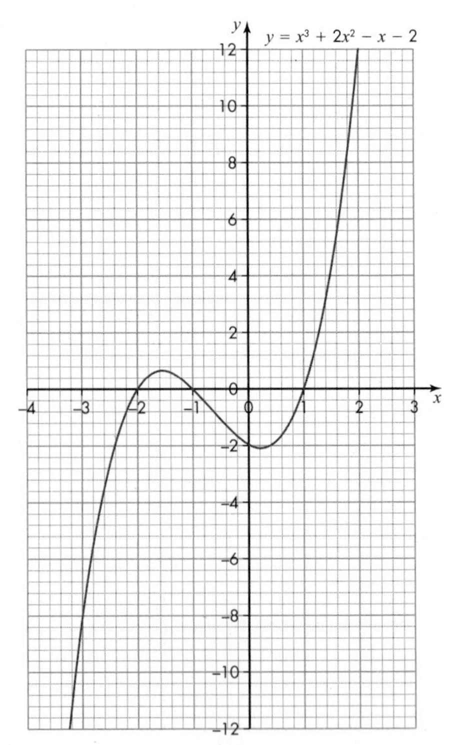

b i (–3, –8)

ii (1.5, 4)

iii When $y = 2x - 2$ the coordinates of the point of intersection are (–3, –8), (0, –2) and (1, 0)

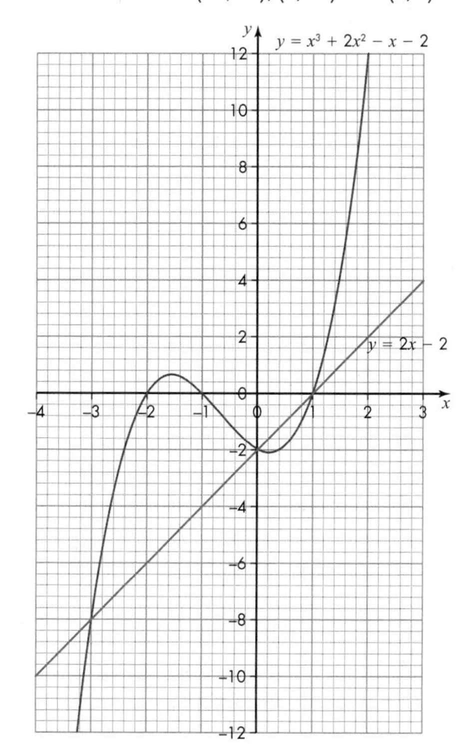

6

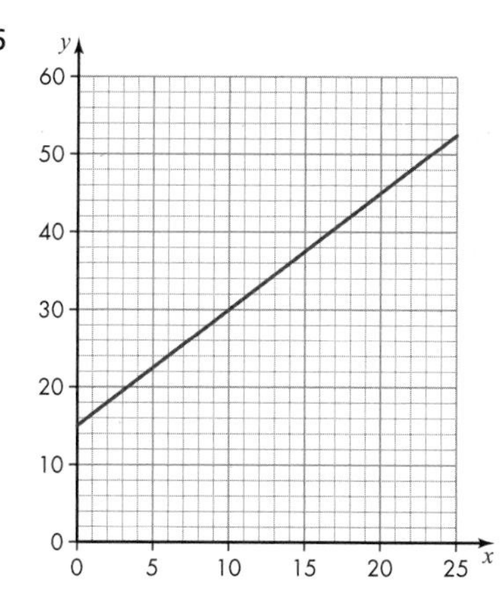

a $y = 15 + 1.50x$ where y is the total cost (£) and x is time in months

b i £18.00

ii £22.50

iii In 10 months

7 a

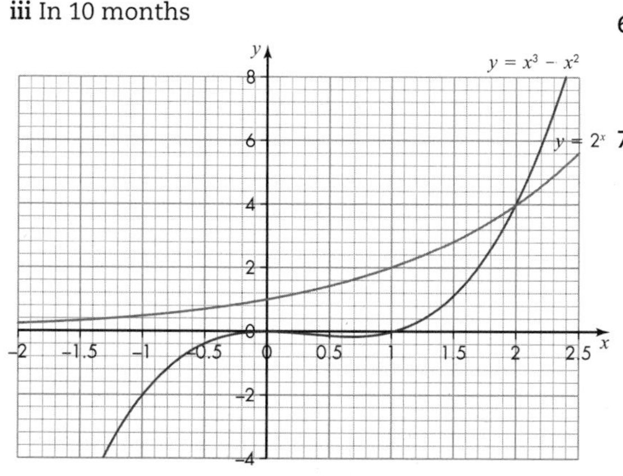

b Cubic and exponential

c 1 solution because there is one point of intersection

d $x = 2$

8 a

x (m)	0	1	2	3	4	5	6
y (m)	0	5	8	9	8	5	0

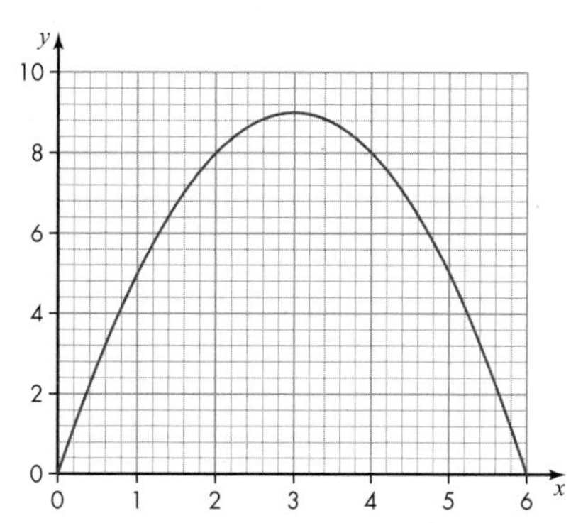

b 2 + 9 = 11 m

c 6 m

d 2 supports which will be 8 m tall

Chapter 12

1 a 60 mph

b 40 mph

c Yes. First section travelled 10 miles in 10 minutes i.e. 60 mph. Last section 15 miles in 10 minutes i.e. 90 mph

2 One graph should be a distance time graph with tangents at the time points with gradients of about 0, 1.4, 2.9, 4.3, 5.8, 7.2, then a velocity time graph showing a fairly straight line graph with gradient of about 3.6. Thus deducing the acceleration due to gravity on Mars to be about 3.6 ms^{-2}.

3 a i 0.13 mmol l^{-1} m^{-1} ii –0.03 mmol l^{-1} m^{-1}

b Gradient is zero, showing the blood sugar level has reached its maximum

c –0.07 mmol l^{-1} m^{-1}

4 a –8 ms^{-1}

b 15 ms^{-1}

c –30 ms^{-1}

d 60 m

5 a 5 ms^{-2}

b 8.75 ms^{-2}

6 a Your own distance-time graph

b when t = 4 seconds

c –6 ms^{-1}

7 One graph should be a distance time graph with tangents at the time points with gradients of about 0, 3.5, 5, 6, 7, 7.7, 8.5, 9.2 then a velocity time graph showing a fairly straight line graph with gradient of about 6. Thus deducing the acceleration due to gravity at 1750 km to be about 6 ms^{-2}

Chapter 13

1 a −0.213
 b 0.599
 c −11.5
 d 3.98
 e −0.394

2 a graph starts at (0, 250) and curves downwards but does not cross the t-axis
 b 81.92 mg
 c 14.4 hours

3 a a = 180
 b 46.4°C
 c 4 minutes 35 seconds

4 a 2.16 kg
 b 101 kg
 c $M = 1 \times 1.08^t$
 d Sketch graph starting at (0, 1) and increasing exponentially
 e Depends on whether there is enough snow and whether the snowball might collapse under its own weight.

5 a 6.79×10^7 bacteria
 b $P = 200 \times 1.7^t$
 c Sketch graph starting at (0, 200) and increasing exponentially

6 a Sketch graph starting at (0, 6) and decreasing exponentially
 b 1.5 litres
 c 2011
 d Yes. Always takes 15 years to reduce it to half its amount. Justification can be by taking values and showing it is true, or by algebraic proof

7 a Sketch graph starting at (0, 60 000) and decreasing exponentially
 b 16 767
 c 2036
 d No. Justification can be by taking values and showing it is false, or by algebraic reasoning using the fact that there is 10 000 to be added.

Index

ACKNOWLEDGEMENTS

The publishers gratefully acknowledge the permission granted to reproduce copyright material in this book. Every effort has been made to contact the holders of copyright material, but if any have been inadvertently overlooked, the publisher will be pleased to make the necessary arrangements at the first opportunity.

cover, YurkaImmortal/Shutterstock, cover, Spectrum Studio/Shutterstock, p. 6, Ververidis Vasilis/Shutterstock, p. 6, Graph adapted from data sourced from Statistical bulletin:Internet Access - Households and Individuals: 2015, Office for National Statistics licensed under the Open Government Licence v.3.0., p. 7, Diagram: Mulder, Steve: Yaar, Ziv, User is Always Right, The: A Practical Guide to Creating and Using Persons for the Web, 1st ed., © 2007. Reprinted by permission of Pearson Education, Inc., New York, New York, p. 8, Dora Zett/Shutterstock, p. 11, Phansak/Shutterstock, p. 12, huyangshu/Shutterstock , p. 12, Graph adapted from data sourced from Energy consumption and greenhouse gas emissions increase in 2012 mainly due to temperature drop Part of UK Environmental Accounts, 2014 Release, Office for National Statistics licensed under the Open Government Licence v.3.0., p. 14, Christopher Elwell / Shutterstock.com, p. 17, aastock/Shutterstock, p. 21, Reprinted from Centers for Disease Control and Prevention, WHO Child Growth Standards, http://www.who.int/childgrowth/en/, Copyright © WHO 2016, p. 41, OlgaLis/Shutterstock, p. 44, Vasabii/ Shutterstock, p. 45, AzriSuratmin/Shutterstock, p. 48, Ollyy/Shutterstock, p. 49, 1000 words/ Shutterstock, p. 53, Ewelina Wachala, p. 56, Brian A Jackson/ Shutterstock, p. 60, scyther5/ Shutterstock, p. 67, restyler/ Shutterstock, p. 69, With permission from Statista Ltd, p. 70, Tax and tax credit rates and thresholds for 2015-16. HM Treasury, licensed under the Open Government Licence v.3.0., p. 71, Tax and tax credit rates and thresholds for 2015-16. HM Treasury, licensed under the Open Government Licence v.3.0., p. 72, RTimages/ Shutterstock, p. 74, corlaffra/ Shutterstock, p. 76, Casper1774 Studio/ Shutterstock, p. 79, Pressmaster/ Shutterstock, p. 81, Mostovyi Sergii Igorevich/Shutterstock, p. 83, PhotoTodos/Shutterstock , p. 87, Dmitry Yashkin/Shutterstock, p. 88, Typical Stopping Distances from The Highway Code, Department for Transport, licensed under the Open Government Licence v.3.0., p. 89, Adapted from data sourced from Typical Stopping Distances from The Highway Code, Department for Transport, licensed under the Open Government Licence v.3.0., p. 90, Nicku/Shutterstock, p. 94, Maatman/Shutterstock , p. 95, Ant Clausen/Shutterstock , p. 98, Martin Christopher Parker/Shutterstock, p. 100, Lone Wolf Photography/Shutterstock, p. 103, gst/Shutterstock, p. 105, Anneka/Shutterstock, p. 106, lazyllama/Shutterstock," p. 106, Open Parliament Licence v3.0. From Oral evidence: The economic and financial costs and benefits of UK membership of the EU, HC 499. Tuesday 8 March 2016,", p. 107, Everett Historical/Shutterstock, p. 109, Alexander Raths/Shutterstock , p. 113, Imran's Photography/Shutterstock, p. 115, Claudio Divizia/Shutterstock, p. 115, nanka / Shutterstock.com, p. 118, bikeriderlondon/ Shutterstock, p. 121, Green Apple/Shutterstock, p. 122, welcomia/Shutterstock, p. 125, bikeriderlondon/Shutterstock, p. 126, Marc Pinter/ Shutterstock, p. 128, blackzheep/Shutterstock, p. 128, "On social media, mom and dad are watching" Pew Research Center, Washington, DC (April, 2015) http://www.pewresearch.org/fact-tank/2015/04/10/on-social-media-mom-and-dad-are-watching/, p. 128, "6 new facts about Facebook" Pew Research Center, Washington, DC (April, 2014) http://www.pewresearch.org/fact-tank/2014/02/03/6-new-facts-about-facebook/, p. 130, amasterphotographer/Shutterstock, p. 132, Andrey_Popov/Shutterstock , p. 133, HarperCollinsPublishers with permission, p. 135, Vinogradov Illya/ Shutterstock, p. 137, alphaspirit/ Shutterstock, p. 144, Jino/Shutterstock, p. 147, Robert Kneschke/Shutterstock, p. 148, Ollyy/ Shutterstock, p. 149, s_bukley / Shutterstock.com, p. 149, Featureflash Photo Agency / Shutterstock.com, p. 151, Jens Goepfert / Shutterstock.com, p. 152, Adapted from data sourced from Office for National Statitstics, licensed under the Open Government Licence v.3.0., p. 153, Rawpixel.com/Shutterstock, p. 154, Photobank gallery/ Shutterstock, p. 155, Giideon/ Shutterstock, p. 155, With permission from Bill Bryson, p. 156, Minerva Studio/Shutterstock, p. 158, Adapted from data sourced from Confidence intervals for the 2011 Census, Office for National Statistics licensed under the Open Government Licence v.3.0., p. 158, © Telegraph Media Group Limited 2016, p. 159, bbernard/ Shutterstock, p. 165, silabob/Shutterstock, p. 166, Adapted from data sourced from Facts & Figures, http://media.ofcom.org.uk/facts/, Copyright Ofcom , p. 166, UK residential broadband connections, by headline speed, Source: Ofcom, based on data provided by the UK's largest ISPs (representing over 90% of the total market), Copyright Ofcom , p. 167, Morrowind/Shutterstock, p. 169, Chapter opener quote from http://www.tylervigen.com/about, Hi, I'm Tyler Vigen!, Licensed under Creative Commons Attribution 4.0 International Public License https://creativecommons.org/licenses/by/4.0/, p. 169, Figures adapted from Total US crude oil imports and the number of honey producing bee colonies in the US/ Online revenue on Black Friday and the number of visitors to Disney World's Animal Kingdom, http://tylervigen.com/ Licensed under Creative Commons Attribution 4.0 International Public License https://creativecommons.org/licenses/by/4.0/, p. 169, Murty/Shutterstock, p. 171, Kingarion/Shutterstock , p. 173, Figure adapted from graph showing Uranium stored at Nucear Power Plants / Maths Doctorates awarded, http://tylervigen.com/ Licensed under Creative Commons Attribution 4.0 International Public License https://creativecommons.org/licenses/by/4.0/, p. 175, Kharkhan Oleg/Shutterstock , p. 176, Figure adapted from graph People who literally worked themselves to death/ Physical copies of video games sold in the UK, http://tylervigen.com/ Licensed under Creative Commons Attribution 4.0 International Public License https://creativecommons.org/licenses/by/4.0/, p. 177, gashgeron/Shutterstock, p. 181, ilolab/Shutterstock, p. 185, JAGS Information Systems Limited with permission, p. 185, Maxisport/Shutterstock, p. 186, luisrsphoto/Shutterstock, p. 188, Anton Gvozdikov/Shutterstock.com, p. 188, Way of Nature Corporation / Wikimedia Commons / Public Domain, p. 188, © Telegraph Media Group Ltd 2015, p. 189, Project Management Institute from "What is Project Management?", Project Management Institute, Inc., 2016. Copyright and all rights reserved. Material from this publication has been reproduced with the permission of PMI., p. 189, From Review of Department of Health: National Programme for IT, Cabinet Office, licensed under the Open Government Licence v.3.0., p. 190, Volt Collection/ Shutterstock, p. 192, Jia Li/Shutterstock, p. 193, With permission from Crossrail, p. 193, The Engineer http://www.theengineer.co.uk/, p. 194, With permission from Crossrail and the photographer CentralPhotography.com, p. 198, John R. Dunlap ed. / Wikimedia Commons / Public Domain, p. 198, Peter Nadolski/Shutterstock, p. 199, Paolo Bona / Shutterstock.com, p. 201, antb/Shutterstock, p. 202, RubinowaDama/ Shutterstock, p. 202, Association for Project Management (2012), The Construction Programme for the London 2012 Olympic and Paralympic Games, APM, Princes Risborough., p. 203, Matt Cardy/GettyImages, p. 205, Alexandru Nika/Shutterstock, p. 205, Nando Machado/ Shutterstock, p. 206, Jack Frog/Shutterstock, p. 207, International Statistical Institute with permission., p. 207, Simon Booth/Shutterstock, p. 208, Chros/Shutterstock, p. 209, Lolostock/Shutterstock, p. 211, Erin Meekhof with permission , p. 211, Ing. Andrej Kaprinay/Shutterstock, p. 212, © Mike Phillips, http://twitter.com/imjustmike, p. 212, Copyright © Stephen Wildish, 2016. Published by arrangement with Summersdale Publishers LTD., p. 213, Joe Seer/Shutterstock, p. 214, antb/Shutterstock, p. 215, Jenn Huls/Shutterstock , p. 217, Featureflash Photo Agency/Shutterstock, p. 218, Matt Benoit/Shutterstock, p. 219, bikeriderlondon/Shutterstock, p. 221, Irina Bg/ shutterstock, p. 223, Monkey Business Images/Shutterstock, p. 225, Arieliona/Shutterstock, p. 228, Peshkova/Shutterstock, p. 229, Copyright © 2014 MathsIsFun.com, p. 229, Sven Hoppe/Shutterstock, p. 231, Brisbane/Shutterstock, p. 231, Ecole Nationale des Ponts et Chausees, p. 232, Antonio Guillem/ Shutterstock, p. 235, Tifonimages/Shutterstock, p. 236, Gerisima/Shutterstock, p. 236, From Controlling fire and explosion risks in the workplace sourced from Health and Safety Executive, licensed under the Open Government Licence v.3.0., p. 238, rangizzz/Shutterstock, p. 239, Jochen Schoenfeld/Shutterstock, p. 241, Petr Malyshev/Shutterstock, p. 243, Roman Prishenko/Shutterstock, p. 245, acceptphoto/ Shutterstock, p. 246, wavebreakmedia/Shutterstock, p. 249, Aitormmfoto/ Shutterstock, p. 249, XiXinXing/Shutterstock, p. 251, Redd Angelo, p. 251, Steve Greenberg with permission , p. 253, Ditty_about_summer/Shutterstock, p. 255, Andrey_Popov/ Shutterstock, p. 258, Gabriele Maltinti/Shutterstock , p. 262, Billion Photos/ Shutterstock, p. 266, Samuel Micut/Shutterstock , p. 268, © Paul Brown / ardea.com, p. 269, Okssi/Shutterstock , p. 273, yuinaya/Shutterstock, p. 275, Oleksandr Lytvynenko/Shutterstock, p. 277, Iasha/Shutterstock , p. 278, testing/ Shutterstock, p. 279, jirapong/Shutterstock , p. 280, Bikeworldtravel/ Shutterstock, p. 282, With permission from Michael L. Pack, Director, CATT Laboratory, p. 282, Andrey Khachatryan, p. 283, iravgustin/ Shutterstock, p. 285, Tan Kian Khoon/Shutterstock, p. 286, Arthimedes/ Shutterstock , p. 288, lukeylukas7/Shutterstock , p. 290-291, Adpted from West Africa: Ebola Outbreak https://data.hdx.rwlabs.org/ebola The Humanitarian. Data Exchange licensed under a Creative Commons Attribution 4.0 International license., p. 295, pio3/Shutterstock, p. 296, Ollyy/Shutterstock , p. 298, Stephen Cullum/Shutterstock , p. 300, fujji/Shutterstock , p. 303, lzf/Shutterstock , p. 305, Andrey Yurlov/ Shutterstock.com, p. 309, HodagMedia/Shutterstock , p. 312, Castleski/Shutterstock , p. 313, adriaticfoto/Shutterstock, p. 314, Reprinted from Centers for Disease Control and Prevention, WHO Child Growth Standards, http://www.who.int/childgrowth/en/, Copyright © WHO 2016, p. 316, mikecphoto/Shutterstock, p. 317, Dan Kosmayer/Shutterstock , p. 317, Krasowit/Shutterstock , p. 322, AdrianNunez/Shutterstock, p. 323, Valeriya Anufriyeva / Shutterstock.com, p. 324, Production Perig/Shutterstock, p. 330, David Crosbie/Shutterstock, p. 331, Gonzalo Aragon/ Shutterstock, p. 332, ajt/Shutterstock , p. 334, ChameleonsEye/Shutterstock, p. 335, beerkoff/Shutterstock, p. 335-336, Age and condition dependent treatment rate for analysis of base-line bed provision, K. J. Davis, and B. Mirkin, March 2004. With permission from Professor Boris Mirkin, p. 338, Andrey_Popov/Shutterstock , p. 347, Data from: Average UK broadband speed continues to rise07 August 2013 / Offcom